T0305832

Phagocytosis of Bacteria and Bacterial Pathogenicity

This book provides up-to-date information on the crucial interaction of pathogenic bacteria and professional phagocytes, the host cells whose purpose is to ingest, kill, and digest bacteria in defense against infection. The introductory chapters focus on the receptors used by professional phagocytes to recognize and phagocytose bacteria, and the signal transduction events that are essential for phagocytosis of bacteria. Subsequent chapters discuss specific bacterial pathogens and the strategies they use in confronting professional phagocytes. Examples include *Helicobacter pylori, Streptococcus pneumoniae,* and *Yersinia,* each of which uses distinct mechanisms to avoid being phagocytosed and killed. Contrasting examples include *Listeria monocytogenes* and *Mycobacterium tuberculosis,* which survive and replicate intracellularly, and actually cooperate with phagocytes to promote their entry into these cells. Together, the contributions in this book provide an outstanding review of current knowledge regarding the mechanisms of phagocytosis and how specific pathogenic bacteria avoid or exploit these mechanisms.

JOEL D. ERNST is the Director of the Division of Infectious Diseases and Professor of Medicine and Microbiology at the New York University School of Medicine.

OLLE STENDAHL is Professor of Medical Microbiology at Linköping University, Sweden.

Published titles

1. *Bacterial Adhesion to Host Tissues.* Edited by Michael Wilson 0521801079
2. *Bacterial Evasion of Host Immune Responses.* Edited by Brian Henderson & Petra Oyston 0521801737
3. *Dormancy in Microbial Diseases.* Edited by Anthony Coates 0521809401
4. *Susceptibility to Infectious Diseases.* Edited by Richard Bellamy 0521815258
5. *Bacterial Invasion of Host Cells.* Edited by Richard Lamont 0521809541
6. *Mammalian Host Defense Peptides.* Edited by Deirdre Devine & Robert Hancock 0521822203
7. *Bacterial Protein Toxins.* Edited by Alistair Lax 052182091X
8. *The Dynamic Bacterial Genome.* Edited by Peter Mullany 0521821576
9. *Salmonella Infections.* Edited by Pietro Mastroeni & Duncan Maskell 0521835046
10. *The Influence of Cooperative Bacteria on Animal Host Biology.* Edited by Margaret J. McFall-Ngai, Brian Henderson & Edward G. Ruby 0521834651
11. *Bacterial Cell-to-Cell Communication.* Edited by Donald R. Demuth & Richard Lamont 0521846387

Over the past decade, the rapid development of an array of techniques in the fields of cellular and molecular biology have transformed whole areas of research across the biological sciences. Microbiology has perhaps been influenced most of all. Our understanding of microbial diversity and evolutionary biology, and of how pathogenic bacteria and viruses interact with their animal and plant hosts at the molecular level, for example, have been revolutionized. Perhaps the most exciting recent advance in microbiology has been the development of the interface discipline of Cellular Microbiology, a fusion of classic microbiology, microbial molecular biology, and eukaryotic cellular and molecular biology. Cellular Microbiology is revealing how pathogenic bacteria interact with host cells in what is turning out to be a complex evolutionary battle of competing gene products. Molecular and cellular biology are no longer discrete subject areas but vital tools and an integrated part of current microbiological research. As part of this revolution in molecular biology, the genomes of a growing number of pathogenic and model bacteria have been fully sequenced, with immense implications for our future understanding of microorganisms at the molecular level.

Advances in Molecular and Cellular Microbiology is a series edited by researchers active in these exciting and rapidly expanding fields. Each volume focuses on a particular aspect of cellular or molecular microbiology and provides an overview of the area, as well as examines current research. This series will enable graduate students and researchers to keep up with the rapidly diversifying literature in current microbiological research.

ADVANCES IN MOLECULAR AND CELLULAR MICROBIOLOGY

Advances in Molecular and Cellular Microbiology 12

Phagocytosis of Bacteria and Bacterial Pathogenicity

EDITED BY
JOEL D. ERNST AND
OLLE STENDAHL

CAMBRIDGE
UNIVERSITY PRESS

University Printing House, Cambridge CB2 8BS, United Kingdom

One Liberty Plaza, 20th Floor, New York, NY 10006, USA

477 Williamstown Road, Port Melbourne, VIC 3207, Australia

314-321, 3rd Floor, Plot 3, Splendor Forum, Jasola District Centre, New Delhi - 110025, India

79 Anson Road, #06-04/06, Singapore 079906

Cambridge University Press is part of the University of Cambridge.

It furthers the University's mission by disseminating knowledge in the pursuit of education, learning and research at the highest international levels of excellence.

www.cambridge.org
Information on this title: www.cambridge.org/9780521845694

First published 2006

A catalogue record for this publication is available from the British Library

ISBN 978-0-521-84569-4 Hardback

Contents

Contributors

Lee-Ann H. Allen
Inflammation Program, and Departments of Medicine and Microbiology, University of Iowa, 2501 Crosspark Rd., MTF D154, Coralville, IA 52241, USA

Eric J. Brown
Program in Host-Pathogen Interactions, University of California San Francisco, Genentech Hall, 600–16th St / N216R, San Francisco, CA 94158, USA

David H. Dockrell
Division of Genomic Medicine, University of Sheffield Medical School, Beech Hill Road, Sheffield S10 2RX, UK

Joel D. Ernst
Departments of Medicine and Microbiology, New York University School of Medicine, 550 First Avenue, NB16S8, New York, NY 10016, USA

Maria Fällman
Dept of Molecular Biology, University of Umeå, 901 87 Umeå, Sweden

Stephen B. Gordon
Division of Genomic Medicine, University of Sheffield Medical School, Beech Hill Road, Sheffield S10 2RX, UK

Sergio Grinstein
Programme in Cell Biology, Hospital for Sick Children, and Department of

Biochemistry, 555 University Ave, University of Toronto, Toronto, Ontario, Canada M5G 1X8

Anna Gustavsson
Department of Molecular Biology, University of Umeå, 901 87 Umeå, Sweden

Wouter L. W. Hazenbos
Program in Host-Pathogen Interactions, University of California San Francisco, Genentech Hall, 600–16th St / N216R, San Francisco, CA 94158, USA

Kassidy K. Huynh
Programme in Cell Biology, Hospital for Sick Children, and Department of Biochemistry, 555 University Ave, University of Toronto, Toronto, Ontario, Canada M5G 1X8

Dominic L. Jack
Division of Genomic Medicine, University of Sheffield Medical School, Beech Hill Road, Sheffield S10 2RX, UK

Robert C. Read
Division of Genomic Medicine, University of Sheffield Medical School, Beech Hill Road, Sheffield S10 2RX, UK

Frederick S. Southwick
Division of Infectious Diseases, University of Florida, PO Box 100277, Gainesville, FL 32610-0277, USA

Olle Stendahl
Division of Medical Microbiology, Department of Molecular and Clinical Medicine, Faculty of Health Sciences, Linköping University, SE–58185 Linköping, Sweden

Andrea Wolf
Departments of Medicine and Microbiology, New York University School of Medicine, 550 First Avenue, NB16S8, New York, NY 10016, USA

CHAPTER 1

Introduction

Olle Stendahl

Through their capacity to recognize, phagocytose and inactivate invading microorganisms, phagocytic cells have a key role in the innate immune response and host defense. During this process there is an intimate interplay between different recognition mechanisms displayed by both the host cells and the microorganisms. Understanding the complex process of phagocytosis requires insight into the mechanisms of receptor function, signal transduction, actin-based movements, membrane and vesicle trafficking, and oxidative activation, as well as how pathogens interfere with and subvert these processes. The complexity is thus in part due to the diversity of receptors capable of stimulating phagocytosis, and in part due to the capacity of different microbes to influence their own fate, as they are recognized and internalized. It is now evident that pathogens are not passive bystanders evading phagocytosis and intracellular killing, but have evolved specific means of subverting the process of phagocytosis through different mechanisms, involving inhibition of opsonization and receptor recognition, inactivation of specific GTPases, dephosphorylation, inhibition of PI-3 kinases, and actin polymerization. Studies of the pathogenicity strategies of bacteria such as *Salmonella, Helicobacter pylori, Streptococcus pneumoniae, Shigella, Mycobacterium tuberculosis, Yersinia pseudotuberculosis,* and *Listeria monocytogenes* have not only shed light on microbial pathogenicity but have also been useful tools for elucidating the phagocytic process *per se*. Understanding how *Listeria* escapes from the phagosome by forming an actin-rich tail has revealed how actin polymerization is initiated and controlled. The role of *Yersinia* YopH protein as a protein phosphatase interfering with signal transduction

Phagocytosis and Bacterial Pathogenicity, ed. J. D. Ernst and O. Stendahl. Published by Cambridge University Press.

and adhesion complexes has also given us an insight into the mechanism of phagocytosis.

A primary challenge for the innate immune system is to discriminate between potential pathogens and self, utilizing a restricted number of phagocyte receptors. This challenge has been met by the evolution of a variety of receptors that recognize conserved motifs on pathogens that are not found in higher eucaryotes. These motifs are essential for the invading agents, and are therefore conserved and not subjected to high mutation rates. These "pathogen-associated molecular patterns" (PAMP) include mannans from yeast, formylated bacterial peptides, lipopolysaccharides, lipoteichoic acid, peptidoglycans, CpG motifs characteristic of microbial DNA, and flagellin of invading microorganisms. Much interest has focused on the role of at least ten different Toll-like receptors (TLR), not only as specific PAMP receptors but also as modulators of the innate immune response and inflammation. An important observation in this respect is that mice resistant to endotoxin and endotoxic shock have a natural mutation in TLR4. In humans a number of polymorphic alleles of TLR4 have also been identified; one of these is associated with increased risk of septic shock, but not of other infections such as meningococcal diseases. The expression of TLR4 has also been linked to susceptibility to urinary tract infections in children. Because TLR drive the transcriptional program associated with a proinflammatory response, they play an integral part in the innate immune response. It was recently shown that TLR not only trigger proinflammatory cytokines, but also regulate phagolysosome maturation and subsequent inactivation of ingested bacteria. On the other hand, phagocytosis of apoptotic cells through phosphatidyl serine receptors does not initiate phagosomal maturation and proinflammatory cytokine production. Whether TLR and other PAMP receptors are engaged in the phagocytic process will thus determine the state of activation of phagocytic cells, and the subsequent inflammatory response.

Several inherited defects in phagocytic cells cause impairment of host defense. These observations have revealed important functions of phagocytic cells in host defense and inflammation. Since the discovery of genetic defects in NADPH-oxidase in chronic granulomatous diseases (CGD), leading to the understanding of the regulation of the respiratory burst and the function of reactive oxygen species, several genetic traits have been characterized linking phagocyte function to host defense. Leukocyte adhesion defects (LAD) led to the discovery of adhesion molecules and the understanding of integrins in adhesion and migration, and novel therapeutic approaches to inflammation. Recently, mutations in the signal transduction genes coding for the GTPase protein Rac1 have been characterized in patients with enhanced susceptibility to infections. These rare but severe phenotypes have deepened our understanding of the function of phagocytic cells, not only in host defense but also in non-infectious inflammatory diseases. In fact, it has become evident that activation of phagocytic cells, leading to generation of reactive oxygen and nitrogen species and to apoptosis, not only plays a vital role in innate immune reaction but forms a link to adaptive immunity as well.

The purpose of this book is to present the current state of understanding of the cellular and molecular mechanisms of phagocytosis and the mechanisms used by pathogenic bacteria to avoid phagocytosis and survive extra- or intracellularly. The book will focus on mechanisms of phagocyte recognition and ingestion, describing receptor-initiated signal transduction, and how certain pathogens interfere with these events. From these reviews it is evident that there is an intimate interplay between phagocytic cell responses and pathogenic microorganisms. A proinflammatory response may be beneficial for both the host and the pathogen, depending on the site and course of infection. Future research must focus on how to control the signaling events and cell responses of neutrophil leukocytes and macrophages interacting with different pathogens. Several animal models targeting specific genes have been very useful in this respect. With new tools of molecular biology it should now be possible in humans to identify genes conferring enhanced susceptibility and resistance to infections. Because reduction of excess inflammation is a major therapeutic goal during treatment of severe infections, and phagocytic cells are important effector cells during inflammation, modulation of the innate immune response in these cells is vital. Understanding the mechanisms of receptor recognition and cross-talk, signal transduction, and intracellular processing, will facilitate new therapeutic approaches to microbe-related inflammatory diseases.

CHAPTER 2

Phagocytosis: receptors and biology

Wouter L.W. Hazenbos and Eric J. Brown

INTRODUCTION

Consumption followed by digestion has developed from a nutrition mechanism in unicellular eukaryotes into a highy regulated and indispensable mechanism of host defense against infection in mammals. *Phagocytosis* of pathogenic microorganisms by *phagocytes*, or "eating cells," is a major host defense mechanism of the innate immune system. The process of phagocytosis was first described at the beginning of the twentieth century by Elie Metchnikoff, who observed ingestion of small particles by cells from starfish larvae. Phagocytosis is generally defined as the internalization of particles with a diameter of at least 0.5 μm, such as bacteria, viruses, parasites, large immune complexes, or apoptotic cells and cell debris. Ingestion of smaller particles, such as small immune complexes or other macromolecules, occurs through a fundamentally distinct mechanism, called *endocytosis*. Phagocytosis and endocytosis are distinguishable by the importance of actin polymerization, which directs membrane motility during phagocytosis, but not endocytosis. Another distinction can be made by the presence of clathrin coats around vacuoles formed during some forms of endocytosis, but not phagocytosis (Greenberg 1986). Recently, one more distinction has been added by showing that endocytosis by IgG Fc receptors (FcγR), but not phagocytosis, requires ubiquitylation (Booth *et al.* 2002). Thus, the specific molecular pathways that direct the process of ingestion depend on the size of the particle. When the target particle is too large to be ingested, a process designated "frustrated phagocytosis" may occur, involving activation of pathways partially similar, but not identical, to those activated during phagocytosis. In contrast

Phagocytosis and Bacterial Pathogenicity, ed. J. D. Ernst and O. Stendahl. Published by Cambridge University Press.

to active entry by invasive pathogens, phagocytosis depends exclusively on molecular mechanisms in the phagocytic cell, while the ingested particle plays an apparently passive role.

Among the first detailed investigations of the mechanisms of phagocytosis were those performed by Cohn and Silverstein in the 1960s and 1970s (Steinman & Moberg 1994; Silverstein *et al.* 1977). Silverstein and colleagues proposed the "zipper hypothesis" to explain particle engulfment during phagocytosis. According to this hypothesis, phagocytosis occurs through initial attachment of a target via specific phagocyte receptors, followed by complete engulfment, which requires sequential and circumferential ("zipper"-like) interactions between ligands distributed around the particle and receptors on the phagocyte (Griffin *et al.* 1975, 1976). Much evidence supports this hypothesis for phagocytosis triggered by IgG receptors. It remains unclear whether complement-receptor-mediated phagocytosis uses a similar mechanism, because it is morphologically distinct and appears to occur through "sinking" of the particle into the phagocyte cytosol. It is possible that additional morphologically distinct mechanisms of phagocytosis also exist.

PHAGOCYTOSIS IN STEPS

The entire process of phagocytosis of microorganisms by phagocytes can be divided into three main steps (see Figure 2.1).

The first step involves the initial *binding* of the target particle to receptors at the phagocyte surface, a recognition process mediated by a limited number of specific receptor–ligand interactions at the contact interface. The multimeric and/or immobile nature of the ligand on the particle causes a local accumulation of relevant receptors on the phagocyte membrane at the contact interface, resulting in an enhanced local receptor concentration, which likely is critical for communication with the relevant intracellular signaling cascades. This local concentration of receptors is referred to as clustering, capping, or multimerization. Receptor clustering is a major mechanism for initiation of signal transduction across the plasma membrane. The second *activation* step involves the interaction of the cytoplasmic tails of clustered receptors with cytosolic molecules, resulting in transmission of a transmembrane signal from the ligated receptor to intracellular signaling pathways, involving both kinases and cytoskeletal proteins. Activation of these signaling components in turn results in membrane motility and initiation of a number of downstream effector functions. In the third step, i.e. the process of *entry*, pseudopod extensions are formed around and closely attached to the target particle (during zippering), or the particle sinks into the ingesting cell, leading

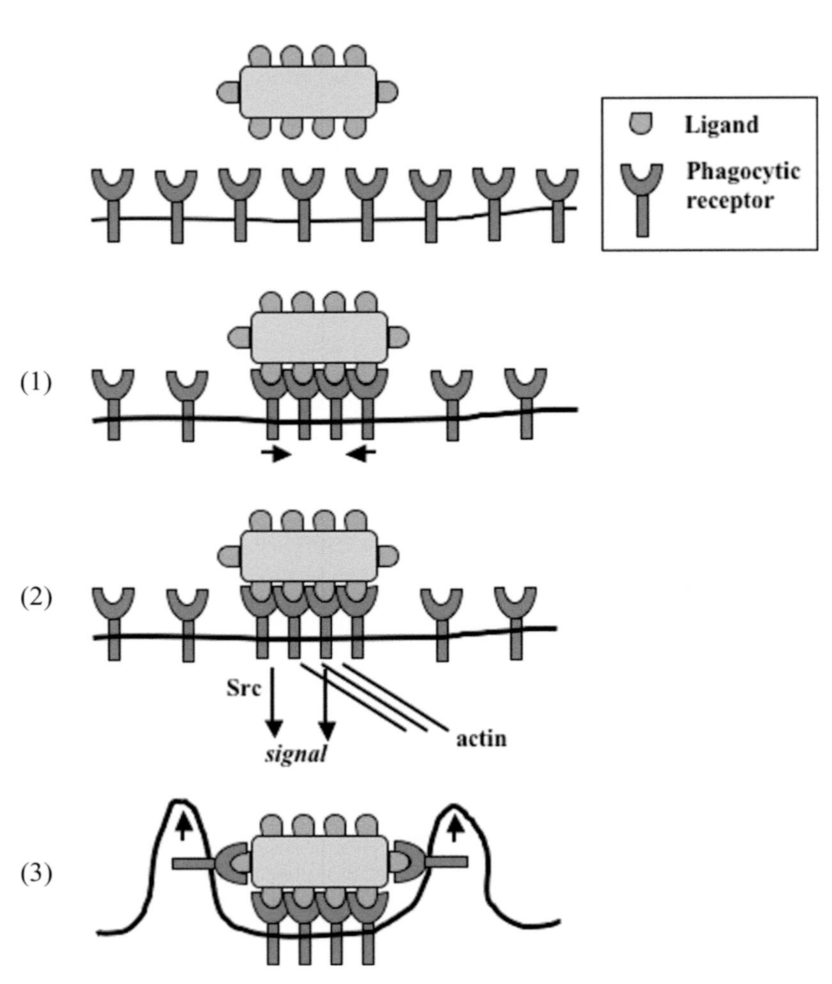

Figure 2.1 Phagocytosis in three steps. In this schematic model, phagocytosis of a ligand-coated particle by a phagocyte occurs through the following three main steps. (1) *Binding.* Unligated phagocytic receptors are normally in monomeric state and unable to signal; binding by a multimeric ligand causes receptor clustering. (2) *Activation.* Clustering of receptors facilitates their interaction with signaling molecules such as the tyrosine kinase Syk or members of the Src family, as well as cytoskeletal components including actin. This leads to cell activation and the initiation of membrane motility. (3) *Entry.* Locally increased membrane motility leads to either complete engulfment by newly formed pseudopods (as shown here) or "sinking" of the particle into the cytoplasm. This is followed by membrane fusion and closure of the phagocytic cup.

to its complete encapsulation by host cell plasma membrane. This is followed by membrane fusion events, allowing the formation of an intracellular vesicle around the particle (the *phagosome*). After a particle binds the phagocyte membrane and successfully initates a transmembrane signal, engulfment can occur quickly, i.e. within a period of one to a few minutes. In general, the process of phagocytosis is accompanied by activation of distinct downstream cellular effector functions, such as phagosome–lysosome fusion, oxidative burst, and release of antimicrobial enzymes. These latter events may facilitate killing and degradation of the ingested pathogenic microorganisms, and eventually processing and presentation of their antigens to the adaptive immune system. In addition, phagocytosis can trigger the release of vasoactive mediators and inflammatory cytokines from phagocytes, which contribute to a general activation of inflammatory and antimicrobial mechanisms at the tissue level.

It should be noted that binding of a particle to the phagocyte membrane *per se* does not necessarily lead to phagocytosis. For example, binding to complement receptors on non-activated cells alone is not sufficient to initiate phagocytosis (Wright & Silverstein 1982; Wright *et al.* 1983). IgG receptors, on the other hand, do not require cell activation prior to triggering phagocytosis. This distinction between IgG- and complement receptors is exemplified by an experiment in which both complement-opsonized pneumococci and IgG-opsonized red blood cells were bound to the surface of the same non-activated macrophages, resulting in IgG-mediated phagocytosis only (Griffin & Silverstein 1974).

PROFESSIONAL AND NON-PROFESSIONAL PHAGOCYTES

Professional phagocytes

Macrophages

Professional phagocytes are hematopoietic cells from the myeloid cell lineage: macrophages, dendritic cells, and neutrophils. Mast cells also have phagocytic potential, although the biological relevance of this is not clear. The “classic” phagocyte is the macrophage. Macrophages are part of a network of phagocytic cells present in most tissues of the body, the main function of which is to efficiently remove potential harmful particles such as microorganisms, as well as dead cells or debris. This cellular network was initially referred to by Aschoff as the reticuloendothelial system. Later, this system was re-defined by Van Furth as the mononuclear phagocyte system (MPS) (Van Furth *et al.* 1972), since it became clear that, although associated with blood vessels, the tissue phagocytes were not of endothelial origin. The MPS concept states that macrophages in all tissues originate from common, non-lymphoid, proliferating precursor cells in the bone marrow, which enter the blood as monocytes, from where they enter the tissues and finally differentiate into macrophages (Van Furth & Cohn 1968). Their main biological function is ingestion and destruction of foreign material and subsequent processing and presentation of antigens to lymphocytes. This is a primary mechanism to link innate and acquired immunity. Tissue macrophages are minimally proliferating cells with a relatively long life-span, which has been estimated to range from four to fifteen days (Van Furth *et al.* 1985). In steady state, i.e. non-inflammatory or non-infectious conditions, there is a constitutive transit of monocytes from the bone marrow to the blood and the tissues, where they develop into macrophages. During infection, this process is significantly accelerated by local inflammatory signals, such as proinflammatory cytokines, produced by tissue macrophages and infiltrating neutrophils in response to microbial products. Inflammatory macrophages usually enter tissue within hours to a few days after the initiation of inflammation.

Dendritic cells

Dendritic cells can be of either myeloid or lymphoid origin. Recently, this cell type has received a lot of attention because of its strong antigen-presenting capacity, making it a potential target for immunotherapy. Myeloid dendritic cells are developmentally and functionally related to macrophages. They originate from myeloid progenitor cells in the bone marrow and are normally resident in most tissues. Upon activation, they migrate to the draining lymph nodes, where they present antigens to lymphocytes. Like macrophages, their main function is ingestion, destruction, and processing of infectious or foreign material, but they are more potent in presenting antigens to lymphocytes to initiate an adaptive immune response. They appear in two functional stages: an immature and a mature stage. "Maturation" of dendritic cells, resembling activation of macrophages, is a prerequisite for these cells to acquire full antigen-presenting capacity. Immature dendritic cells are very efficiently phagocytic and endocytic, but they are inefficient for antigen presentation. Dendritic cell maturation can be induced by diverse stimuli, including microbial Toll-like receptor ligands, such as lipopolysaccharide or peptidoglycan, by IgG-immune complexes, or by ligands for the TNF-receptor family (reviewed in Banchereau & Steinman 1998). Upon maturation, dendritic cells acquire increased capacity to present antigens to T cells; for example, they express high levels of MHC class II and co-stimulatory molecules, while their phagocytic and endocytic capacity weakens compared with that of immature dendritic cells. The reason that this functional separation is necessary remains subject to speculation. Certainly, immature and mature dendritic cells are spatially separated: mature cells migrate to T-cell-rich zones in lymph nodes to present the captured antigen. Perhaps loss of phagocytic capacity is required to prevent them from ingesting the lymphocytes that they activate. In addition, the specificity of the signals that induce maturation of dendritic cells may help them to distinguish self from non-self, preventing presentation of self antigens and consequent autoimmunity.

Neutrophils

Neutrophils are cells with a shorter life span and originate from bone-marrow myeloid precursors that diverged from monocyte precursors. Neutrophils migrate faster and react more aggressively than macrophages, and are therefore more important for the first-line defense against local infection. Whereas neutrophils are more efficient in killing microorganisms than are macrophages, macrophages and dendritic cells more efficiently present pathogen-derived antigens. In response to inflammatory signals, large numbers of neutrophils enter the inflamed tissues quickly, i.e. within minutes to a few hours. Neutrophils harbor a number of cytoplasmic granules containing proteases and antimicrobial peptides. Activation of neutrophils leads to fusion of these granules with phagosomes and with the plasma membrane (degranulation), facilitating both intracellular and extracellular microbial killing. Soon after migration, activation, and killing of pathogens, neutrophils undergo programmed cell death (apoptosis) and are eventually removed by macrophages.

Mast cells

Mast cells are resident in most tissues; they are abundant in the skin and lungs. Their main function is protection against multicellular parasites, through vasoactive mediators such as histamine and serotonin, which are released by degranulation. The aberrant release of these mediators, in response to allergen-mediated crosslinking of IgE bound to mast-cell IgE receptors, is a major trigger of hypersensitivity reactions such as allergy or asthma. Mast cells can ingest particles through IgG-, IgE-, or complement receptors (Vranian *et al.* 1981; Daëron *et al.* 1993; Pierini *et al.* 1996). In addition, interaction with the glycosyl phosphatidyl inositol (GPI)-anchored protein CD48 on mast cells facilitates phagocytosis and killing of *Escherichia coli* (Malaviya *et al.* 1999; reviewed in Shin *et al.* 2000). Not much is known regarding the precise biological significance of phagocytosis by mast cells *in vivo*, because the major contribution of mast cells to the host defense against bacterial infection is through release of the proinflammatory cytokine tumor necrosis factor (TNF)-α, which results in attraction of inflammatory neutrophils (Echtenacher *et al.* 1996; Malaviya *et al.* 1996; Jippo *et al.* 2003).

Non-professional phagocytes

Target binding

A central factor distinguishing professional from non-professional phagocytes is the unique expression of receptors for specific opsonins, e.g. IgG Fc receptors (FcγR) and complement receptors, on professional phagocytes. None the less, the molecular mechanisms involved in the second (activation) and third (entry) steps of phagocytosis are quite generally expressed. This principle has been elegantly demonstrated by transfection studies, in which the introduction of specific receptors into non-professional phagocytes endows them with the capacity to ingest target particles expressing the cognate ligands. For example, Chinese hamster ovary cells were able to ingest IgG-coated *Toxoplasma gondii* parasites efficiently after transfection with full-length FcγRIIB. Cells transfected with FcγRIIB lacking the intracellular tail could not ingest these IgG-coated parasites, showing that binding the target particle by itself is not sufficient and that the intracellular domain of the receptor is essential for transmembrane communication to the relevant signaling components (Joiner *et al.* 1990). Similar results were obtained by using 3T6 fibroblasts and COS cells, which were enabled to ingest IgG-coated red blood cells when transfected with a full-length human FcγRIIA, but not a tail-minus mutant receptor (Tuijnman *et al.* 1992), or FcγRIIA lacking specific cytoplasmic tail tyrosines (Mitchell *et al.* 1994). These and similar experiments are now well accepted as conclusive that non-professional phagocytes possess the relevant machinery for ingestion if they can recognize a target. None the less, they remain at odds to some extent with current paradigms of FcγR function, since it is generally thought that FcγRIIB is primarily an inhibitory rather than a phagocytic receptor (Hunter *et al.* 1998), and FcγRIIA is thought to require Syk kinase for phagocytic function, which is not likely to be present in 3T6 cells. Taken together these data suggest that, although non-professional phagocytes do possess the relevant machinery for ingestion, there are multiple molecular pathways unique to professional phagocytes that markedly increase the efficiency of ingestion, in addition to expression of receptors that broaden the range of pathogens recognized.

Some microorganisms express specific ligands for endogenous surface receptors on non-professional phagocytes, allowing their recognition and ingestion by these cells. These microorganisms include invasive bacteria (reviewed in Cossart & Sansonetti 2004) and some protozoan parasites. For example, *Yersinia* spp. (discussed below, p. 23) and *Shigella* express specific ligands for integrins on epithelial cells. *Streptococcus pneumoniae* can interact with the polymeric IgA receptor at the apical surface of polarized epithelia, leading to its trancytosis and translocation across the mucosal barrier (Zhang *et al.* 2000). Binding of the internalin protein expressed on *Listeria monocytogenes* to E-cadherin on epithelial cells leads to its uptake (Mengaud *et al.* 1996). Thus, direct interaction of some pathogens with surface molecules can promote their internalization by non-professional phagocytes, which in most cases benefits the pathogen rather than the host.

OPSONIZATION

Microorganisms can be recognized by phagocytes either directly, through interaction of a microbial ligand with a receptor(s) on the phagocyte surface, or indirectly, i.e. after first being covered by host-derived soluble serum factors ("opsonization") which act as cognate ligands for receptors on phagocytes, able to trigger the phagocytic process. Opsonization increases the range of pathogens that can be recognized and the efficiency of phagocytosis and intracellular killing. The most common serum factors serving as opsonins are antibodies of the IgG isotype, which interact with IgG-Fc receptors (FcγR), and the complement factor iC3b, which interacts most avidly with complement receptors type 3 and type 4. A number of other host factors have opsonic activity, such as IgA, a ligand for FcαR, or fibronectin, a ligand for β1 integrins (see p. 23). Surfactant protein-A (SP-A) is a major component of lung surfactant and a member of the family of collectins (collagen-like lectins; see also p. 27). SP-A binds to lipopolysaccharide (LPS) on a variety of bacteria, including *E. coli* and *Pseudomonas aeruginosa*, leading to phagocytosis by alveolar macrophages through specific collectin receptors (Van Iwaarden *et al.* 1994; Geertsma *et al.* 1994). SP-A also ligates bacteria to other receptors on lung cells, such as SIRPα (Gardai *et al.* 2003), but the relevance of this for phagocytosis is not known. Serum amyloid P component (SAP) and C-reactive protein (CRP), members of the pentraxin family characterized by a cyclic pentameric structure, can recognize carbohydrates, such as polysaccharides, at microbial surfaces. These acute phase proteins are markers for inflammation since they are released rapidly from the liver into the serum in response to an inflammatory stimulus. Both SAP and CRP enhance phagocytosis by activating the classical pathway of complement and by directly binding to human or murine FcγR (Kaplan & Volanakis 1974; Mold *et al.* 2001). CRP has been shown to use human FcγRIIA as a major receptor (Bharadwaj *et al.* 1999), although there is controversy on this point (Saeland *et al.* 2001).

PHAGOCYTIC RECEPTORS

Immunoglobulin Fc receptors

FcγR

FcγR subclasses

FcγR are prominent inducers of phagocytosis and endocytosis of IgG-coated bacteria, viruses, protozoan parasites, or immune complexes. FcγR-mediated phagocytosis is generally followed by efficient activation of antimicrobial or proinflammatory effector functions such as a respiratory burst and cytokine production. On phagocytic cells, FcγR include three subclasses: FcγRI (CD64), FcγRII (CD32), and FcγRIII (CD16). In humans, FcγRI is encoded by three genes (*FcγRIA*, *-B*, and *-C*); FcγRII is encoded by three genes (*FcγRIIA*, *-B*, and *-C*); and FcγRIII is encoded by two genes (*FcγRIIIA* and *-B*) (reviewed by Daëron, 1997; Ravetch, 2003). In terms of affinity for their ligand, these FcγR can be classified as either high-affinity or low-affinity receptors. The high-affinity FcγRI binds IgG in monomeric form, whereas the low-affinity receptors, FcγRII and FcγRIII, bind IgG only in complexed form, e.g. when bound on a particle such as an invading microorganism, or when aggregated as an immune complex.

FcγR signaling properties

There are four basic mechanisms by which FcγR can transmit signals across the plasma membrane (see Figure 2.2).

1. *Activating γ chain-associated FcγR* (FcγRI and FcγRIIIA). Both FcγRI and FcγRIIIA are single-chain transmembrane receptors, expressed on human and mouse phagocytes, that transmit an activation signal through a γ-chain homodimer containing an immunoreceptor tyrosine-based activation motif (ITAM). The ITAM motif is defined by two YxxL/I sequences interspaced by 7–12 amino acids, and is commonly found among diverse activating immunoreceptors (Reth 1989). Upon multimeric ligand-induced receptor clustering, the ITAM becomes tyrosine-phosphorylated by protein tyrosine kinases of the Src family, rendering it capable of binding other kinases, such as Syk, leading to initiation of activation cascades. Whereas plasma-membrane localization of FcγRIIIA requires its association with the γ-chain, this does not seem the case for

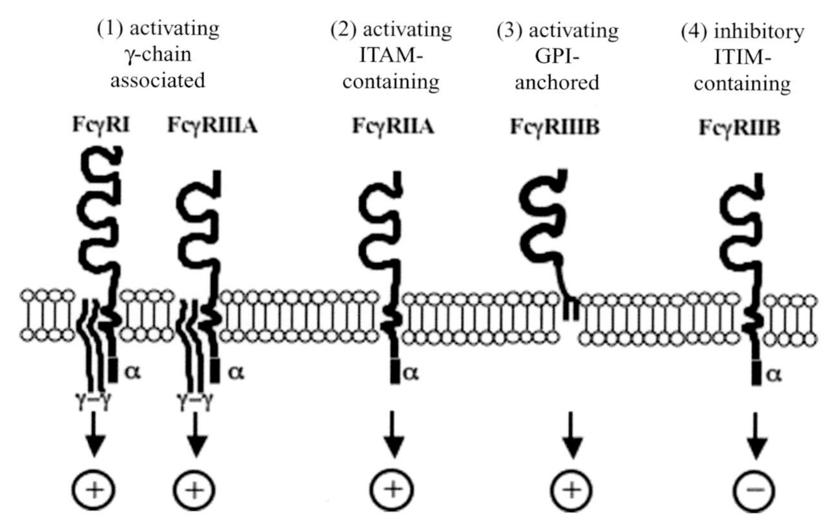

Figure 2.2 Fcγ R signaling mechanisms. Fcγ R can be classified into four categories based on differences in signal transmission mechanisms. (1) *Activating γ-chain associated Fcγ R.* Fcγ RI and Fcγ RIIIA are constitutively associated with the ITAM-containing FcR γ-chain homodimer, which upon ligand-induced receptor clustering becomes tyrosine phosphorylated, leading to downstream activating signals. (2) *Activating ITAM-containing Fcγ R.* Fcγ RIIA has an ITAM motif in its cytoplasmic tail, which becomes tyrosine phosphorylated upon receptor clustering and induces an activation signal. (3) *Activating GPI-anchored Fcγ R.* Fcγ RIIIB lacks a transmembrane domain and is anchored to the membrane through the GPI-glycolipid instead. It mediates activating signals through cooperation with other activatory transmembrane Fcγ R. (4) *Inhibitory ITIM-containing Fcγ R.* Fcγ RIIB contains an inhibitory ITIM motif in its cytoplasmic tail, which upon receptor clustering and tyrosine phosphorylation recruits inhibitory phosphatases, leading to inhibition of signals from co-ligated activating receptors.

Fcγ RI. In the absence of the common γ-chain, Fcγ RI can be expressed but signals poorly, if at all (Van Vugt *et al.* 1996; Barnes *et al.* 2002).

2. *Activating ITAM-containing Fcγ R* (Fcγ RIIA). These are transmembrane receptors expressed on human or other primate but not mouse phagocytes, able to activate cell functions by signaling through their own ITAM-containing intracellular domain. Although, like other activating Fcγ R, it may be found associated with the γ-chain (Masuda & Roos 1993), Fcγ RIIA clearly can function both for phagocytosis and for signal transduction in the absence of this association.

3. *Activating GPI-anchored Fcγ R* (Fcγ RIIIB). This receptor, which is expressed on human but not mouse neutrophils, is attached to the membrane through a GPI-anchor, and is therefore unable to signal by itself, but can function in cooperation with other transmembrane receptors. One potential advantage of the GPI-anchor is that the increased diffusion of this receptor in the plasma membrane increases cellular sensitivity to small immune complexes or minimally opsonized targets.
4. *Inhibitory Fcγ R* (Fcγ RIIB). The intracellular domain of Fcγ RIIB, which is expressed on both human and mouse phagocytes, contains an immunoreceptor tyrosine-based inhibitory motif (ITIM), which is characterized by a single I/VxYxxL/V sequence. Ligand-induced tyrosine phosphorylation of the Fcγ RIIB–ITIM motif creates an SH2 recognition domain that is a docking site for cytosolic tyrosine phosphatases, which are able to block activation signals initiated by co-ligated ITAM-bearing immunoreceptors (reviewed in Vivier & Daeron 1997).

Fcγ R subclasses in phagocytosis

Unlike FcαRI or CR3, Fcγ R do not require cell activation for efficient ligand binding and phagocytosis. Most Fcγ R subclasses expressed on professional phagocytes can promote phagocytosis upon specific ligation. One exception may be Fcγ RIIIB, which is GPI-anchored and thus lacks a transmembrane and intracellular domain. This receptor can promote cell activation, not by itself, but indirectly through cooperation with other transmembrane phagocytic receptors, including Fcγ RIIA and CR3 (see Lipid microdomains, below). When co-ligated with each of these receptors, Fcγ RIIIB synergistically enhances responses such as phagocytosis or respiratory burst (Zhou & Brown 1994; Zhou *et al.* 1995; Edberg & Kimberly 1994; Vossebeld *et al.* 1995).

Whether Fcγ RIIB, which mainly acts as a negative regulator of cell activation, has a significant role in phagocytosis remains controversial. Transfection with Fcγ RIIB has been shown to enable internalization of IgG-coated particles by fibroblasts or mast cells (Joiner *et al.* 1990; Daëron *et al.* 1993), but this may be cell-type-specific, since Fcγ RIIB cannot mediate phagocytosis in COS cells (Indik *et al.* 1994). Mutational analysis demonstrated Fcγ RIIB-mediated phagocytosis to be dependent on two tyrosine-containing sequences in its cytoplasmic domain (Daëron *et al.* 1993). Particles opsonized with $F(ab')_2$ fragments of the anti-Fcγ RII/III monoclonal antibody 2.4G2 could be efficiently ingested by Fcγ RIII-deficient mouse peritoneal macrophages (Hazenbos *et al.* 1996), an indirect indication of the phagocytic capacity of Fcγ RIIB. These observations together indicate that Fcγ RIIB by itself is able to promote phagocytosis, at least in peritoneal macrophages. On the other hand, FcR γ-chain-deficient macrophages, which express FcγRIIB but lack other FcγR, are unable to ingest IgG-coated particles (Takai *et al.* 1994), and FcγRIIB-deficient macrophages displayed enhanced phagocytosis of IgG-coated particles (Clynes *et al.* 1999; Nakamura *et al.* 2002), confirming that FcγRIIB is able to negatively regulate phagocytosis when mediated through other, co-ligated, ITAM-bearing FcγR. Thus, there may be special circumstances in which FcγRIIb can act as a phagocytic receptor, but its more general role on phagocytes is to inhibit signaling through activating FcγR.

Signals triggered by FcγR

During the process of phagocytosis through FcγR, one of the most critical initial molecular events in the activation of an intracellular signal is tyrosine phosphorylation. After ligation of FcγRI or FcγRIII, phosphorylation of the two tyrosines in the ITAM motif of the FcR γ-chain is essential, whereas triggering phagocytosis through FcγRIIA involves phosphorylation of tyrosine residues in its own cytoplasmic tail (Greenberg *et al.* 1993). During phagocytosis, these phosphorylation events are mediated largely by tyrosine kinases of the Src-family, and they are followed by recruitment and activation of the downstream tyrosine kinase Syk (Greenberg *et al.* 1993; Indik *et al.* 1995; Suzuki *et al.* 2000; Strzelecka-Kiliszek *et al.* 2002). Although Src-family kinases are involved in phagocytosis and Syk activation, they are not absolutely indispensible for these processes. Macrophages from mice deficient in the three Src kinases expressed in these cells, i.e. Lyn, Hck and Fgr, were only partly defective in phagocytosis of IgG-coated particles; among these three, Lyn appeared to play a major role (Fitzer-Attas *et al.* 2000). The Src kinase Fgr appears to have an inhibitory effect on FcγR-mediated phagocytosis, by associating with the signal regulatory protein (SIRP)-α (Gresham *et al.* 2000) (see Negative regulation of FcγR-mediated phagocytosis, below). Syk is activated at sites of phagocytosis through interaction of its amino-terminal SH2 domains with the phosphorylated ITAM motif in the Fc receptor complex and has been shown functionally and genetically to be indispensible for phagocytosis through FcγR (Indik *et al.* 1995; Matsuda *et al.* 1996; Crowley *et al.* 1997; Kiefer *et al.* 1998). The main signaling pathways downstream of Syk activation, eventually resulting in activation of phagocyte effector functions, include: (1) activation of phospholipase C (PLC)-γ, leading to generation of diacylglycerol and inositol triphosphate (IP3), which in turn induce Ca^{2+} release from cytoplasmic stores and protein kinase C (PKC) activation; (2) activation of phosphatidylinositol 3-kinase (PI3K), which can generate phosphatidylinositol 3,4,5-triphosphate (PIP3) from phosphatidylinositol

4,5-biphosphate (PIP2), which leads to many effects, including actin polymerization, pseudopod extension, and activation of the Rho family of small GTPases (see 'Cytoskeleton', below); and (3) activation of the Ras GTPase, leading to activation of mitogen-activated protein (MAP) kinases and transcription factors. In order to initiate this process efficiently, ligand-bound, clustered Fcγ R associate with specialized membrane domains, where Src family kinases and other signaling molecules apparently concentrate (see 'Lipid microdomains', below).

Negative regulation of Fcγ R-mediated phagocytosis

Phagocytosis mediated by activating Fcγ R can be negatively regulated by co-ligation of Fcγ RIIB, which presumably occurs naturally whenever immune complexes or opsonized bacteria bind to a professional phagocyte. Ligand-induced phosphorylation of the ITIM motif of Fcγ RIIB by Lyn creates an SH2-recognition domain, which is a binding site for the inhibitory phosphatase SH2-containing inositol 5-phosphatase (SHIP) (Ono *et al.* 1996; reviewed in Coggeshall *et al.* 2002). This phosphatase hydrolyzes the PI3K-generated lipid PIP3, thereby suppressing ITAM-mediated actin polymerization, Rho family activation, and subsequent function of the p21-activated kinase (PAK), which regulates a number of effects of Fcγ R ligation, including activation of the NADPH oxidase. Indeed, macrophages from SHIP-deficient mice exhibit enhanced phagocytosis mediated by ITAM-bearing Fcγ R (Cox *et al.* 2001; Nakamura *et al.* 2002). In addition to Fcγ RIIB, macrophage SIRP-α, which is a ligand for the integrin-associated protein (IAP) or CD47, and which has four ITIM (YxxL/V) motifs, is also able to inhibit ITAM-mediated activation in macrophages. Upon ligation, the SIRP-α ITIM motifs are tyrosyl-phosphorylated, creating a recognition site for the inhibitory phosphatase Src homology 2 domain-containing protein tyrosine phosphatase (SHP)-1. Co-ligation of SIRP-α has been shown to inhibit Fcγ R-mediated phagocytosis (Oldenborg *et al.* 2001; Gresham *et al.* 2000); SHP-1 deficient mice and macrophages have defects resembling those described above for SHIP-deficiency (Coggeshall *et al.* 2002).

Fcγ R knock-out mice

The development of knock-out mouse models, lacking individual Fcγ R subclasses, has significantly advanced our understanding of functions of these receptors *in vitro* and *in vivo*. A knock-out mouse strain deficient in the FcR γ-chain demonstrated the crucial importance of Fcγ R in a wide variety of IgG-induced inflammatory responses, including anaphylaxis and autoimmune diseases, *in vivo*. Macrophages from this mouse exhibited profound defects in phagocytosis of IgG-coated particles *in vitro* (Takai *et al.* 1994). Anticryptococcal IgG antibodies failed to protect FcR γ-chain-deficient mice against *Cryptococcus neoformans* infection; the failure was attributed to a defect in FcγR-mediated phagocytosis of these fungi (Yuan *et al.* 1998). A mouse strain deficient in the ligand-binding chain of FcγRIII showed this FcγR subclass to be the key receptor in diverse IgG-mediated inflammatory reactions *in vivo* (Hazenbos *et al.* 1996; Miyajima *et al.* 1997; Clynes *et al.* 1999; Chouchakova *et al.* 2001). The use of this mouse identified FcγRIII as the principal receptor mediating phagocytosis and other immune effector functions triggered by murine IgG1 (Hazenbos *et al.* 1996, 1998). *In vivo* relevance of this finding was demonstrated by resistance of FcγRIII-deficient mice to IgG1-triggered autoimmune hemolytic anemia, a hematological abnormality caused by phagocytosis of red blood cells coated with autoreactive antibodies (Hazenbos *et al.* 1998; Meyer *et al.* 1998). Mice deficient in FcγRI showed this receptor to be central in phagocytosis and endocytosis triggered by IgG2a. In addition, FcγRI-deficient mice showed impaired IgG-induced antigen presentation and inflammation, and reduced protection against a respiratory *Bordetella pertussis* infection (Ioan-Facsinay *et al.* 2002; Barnes *et al.* 2002). Mice deficient in FcγRII confirmed its negative regulatory role in IgG-mediated phagocytosis, and revealed inhibitory functions in diverse processes *in vivo* such as antibody production, anaphylaxis, alveolitis, and autoimmune reactions (Takai *et al.* 1996; Clynes *et al.* 1999; reviewed in Nakamura & Takai 2004). Of course, these conclusions may be modified in humans and other primates that express two additional activating FcγR, FcγRIIA and FcγRIIIB.

FcγR polymorphism

Some FcγR display a genetic polymorphism, which greatly affects their affinity for IgG and consequently affects host defense against specific infections (reviewed in Van der Pol & Van de Winkel 1998). A G/A polymorphism in the FcγRIIA gene results in either an arginine (R) or a histidine (H) at position 131 (allotypes IIA-R131 or IIA-H131). These allotypes determine the capacity of FcγRIIA to interact with the IgG2 isotype: particles coated with IgG2 can be phagocytosed by neutrophils isolated from individuals homozygous for FcγRIIA-R131, but not FcγRIIA-H131 (Bredius *et al.* 1993). IgG2 deficiency causes enhanced susceptibility to pneumococcal disease (Jefferis & Kumararatne 1990), confirming the essential role of IgG2 in host defense against encapsulated bacteria such as pneumococci and meningococci. Recent epidemiologic data suggest that, whereas R131 homozygous individuals are not more susceptible to acquiring meningococcal infection, the disease is significantly more severe in these individuals (reviewed in Domingo *et al.* 2002). Fcγ RIIA allotype is thus a significant risk factor in human infections with encapsulated bacteria.

FcαR

FcαRI

The low-affinity IgA receptor FcαRI (CD89) is another primate-specific immunoglobulin receptor that, like Fcγ R, associates with the FcR γ-chain, which is essential for signaling and downstream effector functions (reviewed in Monteiro & Van de Winkel 2003). FcαRI appears only on primate leukocytes and is expressed on monocytes, macrophages, neutrophils, eosinophils, and immature dendritic cells; it binds monomeric IgA with low and complexed IgA with high avidity. *In vitro*, FcαRI has been shown to induce a variety of leukocyte effector functions including phagocytosis of IgA-opsonized particles, stimulation of a respiratory burst and release of inflammatory cytokines (Weisbart *et al.* 1988; Shen *et al.* 1989; Van Spriel *et al.* 1999; Van der Pol *et al.* 2000). Efficient FcαRI-mediated phagocytosis requires cell activation, e.g. by LPS or by cytokines such as GM-CSF or IL-8 (Weisbart *et al.* 1988). In a transgenic mouse model, human FcαRI on Kupffer cells was shown to induce *in vivo* phagocytosis of *E. coli* when opsonized with serum IgA, but not with secretory IgA (Van Egmond *et al.* 2000), suggesting that it is more significant in systemic than in to mucosal defense against infection. Targeting of *B. pertussis* to transgenic FcαRI, using bispecific antibodies, promoted elimination of these bacteria during *in vivo* respiratory infection in these mice (Hellwig *et al.* 2001a).

Fcα/μR

Recently, a novel receptor for IgA and IgM, Fcα/μR, has been identified in the mouse; it also has a human homolog. This receptor is expressed on macrophages and B lymphocytes but not neutrophils, and does not appear to associate with the FcR γ-chain. The Fcα/μR binds IgM and IgA with intermediate affinity, and mediates internalization of IgM-opsonized latex beads and *Staphyloccus aureus* (Shibuya *et al.* 2000; Sakamoto *et al.* 2001).

Fcε R

The low-affinity IgE Fc receptor (Fcε RII, CD23) can mediate endocytosis and phagocytosis (Yokota *et al.* 1999). The high-affinity IgE receptor, Fcε RI, on mast cells has also been shown to trigger phagocytosis of IgE-coated particles (Pierini *et al.* 1996). The biological relevance of phagocytosis of IgE-opsonized targets through either IgE receptor is not clear, because for most antigens IgE is a minor fraction of the antibody response. None the less, IgE is a prominent component of the response to worms, and it may be involved in killing of these parasites, induction of inflammatory responses, or antigen presentation to T lymphocytes (Maurer *et al.* 1995).

Complement receptor type 3

Integrins are heterodimeric adhesion molecules composed of a unique α chain and a common β chain (reviewed by Springer 1990). The complement receptor type 3 (CR3) is so named because it recognizes a fragment of the complement component C3, called iC3b, found in abundance on surfaces wherever complement is activated. CR3, also called αMβ2, CD11b/CD18, or Mac-1, is a member of the β2/CD18 integrin family and is expressed as a transmembrane heterodimer, composed of a ligand-specific αM/CD11b chain and a common β2/CD18 chain. CR3 is expressed at the surface of myeloid cells, i.e. monocytes, macrophages, neutrophils, eosinophils, dendritic cells, and mast cells, and is also a major membrane component of the secretory granules of neutrophils. It is also present on NK cells, where it has a role in target recognition. On myeloid leukocytes, CR3 plays a key role in: adhesion of myeloid cells to endothelial cells, epithelial cells, and cells of the interstitium; migration of myeloid cells to sites of inflammation; leukocyte activation; and phagocytosis of some non-opsonized and all complement-opsonized particles, including bacteria. Neutrophils from mice genetically deficient in CR3 exhibit impaired spreading (an initial step during extravasation of inflammatory neutrophils), ingestion of complement-opsonized particles, oxygen radical generation, and apoptosis (Coxon *et al.* 1996). In addition to complement iC3b, ligands for CR3 include intercellular adhesion molecule (ICAM)-1, Factor X of the clotting cascade, the extracellular matrix protein fibrinogen, and perhaps some proteoglycans (Diamond *et al.* 1995). In addition, some microbial products serve as direct ligands for CR3, such as the filamentous hemagglutinin (FHA) of *B. pertussis*, lipopolysaccharide from *E. coli* or *P. aeruginosa*, and *Leishmania* lipophosphoglycan (reviewed in Agramonte-Hevia *et al.* 2002). If this array of apparently unrelated ligands is not sufficiently bewildering, many other proteins have been proposed as CR3 ligands; this has led to the proposal that CR3 may serve as a receptor for denatured proteins in general. However, in most cases, the primary evidence for interaction with CR3 is inhibition by anti-CR3 antibodies of neutrophil adhesion to ligand-coated surfaces. As has been discussed previously (Brown 1991), this is not sufficient evidence for a direct interaction with CR3; therefore, most of these reports must be viewed with some degree of skepticism.

Activation of CR3

In circulating phagocytes unperturbed by inflammatory stimuli, CR3 exists in an inactive form, unable to bind its ligand efficiently or induce phagocytosis (Wright & Silverstein 1982; Wright *et al.* 1983). In order to be able to bind ligand firmly and trigger phagocytosis, CR3 needs to be activated. This appears to be a general property of leukocyte integrins; the cellular process of inducing integrin ligand binding is called "inside-out signaling." In neutrophils, chemoattractants, bacterial products, leukotrienes, growth factors, or other inflammatory stimuli can lead to a rapid increase in receptor expression at the plasma membrane as the result of secretion of internal stores of the molecule. However, it is clear that increased surface expression of the receptor is neither necessary nor sufficient for activation of ligand binding and phagocytosis by CR3. Instead, CR3 activation most likely involves a conformational change leading to increased affinity of individual receptors for the ligand, and multimolecular clustering leading to increased avidity of the cell for iC3b-opsonized targets (Hogg *et al.* 2002). The current model for induction of the conformational change in β2 integrins as a result of inside-out signaling is that it is initiated by an activation-signal-induced separation of the α and β cytoplasmic tails, which is transmitted through the membrane, resulting in unfolding of the integrin (this has been compared to opening of a switchblade knife) and exposure of the ligand-binding pocket (Kim *et al.* 2003; Brown 2005). The mechanism inducing CR3 clustering has not been clarified. Activation of CR3 can be induced by some soluble stimuli, e.g. phorbol esters, or cytokines such as IL-5, TNF-α, M-CSF, GM-CSF and platelet-activating factor, or by ligation of FcγR, the latter involving PI3K (Jones *et al.* 1998) and protein kinase A (Wang & Brown 1999). Upon activation, CR3 is able to bind its ligand, but additional signaling may be required to initiate phagocytosis (Pommier *et al.* 1984). CR3 also can be activated by "integrin crosstalk," i.e. a two-step mechanism initiated by ligation of another integrin on the same cell. For example, adherence to immobilized fibronectin or monoclonal antibodies against the α5β1 integrin, a major fibronectin receptor on phagocytes, enables them to efficiently bind or ingest particles coated with CR3 ligands (Pommier *et al.* 1983, 1984; Hazenbos *et al.* 1993, 1995; Van den Berg *et al.* 1999; Burke-Gaffney *et al.* 2002). The αvβ3 integrin has also been reported to activate CR3 (Van Strijp *et al.* 1993; Ishibashi *et al.* 1994; Capo *et al.* 1999). The biological relevance of this two-step mechanism for the host is probably to separate β2 integrin-dependent leukocyte migration through extracellular matrix, which has ligands for β1 or β3 integrins, from activation for phagocytosis and its accompanying release of toxic oxygen metabolites and hydrolytic enzymes. Thus, tissue damage along the path of phagocyte migration is minimized.

Cell activation and signaling by CR3

The signaling pathways that are induced by CR3 ligation, also referred to as "outside-in signaling," and that promote CR3-mediated phagocytosis, have not been clarified completely. There is evidence for activation of tyrosine kinases, including Syk, upon ligation of integrins including CR3 (Woodside *et al.* 2001; Miranti *et al.* 1998; Chen *et al.* 2003). However, unlike FcγR-mediated phagocytosis, CR3-mediated phagocytosis is normal in Syk-deficient macrophages, indicating that Syk is not an indispensible factor (Kiefer *et al.* 1998). There is evidence of a requirement for other kinases, including phosphatidylinositol-3 kinase (PI3K), and protein kinase C (PKC) in CR3-mediated phagocytosis. PKC likely is activated through a pathway involving phospholipase D during CR3-mediated ingestion, rather than through phospholipase C, as occurs after activation of G-protein-coupled receptors or FcγR (Fallman *et al.* 1992, 1993; Allen & Aderem 1996).

Compared with phagocytosis through FcγR, CR3-mediated phagocytosis is a less potent stimulus for activation of a respiratory burst, which is an important intracellular killing mechanism (Wright & Silverstein 1983; Berton *et al.* 1992; Zhou & Brown 1994). It also has been reported that, during phagocytosis mediated by CR3, fusion of the phagosome with secretory vesicles containing defensins and proteolyic enzymes is less efficient than when mediated by FcγR (Joiner *et al.* 1989). This poor coupling to antibacterial effector mechanisms may explain why some pathogens enter phagocytes through CR3-mediated ingestion, to increase their intracellular survival. Examples of organisms that appear to use CR3 to enter host cells "silently" are non-opsonized or iC3b-opsonized *Leishmania* (Mosser & Edelson 1985, 1987; Russell & Wright 1988) or mycobacteria (Peyron *et al.* 2000), non-opsonized *B. pertussis* (Relman *et al.* 1990; Saukkonen *et al.* 1991), and *Histoplasma capsulatum* (Bullock & Wright 1987). Establishing a benign intracellular environment, where these microorganisms are separated from extracellular immune mechanisms such as complement and antibodies yet also less subject to intracellular killing mechanisms, may thus contribute to the persistence of some infections. For example, when *B. pertussis* is phagocytosed by macrophages through CR3, live bacteria can persist intracellularly long after resolution of clinically apparent infection, providing an intracellular niche for bacterial persistence (Saukkonen *et al.* 1991; Hellwig *et al.* 1999). This is consistent with the observation that bispecific antibody-mediated targeting of *B. pertussis* to FcγR *in vivo* protects against infection, whereas targeting to CR3 does not (Hellwig *et al.* 2001b). CR3-dependent bacterial persistence has clinical significance, not only for the resolution of infection, but also for the persistence of bacterial antigens and cell wall products that may contribute to the pathogenesis of a variety of inflammatory diseases. Both FcγR-mediated phagocytosis and CR3-mediated phagocytosis can be regulated by additional interactions between the phagocytic target and the ingesting cell. For example, both are inhibited by co-ligation of macrophage SIRP-α (Oldenborg *et al.* 2001; Gresham *et al.* 2000) (see Immunoglobulin Fc receptors, above); these inhibitory effects potentially help suppress immune damage to host tissues during infection.

Integrin crosstalk induced by pathogens

Some pathogenic microorganisms have adapted the multi-step mechanism of integrin crosstalk to induce their own uptake by CR3. These microorganisms interact with β1 or β3 integrin receptors, which then activate CR3 via crosstalk to facilitate their uptake. For example, *B. pertussis* expresses filamentous hemagglutinin (FHA) at its surface, which serves as a ligand for CR3 on phagocytes (Relman *et al.* 1990), and which binds stably only to activated CR3. *Bordetella pertussis* expresses ligands for two other integrins on phagocytes: the minor fimbrial subunit FimD, which binds the integrin α5β1 (Hazenbos *et al.* 1995), and FHA, which recognizes αvβ3 (Ishibashi *et al.* 1994). Binding of *B. pertussis* to either α5β1 or αvβ3 induces CR3 activation, leading to stable binding of FHA to CR3 (Hazenbos *et al.* 1993, 1995; Ishibashi *et al.* 1994), thereby promoting CR3-mediated ingestion and intracellular survival by macrophages (Saukkonen *et al.* 1991). A similar two-step mechanism has been proposed for *Mycobacterium avium*, which binds αvβ3 and thereby facilitates recognition of surface-bound complement by CR3 (Hayashi *et al.* 1997). *Coxiella burnetii* is also internalized through the dual interaction with αvβ3 and CR3, a process that is inhibited by virulence factors of these bacteria (Capo *et al.* 1999). In addition to promoting CR3-mediated phagocytosis, αvβ3 ligation suppresses α5β1-mediated phagocytosis, a negative regulatory integrin crosstalk mechanism, through inhibition of the calcium/calmodium-dependent protein kinase II (Blystone *et al.* 1994, 1999).

Cooperation of CR3 with other phagocytic receptors

In addition to stimulating phagocytosis by itself, CR3 is indispensable for normal functions of several other phagocytic receptors (reviewed in Jones & Brown 1996). FcγR-mediated phagocytosis and production of the chemoattractant LTB_4 are markedly reduced in neutrophils from patients with leukocyte adhesion deficiency, which lack the common β2-integrin chain, or in normal neutrophils by blocking anti-CR3 antibodies (Arnaout *et al.* 1983; Gresham *et al.* 1991; Graham *et al.* 1993). These studies indicate the critical requirement for CR3 in phagocytosis through FcγR. CR3 was also found to be required for other FcγR functions, including respiratory burst and antibody-mediated cellular cytotoxicity *in vitro* and antibody-mediated antimelanoma immunity *in vivo* (Kushner & Cheung 1992; Zhou & Brown 1994; Van Spriel *et al.* 2001, 2003). In addition, binding of secretory IgA, but not serum IgA, to FcαRI and the subsequent activation of a respiratory burst depends on the presence of CR3 (Van Spriel *et al.* 2002).

β1 Integrins

Integrins of the β1 family are ubiquitously expressed on both professional and non-professional phagocytes. Many ligands are extracellular matrix molecules such as fibronectin (recognized by α4β1 and α5β1), laminin (α1β1, α2β1, and α6β1), or collagen (α1β1, α2β1, and α3β1). The biological function of these interactions is promoting cell adhesion to matrix, cell migration, cell–cell interactions and signaling (reviewed in Springer 1990). The α4β1 receptor, expressed poorly on neutrophils but otherwise a major integrin of myeloid leukocytes, also recognizes both fibronectin and the endothelial molecule VCAM-1; this latter interaction is important in transendothelial migration of a variety of blood cells, including monocytes, lymphocytes, and eosinophils. Many β1 integrins are capable of endocytosis and phagocytosis; indeed, some are constitutively recycling molecules (Bretscher 1992). As described below, some pathogens have taken advantage of this property of integrins to enhance their own uptake during establishment of infection.

Invasin–β1 integrin interactions

Yersinia spp. express a surface protein, invasin, which binds directly to several β1 integrins, which in turn efficiently triggers the uptake of these bacteria by non-professional phagocytes (Isberg & Leong 1990; reviewed in Isberg *et al.* 2000). This is relevant for the translocation of enteropathogenic *Yersinia* through β1-expressing M cells overlying Peyer's patches, an essential step in efficient host infection. Binding of *Yersinia* invasin to α5β1 has a higher affinity than binding of the natural ligand, fibronectin. Both ligands bind to the same domain in α5β1 (Tran Van Nhieu & Isberg 1991). Whereas fibronectin binding to α5β1 requires its Arg–Gly–Asp sequence, the invasin sequence does not have Arg–Gly–Asp, although an invasin Asp residue is important for binding α5β1 (Leong *et al.* 1995). Invasin is more potent in triggering phagocytosis than fibronectin, likely because of its higher affinity for α5β1 (Tran Van Nhieu & Isberg 1991, 1993). Interestingly, invasin-mediated uptake appears to occur through the "zipper mechanism" (discussed above), because it requires receptor–ligand interactions around the entire surface of the bacterium (Isberg *et al.* 2000). In addition to promoting β1 integrin-mediated phagocytosis through invasin, *Yersinia* secretes proteins designated Yops, which are injected into the host cell by a Type III secretion pathway, and which efficiently block IgG-mediated phagocytosis and intracellular killing by professional phagocytes (China *et al.* 1994; Bliska & Black 1995; Visser *et al.* 1995; Fallman *et al.* 1995) by interfering with signal transduction at key steps (Persson *et al.* 1997. Enteropathogenic *Yersinia* species may thus avoid being killed by professional phagocytes through the expression of Yops, while facilitating invasin-mediated translocation across β1 integrin-expressing M cells.

Extracellular matrix protein binding by pathogens

A number of other pathogens, such as mycobacteria (Schorey *et al.* 1995, 1996), *Trypanosoma cruzi* (Wirth & Kierszenbaum 1984; Ouaissi *et al.* 1984), *Leishmania* (Wyler *et al.* 1985), *Streptococcus pyogenes* (Talay *et al.* 1992), and *S. aureus* (Kuusela 1978), may also use β1 integrins to interact with host cells, not by binding integrins directly through a bacterially encoded protein like invasin, but indirectly through a molecular bridge of fibronectin, collagen, or laminin. For example, mycobacteria, including the pathogenic strains *M. tuberculosis, M. leprae,* and *M. avium,* express fibronectin-binding proteins (Thole *et al.* 1992; Ratliff *et al.* 1993; Schorey *et al.* 1996). These fibronectin-binding proteins may facilitate β1-mediated uptake by host cells. Blockade of fibronectin binding to *M. leprae* impairs uptake by Schwann cell and epithelial cell lines (Schorey *et al.* 1995), which are main target cells during the pathogenesis of leprosy. There is strong evidence that *M. leprae* binding to laminin has a significant role in its uptake by Schwann cells (Rambukkana 2001). In this case, the Schwann cell receptor mediating bacterial invasion is less clear, but β1 and β4 integrins may be involved (Rambukkana *et al.* 1997).

For these organisms, although extracellular matrix binding clearly contributes to pathogenesis, the precise biological significance of ECM-binding proteins for interaction with professional phagocytes remains to be clarified. In addition to inducing β1-mediated entry, the interaction with β1-integrins may activate CR3 through integrin crosstalk (see paragraph Complement receptor type 3, above) and thus facilitate entry through the latter integrins.

Toll-like receptors

In the past 5–10 years, it has become apparent that the family of Toll-like receptors (TLR) is very important in mammalian innate immunity, including a significant role in phagocytosis. Toll itself was originally identified as required for normal *Drosophila* development and then was shown to be necessary for antifungal defense. Toll has an intracytoplasmic domain homologous to the mammalian IL-1 receptor, and mammalian TLR were identified based on their homology with *Drosophila* Toll. Mammalian TLR are prominently expressed on leukocytes including phagocytes, and many if not all are able to recognize "pathogen-associated molecular patterns" (PAMPs), directly, without a host-derived opsonin as an intermediary. A variety of cell biological approaches identified the PAMPs recognized by different TLRs, and mouse strains that lacked specific individual members of the TLR family revealed the aspects of host defense dependent on each of the TLRs (reviewed by Akira, 2003). For example, TLR type 4 (TLR4) was identified as the long-sought main LPS receptor transmitting activation signals triggered by LPS; the "classical" LPS receptor, the GPI-anchored protein CD14, was found to be merely a binding partner without intrinsic signaling capacity. TLR2 was identified as a major receptor for peptidoglycan on Gram-positive bacteria. Members of the TLR family are potent triggers of signaling pathways leading to multiple downstream cellular effector mechanisms, such as the release of inflammatory cytokines or upregulation of co-stimulatory molecules for antigen presentation. Investigation of the roles for TLRs in phagocytosis has only recently begun. TLRs are found in phagosomes in macrophages, even when the target particle has no TLR ligand (Underhill *et al.* 1999). This has been interpreted to suggest that TLRs are "sampling" the contents of phagosomes to determine whether bacterial PAMPs are present. In a recent report, TLR2 and TLR4 were shown to determine the rate and extent of phagocytosis of non-opsonized bacteria, *E. coli*, *Salmonella typhimurium*, and *S. aureus* by macrophages and also to enhance the rate at which the phagosomes fused with lysosomes (Blander & Medzhitov 2004). These early data demonstrate that TLRs will be very important in severals steps of phagocytosis; further investigations are necessary to generate a more complete picture of their roles.

C-type lectins

There are several receptors on phagocytic cells that demonstrate Ca^{2+}-dependent binding of carbohydrate ligands, so called C-type lectins.

Among the most studied are the selectins, which are involved in leukocyte transendothelial migration. However, phagocytes also express C-type lectins, distinct from selectins, that are capable of recognizing these carbohydrates on intact bacteria as well as on molecules shed by bacteria during their proliferation in a mammalian host. Although there are many of these, the ones that have received the most attention with respect to phagocytic host defense are DC-SIGN, the mannose receptor, and Dectin-1 (reviewed in Geijtenbeek *et al.* 2004). These receptors recognize various carbohydrate structures, principally those including mannoses, but also other sugar groups such as fucose.

DC-SIGN

DC-SIGN, which is expressed on dendritic cells, can recognize the mannose-containing lipoarabinomannan on *M. tuberculosis*, lipophosphoglycan of *Leishmania mexicana*, and *Candida albicans* (Cambi *et al.* 2003), which facilitates their internalization (Colmenares *et al.* 2002; Appelmelk *et al.* 2003). In addition, DC-SIGN was found to be a major receptor for HIV or Ebola virus on dendritic cells (Geijtenbeek *et al.* 2000; Cambi *et al.* 2004). Thus, DC-SIGN recognizes a wide variety of bacterial, viral, protozoan, or fungal pathogens through mannose-containing carbohydrates that decorate their surfaces, and it mediates their phagocytosis by dendritic cells, an essential component of innate immunity.

Macrophage mannose receptors

Tissue macrophages express another mannose-binding lectin, called the macrophage mannose receptor (MMR). A role for MMR in opsonin-independent phagocytosis of yeast was suggested by a number of blocking experiments (Marodi *et al.* 1991). However, mice lacking MMR have unaltered resistance to fungal infections with *C. albicans* or *Pneumocystis carinii* (Lee *et al.* 2003; Swain *et al.* 2003). Macrophages isolated from MMR-deficient mice were normally able to ingest non-opsonized *P. carinii* (Steele *et al.* 2003). Thus, the biochemical and genetic studies of the mannose receptor in phagocytosis lead to different conclusions, and the physiologic role of the receptor in phagocytic host defense remains unclear. Because MMR expression is downregulated by inflammation (Chroneos & Shepherd 1995), it is possible that it plays a primary role in tissue homeostasis rather than in host defense.

Dectin-1

A major receptor for β-glucans is Dectin-1, a relatively small (*c.*28 kDa) member of the C-type lectin family. Dectin-1 is a major phagocytic receptor on macrophages recognizing yeast (Brown & Gordon 2001), and has been shown to mediate phagocytosis of *P. carinii* as well (Steele *et al.* 2003). Like FcγRs, but unlike CR3, Dectin-1 also efficiently integrates phagocytosis with proinflammatory responses. Thus, Dectin-1 appears to be a major host defense C-type lectin. However, it is important to emphasize that *in vivo* genetic tests of this hypothesis have not yet been reported.

C-type lectin opsonins

The C-type lectin carbohydrate recognition motif is found not only on receptors on host defense cells, but also in several secreted molecules that can bind both to the pathogen carbohydrate PAMP and to phagocytic cells and thus act as classic opsonins. This family of opsonins is called collectins and is characterized not only by their lectin domain but also by a collagen-like tail that is recognized by phagocyte receptors. Mannose binding protein in plasma and surfactant protein A in alveolar secretions are both in this family, and both have been shown to be able to signal phagocytosis of opsonized targets and to have a role in host defense. The molecular identities of phagocyte receptors for the collectins remain controversial; candidates include calreticulin, the globular C1a receptor (gC1qR), signal regulatory protein alpha, CD93, and the mannose receptor itself (reviewed in Holmskov *et al.* 2003).

CEACAM receptors

Receptors of the carcino-embryonic antigen-related cellular adhesion molecule (CEACAM; CD66) family are members of the immunoglobulin superfamily that were originally identified on tumors and later found to have a more ubiquitous cellular distribution, including phagocytes (reviewed by Hammarstrom 1999). Four of the seven known members of the family, including two transmembrane molecules (CEACAM1 and CEACAM3) and two GPI-anchored receptors (CEACAM5 and CEACAM6), bind opacity-associated (Opa) proteins of *Neisseria* and related genera of bacteria. Two of these four (CEACAM3 and CEACAM6) are expressed on neutrophils or monocytes and can mediate phagocytosis of non-opsonized *Neisseria meningitidis, N. gonorrhoeae, Moraxella catarrhalis,* and *Haemophilus influenzae* (Billker *et al.* 2002; Schmitter *et al.* 2004), leading to bacterial killing. Ingestion through CEACAM3 was found to depend on Src tyrosine kinases and the GTPase Rac (see Cytoskeleton, below) (Schmitter *et al.* 2004). The CEACAM receptors are thus potentially involved in host defense against these Gram-negative pathogens.

LIPID MICRODOMAINS

Recently it has become clear that the plasma membranes of all mammalian cells contain distinct domains that vary in their lipid composition. One domain that is distinct from the bulk disordered lipid of the majority of the membrane contains lipids that are able to pack more densely, in a so-called "liquid ordered" state. These separate lipid microdomains, also referred to as "lipid rafts," are highly and selectively enriched in sphingolipids, cholesterol, and GPI-anchored proteins, and they are characterized by detergent insolubility (Simons & Ikonen 1997). Hence, they are sometimes also called *d*etergent-*i*nsoluble *g*lyco*s*phingolipid rich domains or DIGS. At the inner leaflet of the membrane, these microdomains concentrate components of the intracellular signal transduction machinery, such as myristoylated and/or palmitoylated signaling proteins like kinases from the Src protein tyrosine kinase family and heterotrimeric G proteins, some prenylated proteins like the small GTPases of the ras and rho families, and the cytoskeletal protein actin. These lipid rafts are thought to be essential for the function of several immunoreceptors, such as the T cell antigen receptor, the B cell antigen receptor, and immunoglobulin receptors including FcγR, FcϵR, and FcαR (Langlet *et al.* 2000; Dykstra *et al.* 2003). In the absence of ligand, these transmembrane receptors remain excluded from lipid rafts. After being bound and clustered ("crosslinked") by a multimeric ligand, these receptors associate with lipid rafts by an unknown mechanism. Association with lipid microdomains allows these receptors to come into proximity to cytosolic kinases at the inner leaflet, to initiate the intracellular signal necessary for their effector functions. Raft association is thus an essential step in the signaling pathway of various receptors. This is supported by experimental disruption of these domains, for example by cholesterol depletion, which results in reduced or abolished receptor function. In this way, lipid rafts provide a spatially segregated and dynamic mechanism for the initiation of transmembrane signals related to many essential functions of the immune system. For this reason, rafts have been proposed to function as "signaling platforms."

Functional cooperation of FcγR and CR3 with GPI-anchored proteins

GPI-anchored proteins are anchored to the outer membrane leaflet and are highly concentrated in lipid rafts (Friedrichson & Kurzchalia 1998; Sharma *et al.* 2004). Because these proteins lack a transmembrane domain, they have no obvious mechanism for efficient signaling to the cytoplasm

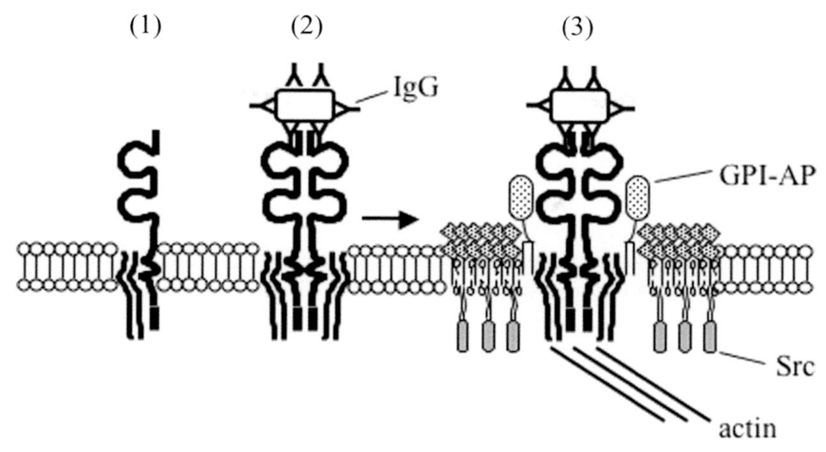

Figure 2.3 Hypothesis for FcγR association with membrane lipid microdomains. (1) Non-ligand-bound FcγR appears in monomeric state, excluded from rafts, and not able to signal. (2) Binding of a multimeric ligand, such as an IgG-containing immune complex or IgG-coated microorganism, induces FcγR clustering. (3) Clustering enhances association of FcγR with lipid microdomains, which concentrate GPI-anchored proteins (GPI-AP), cholesterol, and sphingolipids. GPI-AP are essential for normal FcγR signaling (Hazenbos *et al.* 2004), perhaps reflecting a stable FcγR–microdomain association, possibly mediated by interaction between FcγR and GPI-AP. In lipid microdomains, the clustered FcγR can communicate with protein tyrosine kinases of the Src family, i.e. Lyn, Hck, and Fgr, which are associated with these microdomains at the inner leaflet because of their acylation. Actin polymerized by FcγR as a result of tyrosine kinase activation tends to remove FcγR from lipid microdomains, which may be necessary for termination of signaling and subsequent steps in phagocytosis.

upon ligation. However, several studies showed that co-crosslinking with GPI-anchored proteins synergistically enhances signaling by transmembrane FcγR, indicating a functional cooperation (Edberg & Kimberly 1994; Salmon *et al.* 1995; Green *et al.* 1997). In a myeloid-specific GPI-anchor-deficient mouse strain, GPI-anchored proteins were found to be required for several FcγR functions in response to IgG complexes, including FcR γ-chain tyrosine phophorylation, TNF-α release, antigen presentation, and *in vivo* inflammation (Hazenbos *et al.* 2004). Some GPI-anchored proteins have been shown to associate physically with FcγR (Chuang *et al.* 2000; Pfeiffer *et al.* 2001; Ding & Shevach 2001). It is possible that GPI-anchored proteins are involved in stabilization of the interaction of ligand-clustered FcγR with rafts, allowing a prolonged contact of FcγR with raft-associated intracellular kinases (see Figure 2.3). In addition to FcγR, CR3 also has been shown to associate physically with several GPI-anchored proteins (Zhou *et al.* 1993; Xue *et al.* 1994; Bohuslav *et al.* 1995; Pfeiffer *et al.* 2001); crosslinking with GPI-anchored proteins synergistically enhances CR3 function (Zhou & Brown 1994; Zhou *et al.* 1995; Edberg & Kimberly 1994). Among the different GPI-anchored proteins, the interaction between the urokinase/plasminogen activator receptor (uPAR) and CR3 has been best characterized. Although it is possible that GPI-anchored proteins and transmembrane proteins interact through the GPI anchor itself, uPAR and CR3 demonstrate a protein–protein interaction on the beta propeller motif of the CR3 α-chain (Simon *et al.* 2000). Downmodulation of uPAR by monoclonal antibodies, antisense oligonucleotides, or gene disruption strongly suppresses CR3 functions (Sitrin *et al.* 1996; Gyetko *et al.* 2000; Xia *et al.* 2002), presumably because CR3 association with rafts is diminished in the absence of uPAR. Since both FcγR and CR3 can associate with GPI-anchored proteins, these molecules together may form a functional signaling complex in the membrane. The possible role of these cooperative interactions in phagocytosis remains to be clarified.

Lipid microdomains and phagocytosis

There are several reports of a requirement for lipid rafts in the phagocytosis of pathogens. Mycobacteria may use lipid microdomains for their entry into macrophages: disruption of lipid rafts by treating cells with cholesterol-depleting agents inhibits uptake of non-opsonized *M. tuberculosis, M. bovis* BCG, or *M. kansasii* (Gatfield & Pieters 2000; Peyron *et al.* 2000). Cholesterol depletion also inhibits non-opsonic phagocytosis of FimB-expressing *E. coli* bacteria, which use the raft-associated GPI-anchored protein CD48 as a phagocytic receptor (Shin & Abraham 2001). Evidence for lipid-raft-mediated uptake of *Brucella abortis, Clostridium perfringens, Chlamydia trachomatis, Plasmodium falciparum, Campylobacter jejuni, Salmonella typhimurium, Listeria monocytogenes,* and many viruses, including HIV, has also been reported. The mechanism for raft requirement in internalization may involve recruitment of the lipid-raft-associated protein flotillin to the nascent phagosome. However, the hypothesis that phagocytosis requires initiation in lipid rafts is not proven, because the techniques used in most of these studies are not very precise. Cholesterol depletion, the primary method used, obviously has many additional effects on cellular function (Munro 2003). Phagocytosis of beads (6 μm in diameter) by mast cells through Fcε RI, which uses lipid rafts for a variety of other signals, is apparently raft-independent (Pierini *et al.* 1996). Preliminary experiments in our laboratory indicate specific and rapid localization, i.e. within 2 minutes, of lipid rafts around FcγR-induced phagocytic cups during internalization of beads (4 μm in diameter) by dendritic cells, but we have found no requirement for GPI-anchored proteins in phagocytosis. We believe that the role of lipid rafts in phagocytosis remains controversial and that new techniques for perturbing raft function will be required before definitive analysis is possible.

CYTOSKELETON

Cytoskeletal proteins provide the driving force for the membrane changes that follow the initial contact between the target particle and membrane receptors and that allow formation of a phagocytic cup and membrane closure. Cytoskeletal activity during phagocytosis can be divided into five steps, as follows.

(1) Filamentous (F)-actin nucleation and polymerization

One of the initial events in cytoskeletal activation is actin nucleation, i.e. the formation of an actin trimer, which serves as a starting point for further polymerization. F-actin bundles are formed by polymerization of monomeric globular (g)-actin. The specific mechanisms for initiation of actin polymerization during Fcγ R- and CR3-mediated phagocytosis are thought to be distinct. Fcγ R-induced Syk activation (see Immunoglobulin Fc receptors, above) leads to recruitment of Vav1, a GTPase exchange factor (GEF) for the GTPase Rac, and Cdc42, both of which promote actin polymerization. In contrast, CR3-induced actin polymerization is mediated not by Rac and Cdc42, but by Rho (Caron & Hall 1998). Interestingly, Rac and Cdc42 are known to control actin polymerization during membrane protrusions, while Rho controls the assembly of contractile actomyosin filaments, which fits well with the "zippering" versus "sinking" hypothesis for Fcγ R- versus CR3-mediated phagocytosis (Ridley *et al.* 1992; Kozma *et al.* 1995; Nobes & Hall 1995; Caron & Hall 1998) discussed above. Although distinct rho-family GTPases are involved in Fcγ R- and CR3-mediated phagocytosis, ingestion through both pathways none the less requires the actin-binding and nucleating proteins of the actin-related protein 2/3 (Arp2/3) complex. Arp2/3 can be recruited and activated by the Wiskott–Aldrich syndrome protein (WASP); by using WASP-deficient phagocytes from patients with Wiskott–Aldrich syndrome, WASP was also shown to be required for Fcγ R-mediated phagocytosis (Lorenzi *et al.* 2000). WASP recruitment of Arp2/3 requires Cdc42, thus potentially explaining the requirement for this GTPase in Fcγ R-mediated ingestion. How Arp2/3 is recruited to the phagocytic cup during CR3-mediated ingestion, which has no Cdc42 requirement, remains unclear. Coronin, another actin-binding protein, is required at least in part for phagocytosis by *Dictyostelium* species (Maniak *et al.* 1995), although uptake of *Mycobacterium marinum* by *Dictyostelium* is unaffected by deletion of coronin (Solomon *et al.* 2003). Actin polymerization is enhanced by severing of actin filaments, which increases the local concentration of barbed filament ends. Gelsolin is a major actin-severing protein in phagocytes; it localizes to phagosomes and is required for Fcγ R- but not CR3-mediated phagocytosis (Serrander *et al.* 2000). Another actin-severing protein is cofilin, which is necessary for phagocytosis of serum-opsonized zymosan (Nagaishi *et al.* 1999).

(2) Interaction between F-actin and the membrane

In order to provide the driving force for membrane changes, newly assembled F-actin uses different mechanisms to associate with the plasma membrane. One such mechanism is provided by myristoylated, alanine-rich C kinase substrate (MARCKS) or MARCKS-related protein (MacMARCKS). These actin-crosslinking proteins can associate with the plasma membrane; their crosslinking activity is inhibited by protein-kinase C-mediated phosphorylation and calcium/calmodium binding (Hartwig *et al.* 1992; Li & Aderem 1992). Cells expressing a dominant negative MacMARCKS show impaired phagocytosis (Zhu *et al.* 1995), but macrophages deficient in both MARCKS and MacMARCKS ingest perfectly normally (Underhill *et al.* 1998). This is another example of disagreement between genetic and biochemical experiments in the analysis of phagocytosis, suggesting that there may be multiple overlapping and redundant pathways that mediate this important function.

Another mechanism for actin–membrane interactions can be provided by integrins. This mechanism is likely involved in CR3-mediated phagocytosis, and in the requirement for CR3 in Fcγ R-mediated phagocytosis discussed above. The β-chain of integrins can associate with a variety of actin-binding proteins, such as α-actinin, filamin, talin, and vinculin. Although it is likely that these associations are required for phagocytosis, no experimental tests of this hypothesis have been reported. Fcγ R-mediated phagocytosis induces phosphorylation of cytoskeleton-associated proteins such as paxillin and focal adhesion kinase (FAK), which are concentrated around the phagocytic cup. The same proteins are also phosphorylated at sites of integrin-mediated adhesion ("focal contacts"), suggesting a potential common mechanism for Fcγ R-mediated phagocytosis and focal contact formation. However, whereas Fcγ R-mediated phagocytosis and integrin-dependent focal contact formation are both inhibited by tyrosine kinase blockade, CR3-mediated ingestion is not, emphasizing the distinction between the molecular pathways for internalization initiated by distinct membrane receptors.

(3) F-actin bundling and rearrangement

Actin polymerization occurs primarily at the extending tips of the pseudopod; at the base of the forming phagosome, there is net depolymerization. The loss of polymerized actin at the base of the phagocytic cup may facilitate phagosome–lysosome fusion. During FcγR-mediated phagocytosis, F-actin filaments are crosslinked within the extending pseudopods by actin-bundling proteins, leading to their rearrangement in parallel bundles. However, little is understood about the mechanisms and significance of actin bundling during phagocytosis.

(4) Connecting signaling components to the phagosome

Following receptor ligation, actin rearrangements are required for intracellular signaling. The cytoskeleton may act as a "scaffold" for assembly of signaling cascades, i.e. bringing kinases, phosphatases, their substrates, and adaptor proteins into close proximity with each other and with the ligated phagocytic receptors. Since the early work of Silverstein, it has been clear that signaling involved in phagocytosis is spatially constrained; both cytoskeleton and lipid microdomains may contribute to this essential localization.

(5) Force generation

Three models have been proposed to explain the mechanical force generated by the cytoskeleton to drive pseudopod formation and particle engulfment. First, actin polymerization by itself may be sufficient for membrane extention and phagocytic cup formation, although this cannot explain the forces required for contractility during particle internalization. Second, osmotic changes, in concert with actin polymerization, may help the cytoplasmic gel to form membrane protrusions (Condeelis 1993). It is noteworthy that the Na^+/H^+ exchanger 1 (NHE1) interacts with proteins of the ezrin/radixin/moesin (ERM) family, which is able to induce actin polymerization at phagosome membranes (Defacque *et al.* 2000; Denker *et al.* 2000). Phagocytosis mediated by FcγR and CR3 correlates with NHE1 activation (Fukushima *et al.* 1996), but it remains unclear whether NHE1 is required for this process. Third, myosin may cooperate with actin in force generation. Myosin has been observed at phagocytic cups, and inhibition of myosin activity inhibits phagocytosis (Stendahl *et al.* 1980; Titus 1999; Mansfield *et al.* 2000). Genetic analysis of phagocytosis by *Dictyostelium* has been very important in the developing understanding of the role of myosins in phagocytosis (Titus 2000). Work on this organism has pointed to essential roles for both class I and class VII myosins in phagocytosis. Most phagosomes ultimately traffic from the plasma membrane to a perinuclear location; this inward migration probably is mediated by microtubules and microtubule motors, which otherwise appear to have little role in the phagocytic process.

CELLULAR PROCESSES FOLLOWING PHAGOCYTOSIS

Phagosome–lysosome fusion

After complete engulfment, separation fom the plasma membrane and entry into the cell, ingested material is localized in acidified cytoplasmic vacuoles (phagosomes). In general, these phagosomes acquire proteolytic activity by a series of interactions and fusion with pre-existing organelles ranging from endosomes to lysosomes, which contain antimicrobial enzymes, proteins, and peptides, promoting killing and degradation of ingested microorganisms.

Phagosome maturation is a process of continuous interactions of the nascent vacuole with the endocytic pathway, starting from the early phagosome, via the late phagosome, to the final stage of the phagolysosome (reviewed in Vieira *et al.* 2002). Phagosome maturation can be followed experimentally by tracking acquisition and loss of a variety of markers of specific compartments within the endosomal pathway. Proteins of the Rab family of small GTPases and lysosomal-associated membrane proteins (LAMPs) have been found on phagosomal membranes and implicated in phagosome maturation. In particular, Rab5 seems to be involved in fusion of phagosomes with early endosomes, and Rab7 mediates fusion with late endosomes (Jahraus *et al.* 1998). Phagosome acquisition of Rab7 is generally accompanied by loss of Rab5. Other potentially important factors in phagosome maturation are the soluble *N*-ethylmaleimide-sensitive factor-attachment protein receptors (SNAREs). SNAREs expressed on vesicles can efficiently interact with SNAREs on target membranes and thus promote fusion, and have been implicated in phagosome maturation (Funato *et al.* 1997). The SNAREs have also been implicated in membrane recycling, i.e. the addition of intracellular membranes to the cell surface to compensate for the loss of plasma membrane during phagocytosis. This has been shown to be important in allowing the maximal rate and extent of ingestion by professional phagocytes (Hackam *et al.* 1998).

Many intracellular pathogens can interfere with normal phagosome maturation. There are pathogens that prevent acidification, leading to failure of fusion with late endosomes and lysosomes, pathogens that induce aberrant mechanisms of ingestion so that the phagosome membrane does not have appropriate signals for fusion with the endosomal pathway, and pathogens that escape phagosomes altogether and live free in the cytoplasm of the ingesting cell. Yet other pathogens appear to be able to survive even within the harsh environment of the phagolysosome. Space limitations preclude a full discussion of the interesting virulence mechanisms by which these pathogen strategies are accomplished; the interested reader is referred to several recent reviews (Sinai & Joiner 1997; Meresse *et al.* 1999; Bitar *et al.* 2004; Kim & Weiss 2004; Koul *et al.* 2004).

Proinflammatory and anti-inflammatory functions

The process of phagocytosis may be followed by a general activation of cellular effector functions including degranulation, respiratory burst, synthesis and release of cytokines, chemokines, and leukotrienes, and modulation of expression of surface molecules. As discussed above, the receptor involved in ingestion has a major role in determining whether this will happen; pathogenic bacteria can regulate these responses as well. These effector functions may contribute to inflammation, activation of neighboring cells and migration of cells from distinct tissues, antigen presentation, and eventually to the eradication of a pathogen from infected tissue. Phagocytosis through Fcγ R is a potent inducer of proinflammatory functions. When inappropriately controlled, Fcγ R-induced cellular effector functions can give rise to tissue damage and autoimmunity (reviewed in Hogarth 2002). Phagocytosis of apoptotic cells, on the other hand, induces active suppression of proinflammatory functions (reviewed in Henson *et al.* 2001). Phagocytosis of apoptotic cells does not result in release of inflammatory mediators, but instead induces the release of prominent anti-inflammatory mediators such as transforming growth factor-β, prostaglandin E2, and interleukin-10. These anti-inflammatory effects are likely related to resolution of inflammation after completing the first line of defense against an invading pathogen.

CONCLUDING REMARKS

In the past few years, our knowledge of the receptors involved in phagocytosis and their molecular mechanisms has significantly improved. Several novel phagocytic receptors have been identified. The development of mouse models with genetic deficiencies in specific phagocytic receptors or their signaling components has allowed a deeper understanding of the biological and molecular functions of phagocytic receptors. Receptor cooperation during phagocytosis has been shown to be important not only for ingestion but also for the activation or regulation of subsequent proinflammatory signals. Since many microbes undoubtedly engage several receptors simultaneously, including opsonin receptors and PAMP receptors, this is likely to be of major biological importance. Understanding of the molecular mechanisms involved in the ingestion process itself is also advancing, but it is clear that this can be an area of great complexity. There are at least two distinct mechanisms for ingestion, which also appear to be distinct in their coupling to downstream effector functions. The significant divergence of the conclusions from biochemical and antibody inhibition experiments on one hand and genetic experiments on the other is probably a manifestation of this complexity. None the less, this is the arena in which further investigation is likely to lead to significant advances in understanding the cell biology of phagocytes, with the potential to lead to novel therapies for infectious and idiopathic inflammatory diseases.

Note added in proof

During the printing process, a few recent discoveries relevant to this chapter were reported. Two new phagocytic receptors were identified: FcγRIV, which is expressed on murine myeloid cells, is FcR γ-chain associated, and primarily recognizes IgG2a and IgG2b (Nimmerjahn *et al.*, *Immunity* 2005; **23**(1): 41–51); and a new C3b/iC3b receptor named CRIg, which is expressed on human and murine macrophages (Helmy *et al.*, *Cell* 2006; **124**(5): 915–27). In addition, genetic deletion in mice showed that Rac is required for both FcγR- and CR3-mediated phagocytosis, whereas Vav is required only for CR3-mediated phagocytosis (Hall *et al.*, *Immunity* 2006; **24**(3): 305–16).

REFERENCES

Agramonte-Hevia J, González-Arenas A, Barrera D, Velasco-Velásquez, M. 2002. Gram-negative bacteria and phagocytic cell interaction mediated by complement receptor 3. *FEMS Immunol. Med. Microbiol.* **34**: 255–66.

Akira S. 2003. Mammalian Toll-like receptors. *Curr. Opin. Immunol.* **15**(1): 5–11.

Allen LA, Aderem A. 1996. Molecular definition of distinct cytoskeletal structures involved in complement- and Fc receptor-mediated phagocytosis in macrophages. *J. Exp. Med.* **184**(2): 627–37.

Appelmelk BJ, van Die I, van Vliet SJ *et al.* 2003. Cutting edge: carbohydrate profiling identifies new pathogens that interact with dendritic cell-specific ICAM-3-grabbing nonintegrin on dendritic cells. *J. Immunol.* **170**(4): 1635–9.

Arnaout MA, Todd RF III, Dana N *et al.* 1983. Inhibition of phagocytosis of complement C3- or immunoglobulin G-coated particles and of C3bi binding by monoclonal antibodies to a monocyte-granulocyte membrane glycoprotein (Mol). *J. Clin. Invest.* **72**(1): 171–9.

Banchereau J, Steinman RM. 1998. Dendritic cells and the control of immunity. *Nature* **392**: 245–51.

Barnes N, Gavin AL, Tan PS *et al.* 2002. FcgammaRI-deficient mice show multiple alterations to inflammatory and immune responses. *Immunity* **16**: 379–89.

Berton G, Laudanna C, Sorio C, Rossi F. 1992. Generation of signals activating neutrophil functions by leukocyte integrins: LFA-1 and gp150/95, but not CR3, are able to stimulate the respiratory burst of human neutrophils. *J. Cell Biol.* **116**(4): 1007–17.

Bharadwaj D, Stein MP, Volzer M, Mold C, Du Clos TW. 1999. The major receptor for C-reactive protein on leukocytes is fcgamma receptor II. *J. Exp. Med.* **190**(4): 585–90.

Billker O, Popp A, Brinkmann V *et al.* 2002. Distinct mechanisms of internalization of *Neisseria gonorrhoeae* by members of the CEACAM receptor family involving Rac1- and Cdc42-dependent and -independent pathways. *EMBO J.* **21**(4): 560–71.

Bitar DM, Molmeret M, Abu Kwaik Y. 2004. Molecular and cell biology of Legionella pneumophila. *Int. J. Med. Microbiol.* **293**(7–8): 519–27.

Blander JM, Medzhitov R. 2004. Regulation of phagosome maturation by signals from toll-like receptors. *Science* **304**: 1014–18.

Bliska JB, Black DS. 1995. Inhibition of the Fc receptor-mediated oxidative burst in macrophages by the *Yersinia pseudotuberculosis* tyrosine phosphatase. *Infect. Immun.* **63**(2): 681–5.

Blystone SD, Graham IL, Lindberg FP, Brown EJ. 1994. Integrin alpha v beta 3 differentially regulates adhesive and phagocytic functions of the fibronectin receptor alpha 5 beta 1. *J. Cell Biol.* **127**(4): 1129–37.

Blystone SD, Slater SE, Williams MP, Crow MT, Brown EJ. 1999. A molecular mechanism of integrin crosstalk: alpha beta 3 suppression of calcium/calmodulin-dependent protein kinase II regulates alpha 5 beta 1 function. *J. Cell Biol.* **145**(4): 889–97.

Bohuslav J, Horejsi V, Hansmann C *et al.* 1995. Urokinase plasminogen activator receptor, beta 2-integrins, and Src-kinases within a single receptor complex of human monocytes. *J. Exp. Med.* **181**(4): 1381–90.

Booth JW, Kim MK, Jankowski A, Schreiber AD, Grinstein S. 2002. Contrasting requirements for ubiquitylation during Fc receptor-mediated endocytosis and phagocytosis. *EMBO J.* **21**(3): 251–8.

Bredius RG, de Vries CE, Troelstra A *et al.* 1993. Phagocytosis of *Staphylococcus aureus* and *Haemophilus influenzae* type B opsonized with polyclonal human IgG1 and IgG2 antibodies. Functional hFc gamma RIIa polymorphism to IgG2. *J. Immunol.* **151**(3): 1463–72.

Bretscher MS. 1992. Circulating integrins: alpha 5 beta 1, alpha 6 beta 4 and Mac-1, but not alpha 3 beta 1, alpha 4 beta 1 or LFA-1. *EMBO J.* **11**(2): 405–10.

Brown EJ. 1991. Complement receptors and phagocytosis. *Curr. Opin. Immunol.* **3**(1): 76–82.

Brown E. 2005 Complement receptors, adhesion, and phagocytosis. In *Molecular Mechanisms of Phagocytosis.* C. Rosales, Ed., pp. 49–57. Georgetown, TX: Landes Bioscience.

Brown GD, Gordon S. 2001. Immune recognition. A new receptor for beta-glucans. *Nature* **413**(6851): 36–7.

Bullock WE, Wright SD. 1987. Role of the adherence-promoting receptors, CR3, LFA-1, and p150,95, in binding of *Histoplasma capsulatum* by human macrophages. *J. Exp. Med.* **165**(1): 195–210.

Burke-Gaffney A, Blease K, Hartnell A, Helewell PG. 2002. TNF-α potentiates C5a-stimulated eosinophil adhesion to human bronchial epithelila cells: a role for $\alpha5\beta1$ integrin. *J. Immunol.* **168**: 1380–8.

Cambi A, Gijzen K, de Vries JM *et al.* 2003. The C-type lectin DC-SIGN (CD209) is an antigen-uptake receptor for *Candida albicans* on dendritic cells. *Eur. J. Immunol.* **33**(2): 532–8.

Cambi A, De Lange F, Van Maarseveen NM *et al.* 2004. Microdomains of the C-type lectin DC-SIGN are portals for virus entry into dendritic cells. *J. Cell Biol.* **164**(1): 145–55.

Capo C, Lindberg FP, Meconi S *et al.* 1999. Subversion of monocyte functions by *Coxiella burnetii*: impairment of the cross-talk between alphavbeta3 integrin and CR3. *J. Immunol.* **163**(11): 6078–85.

Caron E, Hall A. 1998. Identification of two distinct mechanisms of phagocytosis controlled by different Rho GTPases. *Science* **282**(5394): 1717–21.

Chen H, Mocsai A, Zhang H *et al.* 2003. Role for plastin in host defense distinguishes integrin signaling from cell adhesion and spreading. *Immunity* **19**(1): 95–104.

China B, N'Guyen BT, de Bruyere M, Cornelis GR. 1994. Role of YadA in resistance of *Yersinia enterocolitica* to phagocytosis by human polymorphonuclear leukocytes. *Infect. Immun.* **62**(4): 1275–81.

Chouchakova N, Skokowa J, Baumann U *et al.* 2001. Fc gamma RIII-mediated production of TNF-alpha induces immune complex alveolitis independently of CXC chemokine generation. *J. Immunol.* **166**(8): 5193–200.
Chroneos Z, Shepherd VL. 1995. Differential regulation of the mannose and SP-A receptors on macrophages. *Am. J. Physiol.* **269**(6 Pt 1): L721–6.
Chuang FY, Sassaroli M, Unkeless JC. 2000. Convergence of Fc gamma receptor IIA and Fc gamma receptor IIIB signaling pathways in human neutrophils. *J. Immunol.* **164**: 350–60.
Clynes R, Maizes JS, Guinamard R *et al.* 1999. Modulation of immune complex-induced inflammation *in vivo* by the coordinate expression of activation and inhibitory Fc receptors. *J. Exp. Med.* **189**(1): 179–85.
Coggeshall KM, Nakamura K, Phee H. 2002. How do inhibitory phosphatases work? *Mol. Immunol.* **39**(9): 521–9.
Colmenares M, Puig-Kroger A, Pello OM, Corbi AL, Rivas L. 2002. Dendritic cell (DC)-specific intercellular adhesion molecule 3 (ICAM-3)-grabbing nonintegrin (DC-SIGN, CD209), a C-type surface lectin in human DCs, is a receptor for *Leishmania* amastigotes. *J. Biol. Chem.* **277**(39): 36766–9.
Condeelis J. 1993. Life at the leading edge: the formation of cell protrusions. A. *Rev. Cell Biol.* **9**: 411–44.
Cossart P, Sansonetti PJ. 2004. Bacterial invasion: the paradigms of enteroinvasive pathogens. *Science* **304**(5668): 242–8.
Cox D, Dale BM, Kashiwada M, Helgason CD, Greenberg S. 2001. A regulatory role for Src homology 2 domain-containing inositol 5′-phosphatase (SHIP) in phagocytosis mediated by Fc gamma receptors and complement receptor 3 (alpha(M)beta(2); CD11b/CD18). *J. Exp. Med.* **193**(1): 61–71.
Coxon A, Rieu P, Barkalow FJ *et al.* 1996. A novel role for the β2 integrin CD11b/CD18 in neutrophil apoptosis: a homeostatic mechanism in inflammation. *Immunity* **5**: 653–66.
Crowley MT, Costello PS, Fitzer-Attas CJ *et al.* 1997. A critical role for Syk in signal transduction and phagocytosis mediated by Fcgamma receptors on macrophages. *J. Exp. Med.* **186**(7): 1027–39.
Daëron M. 1997. Fc receptor biology. *A. Rev. Immunol.* **15**: 203–34.
Daëron, M, Malbec O, Latour S *et al.* 1993. Distinct intracytoplasmic sequences are required for endocytosis and phagocytosis via murine Fc gamma RII in mast cells. *Int. Immunol.* **5**: 1393–401.
Defacque H, Egeberg M, Habermann A *et al.* 2000. Involvement of ezrin/moesin in de novo actin assembly on phagosomal membranes. *EMBO J.* **19**(2): 199–212.

Denker SP, Huang DC, Orlowski J, Furthmayr H, Barber DL. 2000. Direct binding of the Na–H exchanger NHE1 to ERM proteins regulates the cortical cytoskeleton and cell shape independently of H(+) translocation. *Mol. Cell* **6**(6): 1425–36.

Diamond MS, Alon R, Parkos CA, Quinn MT, Springer TA. 1995. Heparin is an adhesive ligand for the leukocyte integrin Mac-1 (CD11b/CD1). *J. Cell Biol.* **130**(6): 1473–82.

Ding L, Shevach E. 2001. Inhibition of the function of the FcγRIIB by a monoclonal antibody to thymic shared antigen, a Ly-6 family antigen. *Immunology* **104**: 28–36.

Domingo P, Muniz-Diaz E, Baraldes MA *et al.* 2002. Associations between Fc gamma receptor IIA polymorphisms and the risk and prognosis of meningococcal disease. *Am. J. Med.* **112**(1): 19–25.

Dykstra M, Cherukuri A, Sohn HW, Tzeng SI, Pierce SK. 2003. Location is everything: lipid rafts and immune cell signaling. *A. Rev. Immunol.* **21**: 457–81.

Echtenacher B, Mannel DN, Hultner L. 1996. Critical protective role of mast cells in a model of acute septic peritonitis. *Nature* **381**(6577): 75–7.

Edberg JC, Kimberly RP. 1994. Modulation of Fc gamma and complement receptor function by the glycosyl-phosphatidylinositol-anchored form of Fc gamma RIII. *J. Immunol.* **152**(12): 5826–35.

Fallman M, Gullberg M, Hellberg C, Andersson T. 1992. Complement receptor-mediated phagocytosis is associated with accumulation of phosphatidylcholine-derived diglyceride in human neutrophils. Involvement of phospholipase D and direct evidence for a positive feedback signal of protein kinase. *J. Biol. Chem.* **267**(4): 2656–63.

Fallman M, Andersson R, Andersson T. 1993. Signaling properties of CR3 (CD11b/CD18) and CR1 (CD35) in relation to phagocytosis of complement-opsonized particles. *J. Immunol.* **151**(1): 330–8.

Fallman M, Andersson K, Hakansson S *et al.* 1995. *Yersinia pseudotuberculosis* inhibits Fc receptor-mediated phagocytosis in J774 cells. *Infect. Immun.* **63**(8): 3117–24.

Fitzer-Attas CJ, Lowry M, Crowley MT, 2000. Fcgamma receptor-mediated phagocytosis in macrophages lacking the Src family tyrosine kinases Hck, Fgr, and Lyn. *J. Exp. Med.* **191**(4): 669–82.

Friedrichson T, Kurzchalia TV. 1998. Microdomains of GPI-anchored proteins in living cells revealed by crosslinking. *Nature* **394**: 802–5.

Fukushima T, Waddell TK, Grinstein S *et al.* 1996. Na+/H+ exchange activity during phagocytosis in human neutrophils: role of Fcgamma receptors and tyrosine kinases. *J. Cell Biol.* **132**(6): 1037–52.

Funato K, Beron W, Yang CZ, Mukhopadhyay A, Stahl PD. 1997. Reconstitution of phagosome-lysosome fusion in streptolysin O-permeabilized cells. *J. Biol. Chem.* **272**(26): 16147–51.

Gardai SJ, Xiao YQ, Dickinson M *et al.* 2003. By binding SIRPalpha or calreticulin/CD91, lung collectins act as dual function surveillance molecules to suppress or enhance inflammation. *Cell* **115**(1): 13–23.

Gatfield J, Pieters J. 2000. Essential role for cholesterol in entry of mycobacteria into macrophages. *Science* **288**(5471): 1647–50.

Geertsma MF, Nibbering PH, Haagsman HP, Daha MR, van Furth R. 1994. Binding of surfactant protein A to C1q receptors mediates phagocytosis of *Staphylococcus aureus* by monocytes. *Am. J. Physiol.* **267**(5 Pt 1): L578–84.

Geijtenbeek TB, Kwon DS, Torensma R *et al.* 2000. DC-SIGN, a dendritic cell-specific HIV-1-binding protein that enhances trans-infection of T cells. *Cell* **100**(5): 587–97.

Geijtenbeek TB, van Vliet SJ, Engering A, 't Hart BA, van Kooyk Y. 2004. Self- and nonself-recognition by C-type lectins on dendritic cells. *A. Rev. Immunol.* **22**: 33–54.

Graham IL, Lefkowith JB, Anderson DC, Brown EJ. 1993. Immune complex-stimulated neutrophil LTB4 production is dependent on beta2 integrins. *J. Cell Biol.* **120**: 1509–17.

Green JM, Schreiber AD, Brown EJ. 1997. Role for a glycan phosphoinositol anchor in Fc gamma receptor synergy. *J. Cell Biol.* **139**: 1209–17.

Greenberg S, Silverstein SC. 1993. Phagocytosis. In *Fundamental Immunology*. W Paul, Ed., pp. 941–64. New York: Raven Press.

Greenberg S, Chang P, Silverstein SC. 1993. Tyrosine phosphorylation is required for Fc receptor-mediated phagocytosis in mouse macrophages. *J. Exp. Med.* **177**(2): 529–34.

Gresham HD, Graham IL, Anderson DC, Brown EJ. 1991. Leukocyte adhesion-deficient neutrophils fail to amplify phagocytic function in response to stimulation. Evidence for CD11b/CD18-dependent and -independent mechanisms of phagocytosis. *J. Clin. Invest.* **88**(2): 588–97.

Gresham HD, Dale BM, Potter JW *et al.* 2000. Negative regulation of phagocytosis in murine macrophages by the Src kinase family member, Fgr. *J. Exp. Med.* **191**(3): 515–28.

Griffin FM Jr, Silverstein SC. 1974. Segmental response of the macrophage plasma membrane to a phagocytic stimulus. *J. Exp. Med.* **139**(2): 323–36.

Griffin FM Jr, Griffin JA, Leider JE, Silverstein SC. 1975. Studies on the mechanism of phagocytosis. I. Requirements for circumferential attachment of particle-bound ligands to specific receptors on the macrophage plasma membrane. *J. Exp. Med.* **142**(5): 1263–82.

Griffin FM Jr, Griffin JA, Silverstein SC. 1976. Studies on the mechanism of phagocytosis. II. The interaction of macrophages with anti-immunoglobulin IgG-coated bone marrow-derived lymphocytes. *J. Exp. Med.* **144**(3): 788–809.

Gyetko MR, Sud S, Kendall T *et al.* 2000. Urokinase receptor-deficient mice have impaired neutrophil recruitment in response to pulmonary *Pseudomonas aeruginosa* infection. *J. Immunol.* **165**(3): 1513–19.

Hackam DJ, Rotstein OD, Sjolin C *et al.* 1998. v-SNARE-dependent secretion is required for phagocytosis. *Proc. Natl. Acad. Sci. USA* **95**(20): 11691–6.

Hammarstrom S. 1999. The carcinoembryonic antigen (CEA) family: structures, suggested functions and expression in normal and malignant tissues. *Semin. Cancer Biol.* **9**(2): 67–81.

Hartwig JH, Thelen M, Rosen A *et al.* 1992. MARCKS is an actin filament crosslinking protein regulated by protein kinase C and calcium-calmodulin. *Nature* **356**(6370): 618–22.

Hayashi T, Rao SP, Catanzaro A. 1997. Binding of the 68-kilodalton protein of Mycobacterium avium to alpha(v)beta3 on human monocyte-derived macrophages enhances complement receptor type 3 expression. *Infect. Immun.* **65**(4): 1211–16.

Hazenbos, WLW, van den Berg BM, van Furth R. 1993. Very late antigen-5 and complement receptor type 3 cooperatively mediate the interaction between *Bordetella pertussis* and human monocytes. *J. Immunol.* **151**(11): 6274–82.

Hazenbos WL, van den Berg BM, Geuijen CW, Mooi FR, van Furth R. 1995. Binding of FimD on *Bordetella pertussis* to very late antigen-5 on monocytes activates complement receptor type 3 via protein tyrosine kinases. *J. Immunol.* **155**(8): 3972–8.

Hazenbos WL, Gessner JE, Hofhuis FM *et al.* 1996. Impaired IgG-dependent anaphylaxis and Arthus reaction in Fc gamma RIII (CD16) deficient mice. *Immunity* **5**(2): 181–8.

Hazenbos WL, Heijnen IA, Meyer D *et al.* 1998. Murine IgG1 complexes trigger immune effector functions predominantly via Fc gamma RIII (CD16). *J. Immunol.* **161**(6): 3026–32.

Hazenbos WL, Clausen BE, Takeda J, Kinoshita T. 2004. GPI-anchor deficiency in myeloid cells causes impaired FcγR effector functions. *Blood* **104**(9): 2825–31.

Hellwig SM, Hazenbos WLW, van de Winkel JG, Mooi FR. 1999. Evidence for an intracellular niche for *Bordetella pertussis* in broncho-alveolar lavage cells of mice. *FEMS Immunol. Med. Microbiol.* **26**: 203–7.

Hellwig SM, van Spriel AB, Schellekens JF, Mooi FR, van de Winkel JG. 2001a. Immunoglobulin A-mediated protection against *Bordetella pertussis* infection. *Infect. Immun.* **69**(8): 4846–50.

Hellwig SM, Van Oirschot HF, Hazenbos WLW *et al.* 2001b. Targeting to Fcgamma receptors, but not CR3 (CD11b/CD18), increases clearance of *Bordetella pertussis. J. Infect. Dis.* **183**: 871–9.

Henson PM, Bratton DL, Fadok VA. 2001. Apoptotic cell removal. *Curr. Biol.* **11**(19): R795–805.

Hogarth PM. 2002. Fc receptors are major mediators of antibody based inflammation in autoimmunity. *Curr. Opin. Immunol.* **14**(6): 798–802.

Hogg, N, Henderson R, Leitinger B *et al.* 2002. Mechanisms contributing to the activity of integrins on leukocytes. *Immunol. Rev.* **186**: 164–71.

Holmskov U, Thiel S, Jensenius JC. 2003. Collections and ficolins: humoral lectins of the innate immune defense. *A. Rev. Immunol.* **21**: 547–78.

Hunter S, Indik ZK, Kim MK *et al.* 1998. Inhibition of Fcgamma receptor-mediated phagocytosis by a nonphagocytic Fcgamma receptor. *Blood* **91**(5): 1762–8.

Indik ZK, Pan XQ, Huang MM *et al.* 1994. Insertion of cytoplasmic tyrosine sequences into the nonphagocytic receptor Fc gamma RIIB establishes phagocytic function. *Blood* **83**(8): 2072–80.

Indik ZK, Park JG, Pan XQ, Schreiber AD. 1995. Induction of phagocytosis by a protein tyrosine kinase. *Blood* **85**(5): 1175–80.

Ioan-Facsinay A, de Kimpe SJ, Hellwig SM *et al.* 2002. FcgammaRI (CD64) contributes substantially to severity of arthritis, hypersensitivity responses, and protection from bacterial infection. *Immunity* **16**: 391–402.

Isberg RR, Leong JM. 1990. Multiple beta 1 chain integrins are receptors for invasin, a protein that promotes bacterial penetration into mammalian cells. *Cell* **60**: 861–71.

Isberg RR, Hamburger Z, Dersch P. 2000. Signaling and invasin-promoted uptake via integrin receptors. *Microbes Infect.* **2**(7): 793–801.

Ishibashi Y, Claus S, Relman DA. 1994. *Bordetella pertussis* filamentous hemagglutinin interacts with a leukocyte signal transduction complex and stimulates bacterial adherence to monocyte CR3 (CD11b/CD18). *J. Exp. Med.* **180**(4): 1225–33.

Jahraus A, Tjelle TE, Berg T *et al.* 1998. *In vitro* fusion of phagosomes with different endocytic organelles from J774 macrophages. *J. Biol. Chem.* **273**(46): 30379–90.

Jefferis R, Kumararatne DS. 1990. Selective IgG subclass deficiency: quantification and clinical relevance. *Clin. Exp. Immunol.* **81**(3): 357–67.

Jippo T, Morii E, Ito A, Kitamura Y. 2003. Effect of anatomical distribution of mast cells on their defense function against bacterial infections: demonstration using partially mast cell-deficient tg/tg mice. *J. Exp. Med.* **197**(11): 1417–25.

Joiner KA, Ganz T, Albert J, Rotrosen D. 1989. The opsonizing ligand on *Salmonella typhimurium* influences incorporation of specific, but not azurophil, granule constituents into neutrophil phagosomes. *J. Cell Biol.* **109**(6 Pt 1): 2771–82.

Joiner KA, Fuhrman SA, Miettinen HM, Kasper LH, Mellman I. 1990. *Toxoplasma gondii*: fusion competence of parasitophorous vacuoles in Fc receptor-transfected fibroblasts. *Science* **249**: 641–6.

Jones SL, Brown EJ. 1996. Functional cooperation between Fcγ receptors and complement receptors in phagocytes. In *Human IgG Fc Receptors,* JGJ Van de Winkel and PJA Capel, Eds., pp. 149–63. Heidelberg: Springer-Verlag.

Jones SL, Knaus UG, Bokoch GM, Brown EJ. 1998. Two signaling mechanisms for activation of alphaM beta2 avidity in polymorphonuclear neutrophils. *J. Biol. Chem.* **273**(17): 10556–66.

Kaplan MH, Volanakis JE. 1974. Interaction of C-reactive protein complexes with the complement system. I. Consumption of human complement associated with the reaction of C-reactive protein with pneumococcal C-polysaccharide and with the choline phosphatides, lecithin and sphingomyelin. *J. Immunol.* **112**(6): 2135–47.

Kiefer F, Brumell J, Al-Alawi N *et al.* 1998. The Syk protein tyrosine kinase is essential for Fcgamma receptor signaling in macrophages and neutrophils. *Mol. Cell Biol.* **18**(7): 4209–20.

Kim K, Weiss LM. 2004. *Toxoplasma gondii*: the model apicomplexan. *Int. J. Parasitol.* **34**(3): 423–32.

Kim M, Carman CV, Springer TA. 2003. Bidirectional transmembrane signaling by cytoplasmic domain separation in integrins. *Science* **301**(5640): 1720–5.

Koul A, Herget T, Klebl B, Ullrich A. 2004. Interplay between mycobacteria and host signalling pathways. *Nature Rev. Microbiol.* **2**(3): 189–202.

Kozma R, Ahmed S, Best A, Lim L. 1995. The Ras-related protein Cdc42Hs and bradykinin promote formation of peripheral actin microspikes and filopodia in Swiss 3T3 fibroblasts. *Mol. Cell Biol.* **15**(4): 1942–52.

Kushner BH, Cheung NK. 1992. Absolute requirement of CD11/CD18 adhesion Molecules, FcRII and the phosphatidylinositol-linked FcRIII for monoclonal antibody-mediated neutrophil antihuman tumor toxicity. *Blood* **76**(6): 1784–90.

Kuusela P. 1978. Fibronectin binds to *Staphylococcus aureus*. *Nature* **276**(5689): 718–20.

Langlet C, Bernard A-M, Drevot P, He H-T 2000. Membrane rafts and signaling by the multichain immune recognition receptors. *Curr. Opin. Immunol.* **12**: 250–5.

Lee SJ, Zheng NY, Clavijo M, Nussenzweig MC. 2003. Normal host defense during systemic candidiasis in mannose receptor-deficient mice. *Infect. Immun.* **71**(1): 437–45.

Leong JM, Morrissey PE, Marra A, Isberg RR. 1995. An aspartate residue of the *Yersinia pseudotuberculosis* invasin protein that is critical for integrin binding. *EMBO J.* **14**(3): 422–31.

Li J, Aderem A. 1992. MacMARCKS, a novel member of the MARCKS family of protein kinase C substrates. *Cell* **70**(5): 791–801.

Lorenzi R, Brickell PM, Katz DR, Kinnon C, Thrasher AJ. 2000. Wiskott–Aldrich syndrome protein is necessary for efficient IgG-mediated phagocytosis. *Blood* **95**(9): 2943–6.

Malaviya R, Ikeda T, Ross E, Abraham SN. 1996. Mast cell modulation of neutrophil influx and bacterial clearance at sites of infection through TNF-alpha. *Nature* **381**(6577): 77–80.

Malaviya R, Gao Z, Thankavel K, van der Merwe PA, Abraham SN. 1999. The mast cell tumor necrosis factor alpha response to FimH-expressing *Escherichia coli* is mediated by the glycosylphosphatidylinositol-anchored molecule CD48. *Proc. Natl. Acad. Sci. USA* **96**: 8110–15.

Maniak M, Rauchenberger R, Albrecht R, Murphy J, Gerisch G. 1995. Coronin involved in phagocytosis: dynamics of particle-induced relocalization visualized by a green fluorescent protein Tag. *Cell* **83**(6): 915–24.

Mansfield PJ, Shayman JA, Boxer LA. 2000. Regulation of polymorphonuclear leukocyte phagocytosis by myosin light chain kinase after activation of mitogen-activated protein kinase. *Blood* **95**(7): 2407–12.

Marodi L, Korchak HM, Johnston RB Jr. 1991. Mechanisms of host defense against *Candida* species. I. Phagocytosis by monocytes and monocyte-derived macrophages. *J. Immunol.* **146**(8): 2783–9.

Masuda M, Roos D. 1993. Association of all three types of Fc gamma R (CD64, CD32, and CD16) with a gamma-chain homodimer in cultured human monocytes. *J. Immunol.* **151**(12): 7188–95.

Matsuda M, Park JG, Wang DC *et al.* 1996. Abrogation of the Fc gamma receptor IIA-mediated phagocytic signal by stem-loop Syk antisense oligonucleotides. *Mol. Biol. Cell* **7**(7): 1095–106.

Maurer D, Ebner C, Reininger B *et al.* 1995. The high affinity IgE receptor (Fc epsilon RI) mediates IgE-dependent allergen presentation. *J. Immunol.* **154**(12): 6285–90.

Mengaud J, Ohayon H, Gounon P, Mege R-M, Cossart P. 1996. E-cadherin is the receptor for internalin, a surface protein required for entry of L. monocytogenes into epithelial cells. *Cell* **84**(6): 923–32.

Meresse S, Steele-Mortimer O, Moreno E *et al.* 1999. Controlling the maturation of pathogen-containing vacuoles: a matter of life and death. *Nature Cell. Biol.* (7): E183–8.

Meyer D, Schiller C, Westermann J *et al.* 1998. FcgammaRIII (CD16)-deficient mice show IgG isotype-dependent protection to experimental autoimmune hemolytic anemia. *Blood* **92**(11): 3997–4002.

Miranti CK, Leng L, Maschberger P, Brugge JS, Shattil SJ. 1998. Identification of a novel integrin signaling pathway involving the kinase Syk and the guanine nucleotide exchange factor Vav1. *Curr Biol.* **8**(24): 1289–99.

Mitchell MA, Huang MM, Chien P *et al.* 1994. Substitutions and deletions in the cytoplasmic domain of the phagocytic receptor Fc gamma RIIA: effect on receptor tyrosine phosphorylation and phagocytosis. *Blood* **84**(6): 1753–9.

Miyajima I, Dombrowicz D, Martin TR *et al.* 1997. Systemic anaphylaxis in the mouse can be mediated largely through IgG1 and Fc gammaRIII. Assessment of the cardiopulmonary changes, mast cell degranulation, and death associated with active or IgE- or IgG1-dependent passive anaphylaxis. *J. Clin. Invest.* **99**(5): 901–14.

Mold C, Gresham HD, Du Clos TW. 2001. Serum amyloid P component and C-reactive protein mediate phagocytosis through murine Fc gamma Rs. *J. Immunol.* **166**(2): 1200–5.

Monteiro RC, Van de Winkel JG. 2003. IgA Fc receptors. *A. Rev. Immunol.* **21**: 177–204.

Mosser DM, Edelson PJ. 1985. The mouse macrophage receptor for C3bi (CR3) is a major mechanism in the phagocytosis of *Leishmania* promastigotes. *J. Immunol.* **135**(4): 2785–9.

Mosser DM, Edelson PJ. 1987. The third component of complement (C3) is responsible for the intracellular survival of *Leishmania major*. *Nature* **327**: 329–31.

Munro S. 2003. Lipid rafts: elusive or illusive? *Cell* **115**: 377–88.

Nagaishi K, Adachi R, Matsui S *et al.* 1999. Herbimycin A inhibits both dephosphorylation and translocation of cofilin induced by opsonized zymosan in macrophagelike U937 cells. *J. Cell Physiol.* **180**(3): 345–54.

Nakamura A, Takai T. 2004. A role of FcgammaRIIB in the development of collagen-induced arthritis. *Biomed. Pharmacother.* **58**(5): 292–8.

Nakamura K, Malykhin A, Coggeshall KM. 2002. The Src homology 2 domain-containing inositol 5-phosphatase negatively regulates Fcgamma receptor-mediated phagocytosis through immunoreceptor tyrosine-based activation motif-bearing phagocytic receptors. *Blood* **100**(9): 3374–82.

Nobes CD, Hall A. 1995. Rho, rac, and cdc42 GTPases regulate the assembly of multimolecular focal complexes associated with actin stress fibers, lamellipodia, and filopodia. *Cell* **81**(1): 53–62.

Oldenborg PA, Gresham HD, Lindberg FP. 2001. CD47-signal regulatory protein alpha (SIRPalpha) regulates Fcgamma and complement receptor-mediated phagocytosis. *J. Exp. Med.* **193**(7): 855–62.

Ono M, Bolland S, Tempst P, Ravetch JV. 1996. Role of the inositol phosphatase SHIP in negative regulation of the immune system by the receptor Fc(gamma)RIIB. *Nature* **383**(6597): 263–6.

Ouaissi MA, Afchain D, Capron A, Grimaud JA. 1984. Fibronectin receptors on *Trypanosoma cruzi* trypomastigotes and their biological function. *Nature* **308**(5957): 380–2.

Persson C, Carballeira N, Wolf-Watz H, Fallman M. 1997. The PTPase YopH inhibits uptake of *Yersinia,* tyrosine phosphorylation of p130Cas and FAK, and the associated accumulation of these proteins in peripheral focal adhesions. *EMBO J.* **16**: 2307–18.

Peyron P, Bordier C, N'Diaye EN, Maridonneau-Parini I. 2000. Non-opsonic phagocytosis of *Mycobacterium kansasii* by human neutrophils depends on cholesterol and is mediated by CR3 associated with glycosylphosphatidylinositol-anchored proteins. *J. Immunol.* **165**(9): 5186–91.

Pfeiffer A, Böttcher A, Orsó E *et al.* 2001. Lipopolysaccharide and ceramide docking to CD14 provokes ligand-specific receptor clustering in rafts. *Eur. J. Immunol.* **31**: 3153–64.

Pierini L, Holowka D, Baird B. 1996. Fc epsilon RI-mediated association of 6-micron beads with RBL-2H3 mast cells results in exclusion of signaling proteins from the forming phagosome and abrogation of normal downstream signaling. *J. Cell Biol.* **134**(6): 1427–39.

Pommier CG, Inada S, Fries LF *et al.* 1983. Plasma fibronectin enhances phagocytosis of opsonized particles by human peripheral blood monocytes. *J. Exp. Med.* **157**(6): 1844–54.

Pommier CG, O'Shea J, Chused T *et al.* 1984. Studies on the fibronectin receptors of human peripheral blood leukocytes. Morphologic and functional characterization. *J. Exp. Med.* **159**(1): 137–51.

Rambukkana A. 2001. Molecular basis for the peripheral nerve predilection of *Mycobacterium leprae. Curr. Opin. Microbiol.* **4**(1): 21–7.

Rambukkana A, Salzer JL, Yurchenco PD, Tuomanen EI. 1997. Neural targeting of *Mycobacterium leprae* mediated by the G domain of the laminin-alpha2 chain. *Cell* **88**(6): 811–21.

Ratliff TL, McCarthy R, Telle WB, Brown EJ. 1993. Purification of a mycobacterial adhesin for fibronectin. *Infect. Immun.* **61**(5): 1889–94.

Ravetch JV. 2003. Fc receptors. In *Fundamental Immunology,* WE Paul, Ed., pp. 685–700. Philadelphia, PA: Lippincott Williams & Wilkins.

Relman D, Tuomanen E, Falkow S *et al.* 1990. Recognition of a bacterial adhesion by an integrin: macrophage CR3 (alpha M beta 2, CD11b/CD18) binds filamentous hemagglutinin of *Bordetella pertussis. Cell* **61**(7): 1375–82.

Reth M. 1989. Antigen receptor tail clue. *Nature* **338**(6214): 383–4.

Ridley AJ, Paterson HF, Johnston CL, Diekmann D, Hall A. 1992. The small GTP-binding protein rac regulates growth factor-induced membrane ruffling. *Cell* **70**(3): 401–10.

Russell DG, Wright SD. 1988. Complement receptor type 3 (CR3) binds to an Arg-Gly-Asp-containing region of the major surface glycoprotein, gp63, of *Leishmania* promastigotes. *J. Exp. Med.* **168**(1): 279–92.

Saeland E, van Royen A, Hendriksen K *et al.* 2001. Human C-reactive protein does not bind to FcgammaRIIa on phagocytic cells. *J. Clin. Invest.* **107**(5): 641–3.

Sakamoto N, Shibuya K, Shimizu Y *et al.* 2001. A novel Fc receptor for IgA and IgM is expressed on both hematopoietic and non-hematopoietic tissues. *Eur. J. Immunol.* **31**(5): 1310–16.

Salmon JE, Millard SS, Brogle NL, Kimberly RP. 1995. Fc gamma receptor IIIb enhances Fc gamma receptor IIa function in an oxidant-dependent and allele-sensitive manner. *J. Clin. Invest.* **95**: 2877–85.

Saukkonen K, Cabellos C, Burroughs M, Prasad S, Tuomanen E. 1991. Integrin-mediated localization of *Bordetella pertussis* within macrophages: role in pulmonary colonization. *J. Exp. Med.* **173**(5): 1143–9.

Schmitter T, Agerer F, Peterson L, Munzner P, Hauck CR. 2004. Granulocyte CEACAM3 is a phagocytic receptor of the innate immune system that mediates recognition and elimination of human-specific pathogens. *J. Exp. Med.* **199**(1): 35–46.

Schorey JS, Li Q, McCourt DW *et al.* 1995. A *Mycobacterium leprae* gene encoding a fibronectin binding protein is used for efficient invasion of epithelial cells and Schwann cells. *Infect. Immun.* **63**(7): 2652–7.

Schorey JS, Holsti MA, Ratliff TL, Allen PM, Brown EJ. 1996. Characterization of the fibronectin-attachment protein of *Mycobacterium avium* reveals a fibronectin-binding motif conserved among mycobacteria. *Mol. Microbiol.* **21**(2): 321–9.

Serrander L, Skarman P, Rasmussen B *et al.* 2000. Selective inhibition of IgG-mediated phagocytosis in gelsolin-deficient murine neutrophils. *J. Immunol.* **165**(5): 2451–7.

Sharma P, Varma R, Sarasij RC *et al.* 2004. Nanoscale organization of multiple GPI-anchored proteins in living cell membranes. *Cell* **116**(4): 577–89.

Shen L, Lasser R, Fanger MW. 1989. My 43, a monoclonal antibody that reacts with human myeloid cells inhibits monocyte IgA binding and triggers function. *J. Immunol.* **143**(12): 4117–22.

Shibuya A, Sakamoto N, Shimizu Y *et al.* 2000. Fc alpha/mu receptor mediates endocytosis of IgM-coated microbes. *Nat. Immunol.* **1**(5): 441–6.

Shin JS, Abraham SN. 2001. Glycosylphosphatidylinositol-anchored receptor-mediated bacterial endocytosis. *FEMS Microbiol. Lett.* **197**(2): 131–8.

Shin JS, Gao Z, Abraham SN. 2000. Involvement of cellular caveolae in bacterial entry into mast cells. *Science* **289**: 785–8.

Silverstein SC, Steinman RM, Cohn ZA. 1977. Endocytosis. *A. Rev. Biochem.* **46**: 669.

Simon DI, Wei Y, Zhang L *et al.* 2000. Identification of a urokinase receptor-integrin interaction site. Promiscuous regulator of integrin function. *J. Biol. Chem.* **275**(14): 10228–34.

Simons K, Ikonen E. 1997. Functional rafts in cell membranes. *Nature* **387**: 569–72.

Sinai AP, Joiner KA. 1997. Safe haven: the cell biology of nonfusogenic pathogen vacuoles. *A. Rev. Microbiol.* **51**: 415–62.

Sitrin RG, Todd RF III, Albrecht E, Gyetko MR. 1996. The urokinase receptor (CD87) facilitates CD11b/CD18-mediated adhesion of human monocytes. *J. Clin. Invest.* **97**(8): 1942–51.

Solomon JM, Leung GS, Isberg RR. 2003. Intracellular replication of *Mycobacterium marinum* within *Dictyostelium discoideum*: efficient replication in the absence of host coronin. *Infect. Immun.* **71**(6): 3578–86.

Springer TA. 1990. Adhesion receptors of the immune system. *Nature* **346**(6283): 425–34.

Steele C, Marrero L, Swain S *et al.* 2003. Alveolar macrophage-mediated killing of *Pneumocystis carinii* f. sp. *muris* involves molecular recognition by the Dectin-1 beta-glucan receptor. *J. Exp. Med.* **198**(11): 1677–88.

Steinman RM, Moberg CL. 1994. Zanvil Alexander Cohn 1926–1993. *J. Exp. Med.* **179**: 1–30.

Stendahl OI, Hartwig JH, Brotschi EA, Stossel TP. 1980. Distribution of actin-binding protein and myosin in macrophages during spreading and phagocytosis. *J. Cell Biol.* **84**(2): 215–24.

Strzelecka-Kiliszek A, Kwiatkowska K, Sobota A. 2002. Lyn and Syk kinases are sequentially engaged in phagocytosis mediated by Fc gamma R. *Immunology* **169**(12): 6787–94.

Suzuki T, Kono H, Hirose N *et al.* 2000. Differential involvement of Src family kinases in Fc gamma receptor-mediated phagocytosis. *J. Immunol.* **165**(1): 473–82.

Swain SD, Lee SJ, Nussenzweig MC, Harmsen AG. 2003. Absence of the macrophage mannose receptor in mice does not increase susceptibility *to Pneumocystis carinii* infection *in vivo. Infect. Immun.* **71**(11): 6213–21.

Takai T, Li M, Sylvestre D, Clynes R, Ravetch JV. 1994. FcR gamma chain deletion results in pleiotrophic effector cell defects. *Cell* **76**(3): 519–29.

Takai T, Ono M, Hikida M, Ohmori H, Ravetch JV. 1996. Augmented humoral and anaphylactic responses in Fc gamma RII-deficient mice. *Nature* **379**(6563): 346–9.

Talay SR, Valentin-Weigand P, Jerlstrom PG, Timmis KN, Chhatwal GS. 1992. Fibronectin-binding protein *of Streptococcus pyogenes*: sequence of the binding domain involved in adherence of streptococci to epithelial cells. *Infect. Immun.* **60**(9): 3837–44.

Thole JE, Schoningh R, Janson AA *et al.* 1992. Molecular and immunological analysis of a fibronectin-binding protein antigen secreted by *Mycobacterium leprae. Mol. Microbiol.* **6**(2): 153–63.

Titus MA. 1999. A class VII unconventional myosin is required for phagocytosis. *Curr. Biol.* **9**(22): 1297–303.

Titus MA. 2000. The role of unconventional myosins in *Dictyostelium* endocytosis. *J. Eukaryot. Microbiol.* **47**(3): 191–6.

Tran Van Nhieu G, Isberg RR. 1991. The *Yersinia pseudotuberculosis* invasin protein and human fibronectin bind to mutually exclusive sites on the alpha 5 beta 1 integrin receptor. *J. Biol. Chem.* **266**(36): 24367–75.

Tran Van Nhieu G, Isberg RR. 1993. Bacterial internalization mediated by beta 1 chain integrins is determined by ligand affinity and receptor density. *EMBO J.* **12**(5): 1887–95.

Tuijnman WB, Capel PJA, Van de Winkel JGJ 1992. Human low affinity IgG receptor FcgammaRIIa (CD32) introduced into mouse fibroblasts mediates EA-phagocytosis. *Blood* **79**: 1651–6.

Underhill DM, Chen J, Allen LA, Aderem A. 1998. MacMARCKS is not essential for phagocytosis in macrophages. *J. Biol. Chem.* **273**(50): 33619–23.

Underhill DM, Ozinsky A, Hajjar AM *et al.* 1999. The Toll-like receptor 2 is recruited to macrophage phagosomes and discriminates between pathogens. *Nature.* **401**(6755): 811–15.

Van der Berg BM, Van Furth R, Hazenbos WLW. 1999. Activation of complement receptor 3 on human monocytes by cross-linking of very-late antigen-5 is mediated via protein tyrosine kinases. *Immunology* **98**(2) 197–202.

Van der Pol W, Van de Winkel JG. 1998. IgG receptor polymorphisms: risk factors for disease. *Immunogenetics* **48**(3): 222–32.

Van der Pol W, Vidarsson G, Vile HA, Van de Winkel JG, Rodriguez ME. 2000. Pneumococcal capsular polysaccharide-specific IgA triggers efficient neutrophil effector functions via FcalphaRI (CD89). *J. Infect. Dis.* **182**(4): 1139–45.

Van Egmond M, van Garderen E, van Spriel AB *et al.* 2000. FcalphaRI-positive liver Kupffer cells: reappraisal of the function of immunoglobulin A in immunity. *Nat. Med.* **6**(6): 680–5.

Van Furth R, Cohn ZA 1968. The origin and kinetics of mononuclear phagocytes. *J. Exp. Med.* **128**(3): 415–35.

Van Furth R, Cohn ZA, Hirsch JG *et al.* 1972. Mononuclear phagocytic system: new classification of macrophages, monocytes and of their cell line. *Bull. World Health Organ.* **47**(5): 651–8.

Van Furth R, Diesselhoff-den Dulk MMC, Sluiter W, Van Dissel JT. 1985. New perspectives on the kinetics of mononuclear phagocytes. In *Mononuclear Phagocytes: Characteristics, Physiology, and Function* (Proceedings of the Fourth Conference on Mononuclear Phagocytes), R Van Furth, Ed., pp. 201–8. The Hague: Martinus Nijhoff.

Van Iwaarden JF, Pikaar JC, Storm J *et al.* 1994. Binding of surfactant protein A to the lipid A moiety of bacterial lipopolysaccharides. *Biochem. J.* **303**(2): 407–11.

Van Spriel AB, van den Herik-Oudijk IE, van Sorge NM *et al.* 1999. Effective phagocytosis and killing of Candida albicans via targeting FcgammaRI (CD64) or FcalphaRI (CD89) on neutrophils. *J. Infect Dis.* **179**(3): 661–9.

Van Spriel AB, Leusen JH, Van Egmond M *et al.* 2001. Mac-1 (CD11b/CD18) is essential for Fc receptor-mediated neutrophil cytotoxicity and immunologic synapse formation. *Blood* **97**(8): 2478–86.

Van Spriel AB, Leusen JH, Vile H, Van de Winkel JG. 2002. Mac-1 (CD11b/CD18) as accessory molecule for Fc alpha R (CD89) binding of IgA. *J. Immunol.* **169**(7): 3831–6.

Van Spriel AB, Van Ojik HH, Bakker A, Jansen MJ, Van de Winkel JG. 2003. Mac-1 (CD11b/CD18) is crucial for effective Fc receptor-mediated immunity to melanoma. *Blood* **101**(1): 253–8.

Van Strijp JA., Russell, DG. Tuomanen E, Brown EJ, Wright SD. 1993. Ligand specificity of purified complement receptor type three (CD11b/CD18, alpha m beta 2, Mac-1). Indirect effects of an Arg-Gly-Asp (RGD) sequence. *J. Immunol.* **151**: 3324–36.

Van Vugt MJ, Heijnen AF, Capel PJ *et al.* 1996. FcR gamma-chain is essential for both surface expression and function of human Fc gamma RI (CD64) *in vivo. Blood* **87**(9): 3593–9.

Vieira OV, Botelho RJ, Grinstein S. 2002. Phagosome maturation: aging gracefully. *Biochem. J.* **366**(3): 689–704.

Visser LG, Annema A, Van Furth R. 1995. Role of Yops in inhibition of phagocytosis and killing of opsonized *Yersinia enterocolitica* by human granulocytes. *Infect. Immun.* **63**: 2570–5.

Vivier E, Daeron M. Immunoreceptor tyrosine-based inhibition motifs. 1997. *Immunol. Today* **18**(6): 286–91.

Vossebeld PJ, Kessler J, von dem Borne AE, Roos D, Verhoeven AJ. 1995. Heterotypic Fc gamma R clusters evoke a synergistic Ca^{2+} response in human neutrophils. *J. Biol. Chem.* **270**(18): 10671–9.

Vranian G Jr, Conrad DH, Ruddy S. 1981. Specificity of C3 receptors that mediate phagocytosis by rat peritoneal mast cells. *J. Immunol.* **126**(6): 2302–6.

Wang J, Brown EJ. 1999. Immune complex-induced integrin activation and L-complex phosphorylation require protein kinase A. *J. Biol. Chem.* **274**: 24349–56.

Weisbart RH, Kacena A, Schuh A, Golde DW. 1988. GM-CSF induces human neutrophil IgA-mediated phagocytosis by an IgA Fc receptor activation mechanism. *Nature* **332**(6165): 647–8.

Wirth JJ, Kierszenbaum F. 1984. Fibronectin enhances macrophage association with invasive forms of *Trypanosoma cruzi*. *J. Immunol.* **133**(1): 460–4.

Woodside DG, Obergfell A, Leng L, *et al.* 2001. Activation of Syk protein tyrosine kinase through interaction with integrin beta cytoplasmic domains. *Curr. Biol.* **11**(22): 1799–804.

Wright SD, Silverstein SC. 1982. Tumor-promoting phorbol esters stimulate C3b and C3b′ receptor-mediated phagocytosis in cultured human monocytes. *J. Exp. Med.* **156**: 1149–64.

Wright SD, Silverstein SC. 1983. Receptors for C3b and C3bi promote phagocytosis but not the release of toxic oxygen from human phagocytes. *J. Exp. Med.* **158**(6): 2016–23.

Wright SD, Craigmyle LS, Silverstein SC. 1983. Fibronectin and serum amyloid P component stimulate C3b- and C3bi-mediated phagocytosis in cultured human monocytes. *J. Exp. Med.* **158**: 1338–43.

Wyler DJ, Sypek JP, McDonald JA. 1985. In vitro parasite-monocyte interactions in human leishmaniasis: possible role of fibronectin in parasite attachment. *Infect. Immun.* **49**(2): 305–11.

Xia Y, Borland G, Huang J *et al.* 2002. Function of the lectin domain of Mac-1/complement receptor type 3 (CD11b/CD18) in regulating neutrophil adhesion. *J. Immunol.* **169**(11): 6417–26.

Xue W, Kindzelskii AL, Todd RF III, Petty HR. 1994. Physical association of complement receptor type 3 and urokinase-type plasminogen activator receptor in neutrophil membranes. *J. Immunol.* **152**: 4630–40.

Yokota A, Yukawa K, Yamamoto A *et al.* 1999. Two forms of the low-affinity Fc receptor for IgE differentially mediate endocytosis and phagocytosis: identification of the critical cytoplasmic domains. *Proc. Natl. Acad. Sci. USA.* **89**(11): 5030–4.

Yuan R, Clynes R, Oh J, Ravetch JV, Scharff MD, 1998. Antibody-mediated modulation of *Cryptococcus neoformans* infection is dependent on distinct Fc receptor functions and IgG subclasses. *J. Exp. Med.* **187**(4): 641–8.

Zhang JR, Mostov KE, Lamm ME *et al.* 2000. The polymeric immunoglobulin receptor translocates pneumococci across human nasopharyngeal epithelial cells. *Cell* **102**(6): 827–37.

Zhou MJ, Brown EJ. 1994. CR3 (Mac-1, alpha M beta 2, CD11b/CD18) and Fc gamma RIII cooperate in generation of a neutrophil respiratory burst: requirement for Fc gamma RIII and tyrosine phosphorylation. *J. Cell Biol.* **125**(6): 1407–16.

Zhou M, Todd RF III, Van de Winkel JG, Petty HR. 1993. Cocapping of the leukoadhesin molecules complement receptor type 3 and lymphocyte function-associated antigen-1 with Fc gamma receptor III on human neutrophils. Possible role of lectin-like interactions. *J. Immunol.* **150**: 3030–41.

Zhou MJ, Lublin DM, Link DC, Brown EJ. 1995. Distinct tyrosine kinase activation and Triton X-100 insolubility upon Fc gamma RII or Fc gamma RIIIB ligation in human polymorphonuclear leukocytes. Implications for immune complex activation of the respiratory burst. *J. Biol. Chem.* **270**(22): 13553–60.

Zhu Z, Bao Z, Li J. 1995. MacMARCKS mutation blocks macrophage phagocytosis of zymosan. *J. Biol. Chem.* **270**(30): 17652–5.

CHAPTER 3

Receptor-initiated signal transduction during phagocytosis

Kassidy K. Huynh and Sergio Grinstein

INTRODUCTION

Phagocytosis, an essential component of the innate immune response, is a complex process whereby extracellular particles of diameter $\geq 0.5\ \mu m$ are internalized into a specialized membrane-bound vacuolar compartment. Following particle engulfment the resulting vacuoles, better known as phagosomes, undergo a maturation sequence involving fusion and fission events with components of the endocytic pathway. The resulting hybrid compartments, termed phagolysosomes, acquire the molecular machinery necessary to destroy the ingested pathogens.

The process of internalization can be envisaged as involving five distinct steps, beginning with particle recognition, receptor signaling, membrane remodeling and actin polymerization allowing for pseudopod extension, and climaxing with particle uptake into the cytosol. Although the mechanism of internalization is largely conserved among phagocytic cell types, the purpose of phagocytosis is diverse. Unicellular organisms such as amoebae engulf bacteria to obtain nutrients. In mammalian cells, neutrophils and macrophages employ phagocytosis to prevent the spread of infectious agents, but the same process is also employed for the clearance of apoptotic bodies.

Phagocytosis is a receptor-mediated event initiated by ligand recognition and particle binding. Multiple receptors recognize and prompt the elimination of the varied army of pathogenic organisms; a different set of receptors is needed to identify and clear apoptotic bodies. Indeed, the former elicit an immune and inflammatory response, whereas the latter do not.

Phagocytosis and Bacterial Pathogenicity, ed. J. D. Ernst and O. Stendahl. Published by Cambridge University Press.

The objective of this review is to summarize succinctly the current knowledge of the events that lead to phagocytosis. An overview of the different types of phagocytic receptor is presented first, followed by a description of the signaling cascades that they initiate.

PHAGOCYTIC RECEPTORS

Non-opsonic phagocytosis

The simplest innate immune response is initiated by a set of receptors termed pathogen-recognition-receptors (PRRs) that recognize conserved pathogen-associated-molecular patterns (PAMPs). PAMPs are unique to pathogens and do not exist in higher organisms. Moreover, because they are required for pathogen survival, PAMPs are conserved through evolution. PAMP motifs include the mannans of the yeast cell wall, formylated peptides of bacteria, and double-stranded RNA, as well as lipopolysaccharide and lipoteichoic acids on the surface of Gram-negative and Gram-positive bacteria (Aderem & Underhill 1999). These ligands are inherent to the pathogens and phagocytic receptors that recognize PAMPs do not require ancillary factors to associate with them. Such receptors are therefore designated as being non-opsonic.

The β-glucan/Dectin-1 receptor

β-Glucans are a group of long chain polysaccharides that are major components of the cell wall of a variety of fungi, plants and bacteria (Brown & Gordon 2003), yet are never found in animal cells. Non-opsonic recognition of β-glucans can be ascribed to a number of receptors, which include CR3, scavenger receptors and Dectin-1. Of these receptors, only Dectin-1 was found to elicit an immune response to β-glucans, stimulating production of proinflammatory cytokines (Figure 3.1).

Dectin-1 is a small (28 kDa) type-II membrane protein having a C-type lectin domain fold and a cytoplasmic tail with an unpaired immunotyrosine-based activation motif (ITAM) (Ariizumi *et al.* 2000). Within the lectin domain is a carbohydrate recognition domain, which contains an amino acid sequence (W^{221}-I^{222}-H^{223}) that is critical for β-glucan interaction and not present in other C-type recognition receptors (Adachi *et al.* 2004). Dectin-1 binds a variety of β(1,3)-linked and β(1,6)-linked glucans (Brown & Gordon 2001).

Dendritic cells were thought to be the sole type of phagocytes to express Dectin-1 (Ariizumi *et al.* 2000). However, in recent studies Dectin-1 was detected also in macrophages isolated from lung, thymus, and liver (Brown

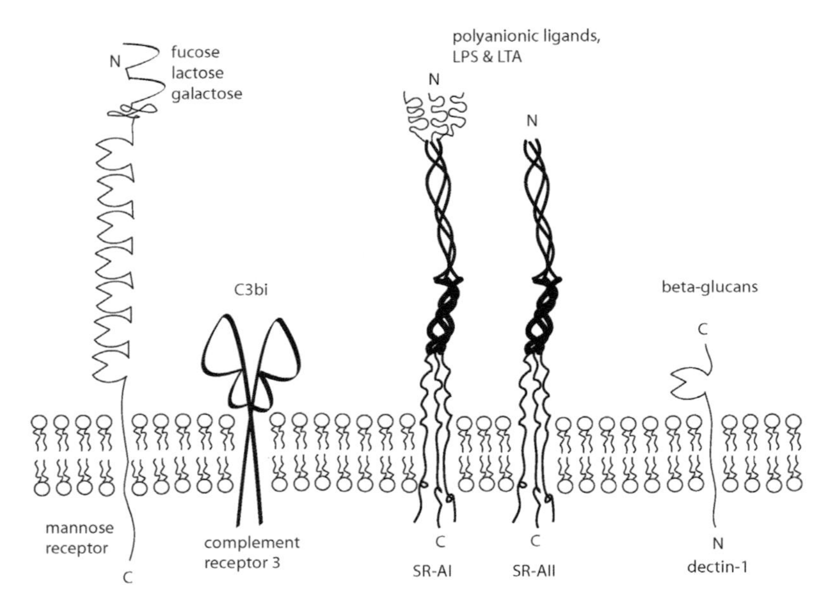

Figure 3.1 Non-opsonic and opsonic phagocytic receptors. Mannose receptors preferentially bind terminal carbohydrate residues that do not exist in mammalian cells and are thus able to differentiate host from invading pathogens. Mannose receptors possess eight tandem C-type lectin domains used to identify pathogens. Dectin-1 binds β-glucans, the major constituent of cell walls of fungi and bacteria. These receptors possess only one C-type lectin domain and are the only group of receptors that upon binding β-glucans can elicit an immune response. SR-A receptors recognize various polyanionic ligands in addition to lipopolysaccharide and lipoteichoic acid, present on Gram-negative and Gram-positive bacteria, respectively. SR-A comprise a collagen-like domain and an α-helical coiled-coil domain. Type I and type II SR-A differ in their N-terminus; only type I contains a cysteine-rich domain. CR3 is an integrin dimer that binds to complement fragment C3bi. Engagement of C3bi initiates "passive" phagocytosis, when the particle sinks into the phagocyte rather than "active" internalization mediated by FcR.

& Gordon 2001). Dectin-1 expressed in NIH3T3 cells was able to bind to *Saccharomyces cerevisiae* and *Candida albicans*, consistent with the expression of β(1,3)- and β(1,6)-linked glucans on the cell walls of these organisms. Dectin-1 can mediate the non-opsonic phagocytosis of heat-killed *C. albicans* and of zymosan, a yeast wall product rich in β-glucans, in an actin-dependent manner (Brown & Gordon 2001) and can elicit a respiratory burst (Brown *et al.* 2003; Gantner *et al.* 2003). Zymosan, which binds to both Dectin-1 and the Toll-like receptor 2 (TLR2) also elicits a proinflammatory-response with activation of NFκB (Brown *et al.* 2003) and production of TNF-α and IL-12 (Brown *et al.* 2003; Gantner *et al.* 2003). Note that, whereas most phagocytic receptors require tyrosine phosphorylation of the ITAM motif, internalization of yeast by macrophages is apparently not dependent on Syk-kinase (Herre *et al.* 2004).

Mannose receptors

The host defense system is armed with a variety of sugar-recognition systems or lectins that are part of the initial innate immune response to infection. Mannose receptors are expressed by a variety of professional phagocytes, including tissue macrophages, dendritic cells, and retinal pigment epithelium, and also in non-professional phagocytes like tracheal smooth muscle cells and kidney mesangial cells. Transfection of the human isoform of the mannose receptor into non-phagocytic COS-1 cells confers the ability to engulf *C. albicans* (Ezekowitz *et al.* 1990). Mannose receptors have been implicated in binding and phagocytosis of additional microorganisms such as Gram-negative and Gram-positive bacteria, fungi, and even protozoa (Linehan *et al.* 2000). All these microorganisms express carbohydrate moieties that interact with mannose receptors to trigger inflammatory responses.

Mannose receptors are 175 kDa type I transmembrane glycoproteins, with an array of eight tandem lectin-like carbohydrate recognition domains (CRDs) and a cytoplasmic domain. The CRDs are thought to facilitate binding of the receptor to oligosaccharides that ornament the surface of pathogens. The specific terminal sugar of oligosaccharides present on the surface of some pathogens determines the affinity of their interaction with the mannose receptor: L-fucose is bound preferentially, followed by D-mannose, with D-*N*-acetylglucosamine and D-galactose displaying comparatively low-affinity interactions (Figure 3.1) (Linehan *et al.* 2000). These terminal sugars are commonly found on the surface of pathogens but are rare in mammalian glycoproteins. This enables mannose receptors to discern self from non-self, invading particles (East & Isacke 2002).

Engagement by mannose receptors in the lung or blood is important for the rapid clearance of microorganisms. This is best illustrated by immunocompromised human hosts that are more susceptible to infection by *Pneumocystis carinii*, a pulmonary pathogen. Alveolar macrophages of such patients display reduced mannose receptor-mediated binding and phagocytosis of these organisms by alveolar macrophages (Koziel *et al.* 1998). Moreover, internalization by macrophages of microspheres coated with lipoarabinomannan, a lipoglycan found on the surface of *Mycobacterium tuberculosis*, is performed predominantly via mannose receptors (Kang & Schlesinger 1998).

Stimulation of mannose receptors mediates release of several pro-inflammatory cytokines, notably IL-1β, IL-6, GM-CSF, TNF-α and IL-12 (Aderem & Underhill 1999). In turn, a variety of cytokines can up- or downregulate the expression and activity of mannose receptors (Aderem & Underhill 1999; East & Isacke 2002; Egan *et al.* 2004; Martinez-Pomares *et al.* 2003; Schreiber *et al.* 1990; Raveh *et al.* 1998), adding an extra layer of complexity to the regulation of the inflammatory response.

Unexpectedly, mannose-receptor-deficient mice were shown to have elevated concentrations of lysosomal hydrolases in their serum (Lee *et al.* 2002). This was interpreted to mean that cell-surface expression and activity of mannose receptors are essential to retrieve secreted lysosomal hydrolases, moderating their extracellular concentration and limiting potential tissue damage.

Scavenger receptors

Scavenger receptors were originally defined by their ability to mediate the uptake of modified lipoproteins such as acetylated low-density lipoproteins. Since their discovery, the family of scavenger receptors has expanded to include several different classes, catalogued as A through F on the basis of structure domain similarities (Gough & Gordon 2000). All scavenger receptors are transmembrane proteins; however, there is no common domain that is shared by all the various classes. Although structurally unrelated, scavenger receptors do share a broad range of polyanionic ligands, including the above mentioned modified low-density lipoproteins. A detailed listing of ligands is beyond the scope of this review, but can be found elsewhere (Gough & Gordon 2000; Platt & Gordon 2001). Most scavenger-receptor ligands studied to date are polyanionic; however, not all polyanions are able to engage these receptors, suggesting that recognition has both charge and structural requirements.

Class A scavenger receptors (SR-A) encompass three splice variants, two of which encode functional isoforms, type I and type II, and the third of which encodes an inert type III protein that is trapped in the endoplasmic reticulum (Gough & Gordon 2000). These receptors are able to bind specifically to microbial cell surface components, particularly the lipopolysaccharide (LPS) of Gram-negative bacteria (Hampton *et al.* 1991) and lipoteichoic acid (LTA) of Gram-positive bacteria (Figure 3.1) (Dunne *et al.* 1994). Administering the soluble form of these substances in excess of pathogen particles can inhibit interactions with SR-A. Collectively, these data indicate that scavenger receptors have the potential to behave as PRRs.

The ability of SR-A to conduct phagocytosis was first observed in studies involving ingestion of apoptotic thymocytes by macrophages (Platt *et al.* 1996). Phagocytic efficiency was greatly decreased when the macrophages were pretreated with a blocking anti-SR-A antibody. Furthermore, SR-$A^{-/-}$ macrophages display a 50% reduction in phagocytosis of apoptotic bodies (Platt *et al.* 1996). Conversely, introduction of SR-A into non-phagocytic cells can confer phagocytosis of apoptotic cells (Platt *et al.* 1998). These data solidify the idea that SR-A can initiate phagocytosis of apoptotic bodies. Studies of SR-A-mediated bacterial phagocytosis, however, reveal a more complicated mechanism for uptake. Whereas expression of the mannose receptor or the complement receptor suffices to reconstitute phagocytosis in nonprofessional phagocytes, expression of SR-A can confer binding of bacteria but fails to promote significant internalization (Peiser *et al.* 2000; Van der Laan *et al.* 1999; Elomaa *et al.* 1995). The participation of SR-A in bacterial phagocytosis was proven definitively when SR-$A^{-/-}$ macrophages were found to display a 40% decrease in the uptake of BCG. These data suggest that, although SR-A is capable of initiating phagocytosis of bacteria, achievement of full phagocytic potential can depend on various factors such as presence of serum proteins and may require the cooperation of other receptors that may not be present in the cells used as recipients of heterologous transfection (Peiser *et al.* 2000).

Although the role of SR-A in phagocytosis remains to be defined precisely, detailed studies of SR-$A^{-/-}$ macrophages revealed that these receptors are essential for innate immunity. SR-A have been implicated in protecting against infection by *Listeria monocytogenes* and herpes simplex virus (Suzuki *et al.* 1997). Moreover, SR-$A^{-/-}$ macrophages are also more sensitive to infection by *Staphylococcus aureus* (Thomas *et al.* 2000) and by *Neisseria meningitidis* (Peiser *et al.* 2002).

Although a role for SR-A in phagocytosis and innate immunity is evident, the specific details of SR-A signaling, and SR-A mediated phagocytosis and bacterial killing, are limited, owing to the lack of SR-A-specific reagents. Nevertheless, it has been reported that engagement of SR-A produces changes in the host cell phosphatidylinositol-4,5-bisphosphate content and fluctuations in intracellular [Ca^{2+}], followed by release of urokinase-type plasminogen activator and nitric oxides and increased transcription of TNF-α via NF-κB (Gough & Gordon 2000).

Macrophage receptor with collagenous structure (MARCO) receptors are also included in the family of Class A receptors and have been implicated to participate in phagocytosis of microbes. MARCO receptors are capable of binding both Gram-positive and Gram-negative bacteria (Van der Laan *et al.* 1999) and latex beads (Palecanda *et al.* 1999). Expression of MARCO receptors is restricted to a subpopulation of macrophages in the spleen and lymph nodes (Van der Laan *et al.* 1999). This expression pattern and the response to microbial surface components suggest that MARCO behaves as a pattern recognition receptor. To further support this contention, antibodies to MARCO were found to inhibit bacterial uptake (Van der Laan *et al.* 1999); however, contradictory data have been presented. MARCO, when heterologously expressed in COS-7 cells, confers only the ability to bind *Escherichia coli* and *S. aureus* (Elshourbagy *et al.* 2000) but cannot induce phagocytosis. Thus, although some of the available data are suggestive of a role for MARCO in pathogen pattern recognition and phagocytosis, other evidence is in apparent disagreement and cannot be ignored. Clearly, further studies are required to validate the role of MARCO as a PRR and its putative function in particle uptake.

CD14

Bacterial endotoxin, a lipopolysaccharide (LPS), is the major structural and highly conserved component found on the surface of most Gram-negative bacteria. LPS is a potent microbial instigator of the proinflammatory signaling cascade. LPS elicits this response by engaging several receptors, including CD14. The latter, which is considered to be one of the primary LPS-receptors, transduces the immune response signals in conjunction with auxiliary components such as Toll-like receptors (TLR), which are discussed below. CD14 is a GPI-anchored protein expressed on the surface of monocytes, macrophages, and neutrophils. LPS associates with LPS-binding protein (LBP) in the extracellular fluid; a ternary complex with CD14 is formed subsequently on the surface of cells.

CD14 also exists in a soluble form (sCD14) in human serum, where receptor levels are especially elevated in septic patients. Like its membrane-bound counterpart, sCD14 can also bind LPS and mediates induction of an inflammatory response. It is noteworthy, however, that the LPS –sCD14 complex can interact with both CD14-positive and CD14-negative cells. This indicates that the LPS–sCD14 complex does not need to interact with CD14 and must engage a different receptor for signal transduction. Indeed, biochemical evidence has been presented documenting the existence of a separate cell surface complex that transduces the signals delivered by LPS–sCD14 (Vita *et al.* 1997).

Although the major ligand for CD14 is LPS, several other bacterial components can engage CD14 and initiate activation of myeloid cells, which is sensitive to blockade by CD14 antibodies. A detailed list of other microbially derived ligands is presented elsewhere (Antal-Szalmas 2000). These alternative ligands are also conserved structural components found on the surface of a variety of pathogens. Thus, CD14 conforms to the requisites of a PRR, having multiple microbial ligand specificity.

The LPS–CD14 complex triggers macrophage activation, culminating in a proinflammatory response involving the production of cytokines such as TNF-α, IL-1, IL-6, IL-8, and IL-12, in addition to mediators including platelet-activating factor, prostaglandins, and nitric oxide. These molecules contribute to the efficient eradication of invading pathogens. Although numerous biochemical and genetic studies have confirmed that CD14 binds LPS, CD14 itself was never observed to directly initiate the signaling cascade. Because, as stated earlier, CD14 lacks a transmembrane, the participation of ancillary components was suspected. CD14 is in fact regarded as a high-affinity conveyor of LPS to signaling receptors. Once coupled to LBP, LPS is attracted to the lipid bilayer by CD14 and is relayed to other signaling components.

The classical model for LPS signal transduction involves a complex including CD14, TLR4, and MD-2. The involvement of TLR4 was suggested from a series of mutational studies (Fujihara *et al.* 2003) and was confirmed when TLR4-null mice were found to be hyporesponsive to LPS challenge (Hoshino *et al.* 1999). TLR2 was also thought to mediate LPS signaling. However, this observation was later proven to be a response to the bacterial lipoproteins and peptidoglycans found in the LPS stocks used, rather than a response to LPS (Fujihara *et al.* 2003; Dobrovolska & Vogel 2002). Although TLR4 is essential for LPS signaling, this receptor is not involved in the initial LPS binding (Dunzendorfer *et al.* 2004) and expression of TLR4 alone in LPS-unresponsive cells (HEK293 and Ba/F3) does not confer LPS sensitivity. Myeloid differentiation protein-2 (MD-2) associates with the extracellular portion of TLR4. Expression of MD-2 alongside TLR4 in LPS-unresponsive cells confers on the cells LPS sensitivity (Shimazu *et al.* 1999), demonstrating a requirement for MD-2 in this activation cluster. The interaction of MD-2 and TLR4 on the surface is thought to prime TLR4 to receive LPS from CD14 and to stabilize the resultant complex (Schmitz & Orso 2002). In addition, MD-2 itself is able to receive LPS molecules and present them to CD14-TLR4 complex (Mancek *et al.* 2002).

The role of CD14 in phagocytosis of microbial particles is not entirely clear. Some evidence suggests that an LBP- and CD14-dependent pathogen clearance mechanism exists. Both the cell-surface and soluble forms of CD14 bind Gram-negative bacteria. Moreover, LBP- and LPS-coated erythrocytes can elicit a phagocytic response from myeloid cells. These data suggest that CD14 may take part in the uptake of Gram-negative bacteria (Antal-Szalmas 2000). Indeed, under non-opsonic conditions, phagocytosis of Gram-negative bacteria was shown to occur via a CD14-mediated pathway (Grunwald *et al.* 1996; Heale *et al.* 2001). To further support a role for CD14 in phagocytosis of microbes, internalization and killing of smooth-type *Salmonella typhimurium* was found to be arrested in CD14-deficient mutants. On the other hand, the same mutants displayed normal phagocytosis and killing of rough-type *S. typhimurium* and *Staphylococcus aureus* (Onozuka *et al.* 1997).

Toll-like receptors (TLR)

Toll receptors were first identified in *Drosophila melanogaster* and shown to be crucial for development. They were subsequently described as essential for the innate immune response against fungal infections. Following the discovery of TLR in mammalian systems, it was thought that each TLR recognizes a specific group of microbial structures and that, collectively, TLR would be able to detect most microbes. To date, eleven TLR have been reported in mammals, where they sense organisms ranging from protozoa to bacteria, fungi, and viruses (Beutler 2004). Detailed compilations of TLR ligands have been presented in recent reviews and will not be presented here (Akira & Takeda 2004; Takeda *et al.* 2003). The primary function ascribed to TLR is to transduce signals leading to stimulation of protein kinases and activation of transcription factors, culminating in proinflammatory responses. The downstream cascade triggered by TLR in macrophages involves the recruitment of the adaptor protein MyD88, followed by association with TNF-α receptor-associated factor-6 (TRAF6) and receptor-interacting protein-2 (RIP-2), an adaptor protein, and activation of the serine kinase interleukin-1 receptor-associated Ser/Thr kinase (IRAK). These molecules signal the translocation of NFκ-B to activate genes encoding proinflammatory cytokines (O'Neill 2002).

Both TLR-signaling and phagocytosis are hallmarks of the immune response, although the relationship between the two is unclear. TLR1 (Ozinsky *et al.* 2000), TLR2 (Ozinsky *et al.* 2000; Underhill *et al.* 1999), TLR4 (Shiratsuchi *et al.* 2004), and TLR6 (Ozinsky *et al.* 2000) are recruited to the phagosome; both TLR2 and TLR4 have been implicated in phagosome maturation (Shiratsuchi *et al.* 2004; Blander & Medzhitov 2004). Interestingly, recent data show that binding of TLR can promote phagocytosis of bacteria that is dependent on MyD88 signaling, causing upregulation of scavenger receptors (Doyle *et al.* 2004). In support of this notion, MyD88-mediated TLR signaling is needed for efficient phagocytosis of bacteria but not of apoptotic bodies or latex beads, especially in the absence of serum (Blander & Medzhitov 2004). This signaling pathway initiates an inducible maturation pathway that results in accelerated delivery of lysosomes compared with the constitutive classical pathway. In agreement with this new role, TLR ligation was reported to modulate actin-driven processes in dendritic cells, including membrane ruffling and macropinocytosis (West *et al.* 2004). Clearly, the contribution of TLR to phagosome formation and maturation is an important notion that requires additional study.

Opsonic phagocytosis

Whereas non-opsonic phagocytosis is dependent on ligands that are inherent to the pathogen, opsonic phagocytosis requires priming of the target with a coat of opsonins, host-derived products that make the particles suitable for phagocytosis. Opsonins are host serum proteins and comprise mainly complement fragments and IgG antibodies that are recognized by specific receptors on the surface of phagocytes.

Complement receptors

Complement fragments C3b and C3bi are formed from the complement component C3 and bind to complement receptors (CR) of leukocytes. There are four types of complement receptors: CR1–CR4. CR1 is a single-pass transmembrane glycoprotein and is thought to facilitate particle binding, but is not by itself competent to initiate phagocytosis; it is thought to act coordinately with Fc receptors to facilitate phagocytosis of mannan-binding lectin and IgG-opsonized bacteria in polymorphonuclear leukocytes (Ghiran *et al.* 2000). Activation of CR1 appears to be dependent on protein kinase C (PKC) and calcium mobilization (O'Shea *et al.* 2005). CR2 is not thought to participate in phagocytosis.

CR3 and CR4 are integrins with heterodimers of differing α-chains, CD11b and CD11c, respectively, and a shared β-chain, CD18. These two complement receptors mediate phagocytosis of C3bi-opsonized particles exclusively (Figure 3.1). However, this requires prior "inside-out" activation of the receptors (Underhill & Ozinsky 2002). Activation can be induced with PKC agonists such as phorbol myristate acetate, which phosphorylate essential serine residues of the β-subunit of CR3, increasing its affinity, and additionally facilitate the receptor clustering needed for CR3-mediated phagocytosis. Activation of CR3 appears to be regulated by the stimulation of Rap1, a small GTPase of the Ras family (Caron *et al.* 2000).

Phospholipase Cγ (PLCγ) (Makranz *et al.* 2004), along with phosphatidylinositol 3-kinase (PI3K) activity (Makranz *et al.* 2004; Lutz & Correll 2003), has recently been proposed to mediate particle internalization via CR3. The role of PI3K is consistent with a previous report showing that SHIP is recruited to CR3-phagoctyic cups during receptor clustering and is able to negatively regulate CR3-mediated phagocytosis by suppressing downstream PI3K-dependent pathways (Cox *et al.* 2001). PKC activity during CR3-mediated phagocytosis has also been reported and is associated with stimulation of phospholipase D (PLD) (Serrander *et al.* 1996; Della Bianca *et al.* 1991), creating a potential positive feedback loop (Fallman *et al.* 1992). During phagocytosis, PLD cleaves phosphatidylcholine, producing phosphatidic acid, which can in turn be further modified to form diacylglycerol (DAG). DAG is thus able to further activate PKC. Recent evidence indicates that both PLD1 and PLD2 contribute in an additive manner to regulate complement-mediated phagocytosis (Iyer *et al.* 2004). Phospholipase A_2 (PLA_2) activation may also contribute to CR3-mediated phagocytosis (Traynor & Authi 1981). Uptake of C3bi-opsonized zymosan particles induced production of arachidonic acid (Fernandez *et al.* 2003; Della Bianca *et al.* 1990) supporting a role for PLA_2, although the opposite effect was reported by others (Aderem *et al.* 1985).

In macrophages, CR-initiated phagocytosis is characterized by the absence of lamellipodia and by its dependence on Rho activation, with little or no involvement of Cdc42 or Rac1. Complement-opsonized particles appear to "sink" into the cell and are destroyed without the production of inflammatory cytokines. In contrast, activation of both Cdc42 and Rac1 has been reported during C3bi-stimulated phagocytosis in neutrophils (Forsberg *et al.* 2003; Caron & Hall 1998). It is unlikely that completely different processes are involved in these closely related types of leukocyte, and further studies are needed to clarify these apparent inconsistencies.

Although CR3-mediated phagocytosis is a non-inflammatory process, inflammatory mediators, such as TNF-α and GM-CSF, or adhesion to laminin- or fibronectin-coated surfaces, can themselves stimulate complement receptor activation, and thus phagocytosis (Pommier *et al.* 1983; Wright *et al.* 1983).

Fcγ receptors

The majority of our knowledge of the signaling events that mediate phagocytosis stems from studies on Fcγ receptors (FcγR). Fcγ receptors are immunoreceptors of the immunoglobulin (Ig) superfamily that possess extracellular Ig domains that interact with IgG-opsonized particles.

IgG molecules, present in serum, bind to foreign particles via their F(ab)2 portion, leaving the Fc moiety exposed and available for interaction with FcγR.

FcγR are categorized into three classes (FcγRI, FcγRII, and FcγRIII) according to the number of extracellular Ig domains they possess. FcγRI isoforms A, B, and C possess three Ig domains, resulting in high-affinity interactions with monomeric IgG. Both FcγRII and FcγRIII have only two IgG domains and interact with IgG with lower affinity. Stable interactions are only obtained by association with multimeric IgG, a consequence of the high avidity of the complex. Three isoforms have also been identified for FcγRII, designated A, B, and C, whereas only two isoforms are reported to exist for FcγRIII, designated A and B. All three classes exhibit different expression profiles, which are detailed elsewhere (Garcia-Garcia & Rosales 2002).

SIGNAL TRANSDUCTION

Although at the macroscopic level phagocytosis induced by different receptors may appear to be similar, there is increasing evidence that considerable differences exist in the signal transduction pathways and even in the mechanical elements involved in the engulfment process. A comprehensive review of the similarities and differences between different types of phagocytosis is not only beyond the scope of this chapter, but would be woefully incomplete, since little is known about most types of phagocytic processes. The exception is FcγR-mediated phagocytosis, which has been studied most extensively and has provided most of our current mechanistic insights (Figure 3.2). The sections below are therefore restricted to the analysis of the signals elicited by FcγR.

Tyrosine phosphorylation

FcγR signal through immunoreceptor tyrosine-based activation motifs (ITAMs), which have the consensus sequence YXXL$(X)_{7/12}$YXXL. Phosphorylation of the ITAM tyrosine residues by Src family kinases is crucial to the successful transduction of signals by FcγR (Ibarrola *et al.* 1997). ITAMs are an intrinsic motif of the cytoplasmic domain of FcγRIIA/C, but are not inherent in the other types of FcγR. FcγRI and FcγRIII require the assistance of ancillary γ- or ζ-chains that contain the ITAM motif needed for signaling. It is noteworthy that FcγRIIB is different from the other FcγR. FcγRIIB possesses an immunoreceptor tyrosine inhibitory motif (ITIM) that negatively regulates ITAM-induced signaling.

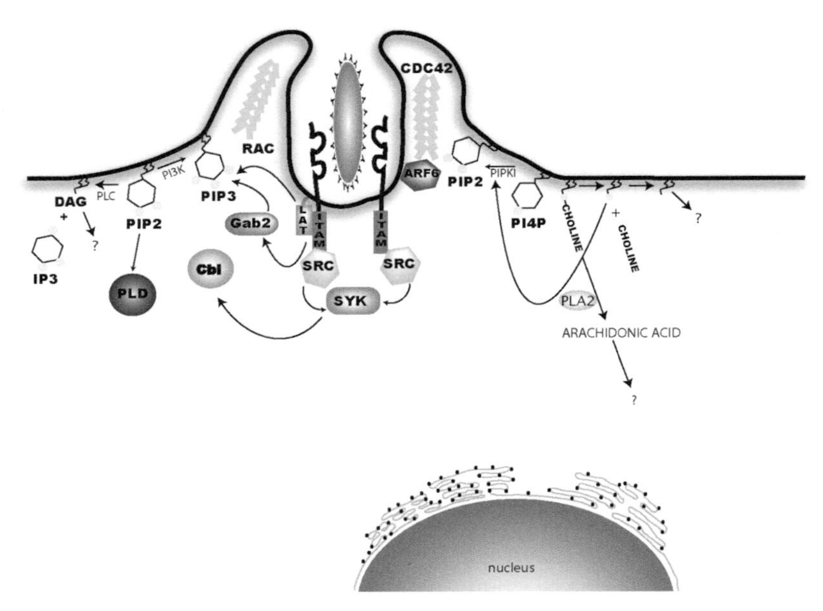

Figure 3.2 Fcγ receptor-mediated signaling. IgG stimulation results in Fcγ receptor clustering followed by phosphorylation of the ITAM motif by Src kinases. Phosphorylated tyrosines serve as docking sites for Syk kinase, initiating a cascade of responses involving adaptor proteins, serine/threonine kinases, phosphoinositides, and proteins that modify these lipids. These molecules generate synergistic signals culminating in pseudopodial extension and particle internalization. See text for details.

Upon addition of IgG-opsonized particles, clustering of FcγR takes place at the phagocytic cup. Cross-linking of FcγR can also be accomplished by using anti-receptor antibodies, resulting in the formation of patches and caps. In both phagocytic cups and receptor caps, FcγR are found to be located in lipid rafts; there is emerging evidence that these rafts may partake in signaling (Kono *et al.* 2002; Barabe *et al.* 2002; Kwiatkowska *et al.* 2003; Bodin *et al.* 2003). It has been reported that dispersal of the rafts prevents FcγR signaling (Bodin *et al.* 2003; Kwiatkowska & Sobota 2001; Katsumata *et al.* 2001), even though receptor clustering is unaffected (Kwiatkowska & Sobota 2001). However, this finding is not universal (Kwiatkowska *et al.* 2003).

Receptor clustering is followed by ITAM tyrosine phosphorylation by activated Src-family kinases. Phagocytes express six members of the Src kinase family: Fgr, Fyn, Hck, Lyn, Yes, and Src (Garcia-Garcia & Rosales 2002). Macrophages isolated from Fgr/Hck/Lyn-deficient mice exhibit a marked reduction in phagocytic efficiency, associated with a decrease in overall tyrosine phosphorylation, implicating these kinases in the activation event (Fitzer-Attas *et al.* 2000). The residual phagocytic activity in these cells could be due to the expression of other Src kinases that may act as a substitute. Dispersal of Lyn from lipid rafts prevents Fcγ R phosphorylation and disrupts downstream signaling (Kwiatkowska *et al.* 2003), further supporting the concept that the activity of Src-family kinases is required.

Phosphorylation of ITAM motifs by Src-family kinases is followed shortly by the recruitment of yet another tyrosine kinase, Syk (Strzelecka-Kiliszek *et al.* 2002; Strzelecta-Kiliszek & Sobota 2002). The phosphotyrosine moieties generated by phosphorylation of the ITAM provide docking sites for the dual SH2 domains present on Syk (Turner *et al.* 2000; Cooney *et al.* 2001; Kiefer *et al.* 1998). After docking onto the receptor Syk is itself phosphorylated, most likely also by Src-family kinases, resulting in its activation (Garcia-Garcia & Rosales 2002). Several lines of evidence indicate that Syk activity is essential for induction of phagocytosis. Co-expression of Syk and Fcγ R in COS cells enhances phagocytic capacity from cells expressing Fcγ R alone (Hunter *et al.* 1999). Moreover, expression of a transmembrane chimera bearing the catalytic domain of Syk is sufficient to induce phagocytosis in COS cells (Greenberg *et al.* 1996). Most importantly, monocytes and macrophages lacking Syk expression are defective in Fcγ R-mediated phagocytosis (Kiefer *et al.* 1998; Crowley *et al.* 1997). Although unable to internalize particles, binding of IgG-opsonized particles is unaffected in these Syk-deficient cells (Kiefer *et al.* 1998; Crowley *et al.* 1997). There is some discrepancy in the literature as to whether Syk is required or dispensable for actin assembly at the phagocytic cup (Crowley *et al.* 1997; Cox *et al.* 1996). The events that are triggered by Syk are not entirely defined, but include the activation of Cbl (Kant *et al.* 2002; Sato *et al.* 1999; Haseki *et al.* 1999), phosphatidylinositol 3-kinase (PI3K), phospholipase C (PLC), and mitogen-activated protein kinases (MAPKs) (Raeder *et al.* 1999).

Negative regulation is required to modulate the phagocytic response and takes place in part by the action of phosphatases. One such regulator is the Src homology 2-containing phosphatase 1 (SHP-1), a protein tyrosine phosphatase. SHP-1 is recruited to sites of phagocytosis, associating with the ITAM of Fcγ RIIA, and becomes activated after receptor clustering (Ganesan *et al.* 2003). SHP-1 also associates with Syk: the net result of these interactions is a decrease in total cellular phosphorylation (Ganesan *et al.* 2003). Dephosphorylation exerts a negative effect on phagocytosis, as evinced by the over-expression of wild-type or constitutively active SHP-1 (Huang *et al.* 2003). SHP-1 activation reduces the interaction between receptors, adaptor proteins, and downstream elements of the signaling cascade.

Adaptor proteins

Adaptor proteins serve to link upstream signaling molecules to downstream effectors. A number of adaptor molecules have been implicated in conveying phagocytic signals, including Gab2/Grb2, Cbl/Nck/Crkl, LAT, Felic (CIP4b), SLP-76, and BLNK/SLP-65. Cbl is a cytosolic protein that is thought to mediate some aspects of Fcγ R signaling. During Fcγ RII clustering Cbl is rapidly tyrosine phosphorylated by Syk (Haseki *et al.* 1999; Matsuo *et al.* 1996). and associates with both Syk itself and with the p85 subunit of PI3K (Haseki *et al.* 1999; Momose *et al.* 2003). Additionally, transforming mutants of Cbl are able to enhance Fcγ R-mediated phagocytosis. Cbl interactions with other adaptor proteins have also been documented in myeloid cells. Interactions between Cbl and Nck, Grb2, and Crkl have been reported and postulated to mediate Fcγ RI signaling (Kyono *et al.* 1998; Park *et al.* 1998; Izadi *et al.* 1998).

Gab2, a scaffolding adaptor, is also essential for phagocytosis. Upon Fcγ R stimulation, Gab2 is recruited to the nascent phagosome and becomes tyrosine phosphorylated by Lyn and/or other Src kinases. Gab2 activation allows for associations with the p85 subunit of PI3K and with SHP-2. The phagocytic defect observed in $Gab2^{-/-}$ macrophages is thought to be a result of interruption of pathways downstream of PI3K, and Gab2 was proposed to control local production of PIP_3 at the phagocytic cup (Gu *et al.* 2003).

LAT, linker for activation of T cells, is a membrane-associated adaptor found primarily in lipid rafts, which is essential for successful signaling by T-cell receptors. LAT has also been implicated in phagocytosis. Fcγ R stimulation results in tyrosine phosphorylation of LAT, allowing for the recruitment of SH2-containing proteins (Tridandapani *et al.* 2000). Unlike other adaptors, LAT was found to be constitutively associated with Fcγ R; upon receptor activation LAT phosphorylation induces recruitment of the p85 subunit of PI3K, PLCγ 1, and Grb2 (Tridandapani *et al.* 2000), consistent with data obtained from T-cells (W. Zhang *et al.* 1998). Studies from LAT-deficient macrophages revealed that LAT serves to enhance Fcγ R-mediated phagocytosis (Tridandapani *et al.* 2000).

Fcγ R signaling is also coupled to phosphorylation of the scaffolding adaptors BLNK/SLP-65 and SLP-76 (Bonilla *et al.* 2000). SLP-76 was observed in a multimeric complex that is recruited to forming phagosomes. This complex consists of proteins that may link Fcγ R-signaling to actin polymerization (Coppolino *et al.* 2001). The significance of SLP-76 in particle engulfment is questionable, since SLP76-deficient macrophages do not display any defects in phagocytic efficiency (Banilla *et al.* 2000). Similarly, macrophages devoid of both SLP-65 and SLP-76 appear to exhibit normal Fcγ R activation and

FcγR-mediated phagocytosis (Nichols *et al.* 2004). Therefore, expression of neither adaptor is essential for successful signal transduction.

Felic/CIP4b, a cytoskeleton scaffolding adaptor protein, is also recruited to the phagocytic cup, where it is believed to link Src kinases to Cdc42 activated pathways, although the underlying mechanisms and significance remain to be clarified (Dombrosky-Ferlan *et al.* 2003).

Lipid metabolism

Signal transduction was traditionally thought to be mediated largely, if not entirely, through protein–protein interactions. However, recent research indicates that lipids play a critical role in signaling actin remodeling and membrane dynamics. During phagocytosis, a series of lipid-modifying enzymes is activated, including kinases, phosphatases and lipases, which can generate, modify, or eliminate lipids that coordinate signal transduction. A brief overview of lipid metabolism in phagocytosis is presented below, with particular emphasis on phosphoinositides (PI), which play a pre-eminent role in the process.

Phosphoinositides

Phosphoinositides are composed of a diacylglycerol backbone, with two acyl chains of variable length, and an inositol ring linked via a phosphodiester bond at position 1′. Only positions 3′ through 5′ of the inositol ring are subject to changes by the actions of phosphatidylinositol kinases or phosphatidylinositol phosphatases, yielding seven possible species of differentially phosphorylated phosphoinositide: phosophatidylinositol-3-phosphate (PI3P), phosphatidylinositol-4-phosphate (PI4P), phosphatidylinositol-5-phosphate (PI5P), phosphatidylinositol-3,4-bisphosphate ($PI(3,4)P_2$), phosphatidyl-3,5-bisphosphate ($PI(3,5)P_2$), phosphatidylinositol-4,5-bisphosphate ($PI(4,5)P_2$), and phosphatidylinositol-3,4,5-trisphosphate (PIP_3). The multiplicity of these chemical species, in addition to their differential ability to interact with proteins, allow for a variety of possible signals and cellular roles. Proteins interact with phosphoinositides via specialized high-affinity binding domains that are specific for the phosphoinositide. These domains serve to recruit and anchor the protein to the membranes bearing the specific inositide. Of the seven species present, only $PI(4,5)P_2$, $PI(3,4)P_2$, and PIP_3 have been documented to be involved in phagocytosis (although PI3P is involved in phagosome maturation).

In resting phagocytes, $PI(4,5)P_2$ is found primarily on the plasma membrane and the Golgi network. The primary source of $PI(4,5)P_2$ is the phosphorylation of PI4P by phosphatidylinositol-4-phosphate 5-kinase type I (PIPKI), also known as PI4P5K. Currently, three isoforms, α, β, and γ, of PIPKI have been cloned. The activity of all three isoforms of PIPKI is regulated by the Rho GTPases, RhoA, Rac and Cdc42 (Weernink *et al.* 2004), as well as by Arf6 (Brown *et al.* 2001) and phosphatidic acid, a product of phospholipase D (PLD) (Skippen *et al.* 2002). Notably, all the molecules that participate in PIPKI regulation are known to contribute to signaling during phagocytosis.

Consumption of $PI(4,5)P_2$ can occur by degradation, through the action of phosphoinositide 5′-phosphatases, by conversion to PIP catalyzed by PI3K, or by cleavage mediated by PLC isoforms to form DAG and IP_3, both secondary signaling molecules themselves.

The role of $PI(4,5)P_2$ in signaling phagocytosis is only now starting to be elucidated. By using the PH domain of PLCδ as a probe, the presence of $PI(4,5)P_2$ at sites of contact with the phagocytic particle was found to be biphasic, initially accumulating locally at the phagocytic cup and extending pseudopods, followed by rapid disappearance at the base of the cup even before internalization is complete (Figure 3.3A) (Botelho *et al.* 2000). The initial phase suggests a local stimulation of $PI(4,5)P_2$ production. Indeed, this idea is supported by the transient accumulation of PIPKIα at the phagocytic cup. The activity of this enzyme also appears to be important, as expression of a catalytically inactive mutant impairs particle internalization (Coppolino *et al.* 2002), consistent with observations that sequestering $PI(4,5)P_2$ also attenuates the process (Botelho *et al.* 2000). $PI(4,5)P_2$ likely participates in phagocytosis by modulating actin assembly. A detailed account of the data supporting this role was recently given (Hilpela *et al.* 2004) and will not be belabored here. Because of the regulatory role of $PI(4,5)P_2$ in actin assembly, the disappearance of the inositide from the forming phagosome is likely to coincide with, and possibly contribute to, the disassembly of actin from the cup, allowing the phagosome to depart from the plasma membrane into the cytosol. The loss of $PI(4,5)P_2$ during this stage is likely due to the combined action of PLC and PI3K.

There are three classes of PI3K, of which only class I has been documented to be involved in phagocytosis, while class III (vps34) is involved in phagosome maturation. The main substrate of class I PI3K is $PI(4,5)P_2$, forming PIP_3. PI3K activity is needed for internalization mediated by all classes of FcγR and seemingly also for complement-mediated phagocytosis. PI3K was first implicated in phagocytosis when it was observed that phagocytic activity is antagonized by wortmannin, an irreversible inhibitor of the enzyme (Ninomiya *et al.* 1994). In addition, expression of a transmembrane chimera

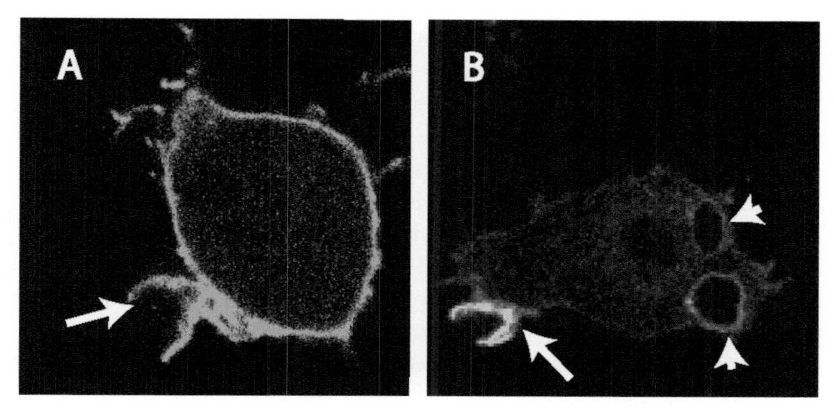

Figure 3.3 Visualization of phosphoinositides during phagocytosis. (A) Macrophage-like RAW264.7 cells expressing a fusion of CFP and the PH domain of PLCδ, a probe for $PI(4,5)P_2$. During engulfment of IgG-opsonized particles, macrophages accumulate fluorescence transiently at the phagocytic cup that is rapidly depleted from sealed phagosomes. (B) Macrophages expressing the PH domain of Akt fused to GFP, a probe for 3′-polyphosphoinositides. Fluorescence accumulates substantially at the phagocytic cup and in recently formed phagosomes, but is absent from more mature phagosomes. Phagocytic cups are indicated by arrows and sealed phagosomes by arrowheads.

containing the p85 subunit of class I PI3K was sufficient to initiate phagocytosis in COS cells (Lowry *et al.* 1998). More recently, products of PI3K activity were found to accumulate markedly at the phagocytic cup (Figure 3.3B) (Marshall *et al.* 2001). The contribution of individual 3′-polyphosphoinositide products of PI3K to phagocytosis is unclear, although some evidence points to a greater importance of PIP3 compared with $PI(3,4)P_2$. As described earlier, overexpression of SHIP, a 5′-phosphatase acting on PI3P, attenuates phagocytosis, resembling the phenotype observed when PI3K is antagonized (Huang *et al.* 2003).

Cells obtained from mice lacking the regulatory p85 subunit of class I PI3K displayed a reduced phagocytic capacity that is more pronounced when larger particles are the phagocytic targets, indicating that the requirement for PI3K is size-dependent (Vieira *et al.* 2001), consistent with previous data demonstrating a size dependence of the effect of wortmannin (Araki *et al.* 1996; Cox *et al.* 1999). Cells treated with this PI3K inhibitor displayed blunted pseudopodial extension during phagocytosis. The resulting short pseudopods sufficed for phagocytosis of small particles but not of larger ones (≥ 3 μm), which consequently failed to be engulfed (Cox *et al.* 1999). Myosin X is recruited to the phagocytic cup in a PI3K-dependent manner and may be the critical factor in pseudopod elongation (Cox *et al.* 2002).

The SH2-domain containing inositol phosphatase 1 (SHIP-1) specifically cleaves the 5′-phosphate of PIP_3. Receptor clustering quickly induces recruitment of SHIP to ITAMs (Maresco *et al.* 1999) and ITIMs (Fong *et al.* 1996; Tridandapani *et al.* 1997), which is accompanied by its phosphorylation and activation. Association of SHIP with sites of phagocytosis is thought to occur for the purpose of limiting and terminating the accumulation of PIP_3 at the phagocytic cup. Accordingly, overexpression of wild-type or constitutively active SHIP was reported to enhance the inhibitory effect of FcγRIIB on FcγRIIA (Cox *et al.* 2001), Hydrolysis of PIP_3 can also be mediated by the 3′-inositol phosphatase PTEN. Recent reports suggests that PTEN also negatively regulates Fc-mediated phagocytosis (Cao *et al.* 2004; Kim *et al.* 2002), although the mechanism of recruitment or activation remain undefined.

Phospholipase C

A phosphoinositide-specific PLC is a critical component of the phagocytic machinery. In mammals, there are eleven genes that code for PLC; their products are categorized into four groups: PLCβ, PLCδ, PLCϵ, and PLCγ. These enzymes catalyze the hydrolysis of a primary substrate, $PI(4,5)P_2$, to form two second messengers: DAG and IP_3. A catalytic core is composed of X and Y domains, and combinations of regulatory domains that mediate recruitment to the vicinity of activators or substrate facilitate this reaction. A comprehensive review of these regulatory domains and their functions appeared recently (Rhee *et al.* 2001).

PLCγ signaling is commonly associated with stimulation of receptor tyrosine kinases by growth factors. However, activation of PLCγ has also been noted during cross-linking of FcγR (Liao *et al.* 1992; Azzoni *et al.* 1992; Dusi *et al.* 1994). During phagocytosis, PLCγ is recruited to the cup, correlating with the disappearance of $PI(4,5)P_2$ and appearance of DAG (Botelho *et al.* 2000) and with the reported increases in cytosolic IP_3 and Ca^{2+} (Della Bianca *et al.* 1991), indicating that PLCγ hydrolyzes $PI(4,5)P_2$. Consumption of $PI(4,5)P_2$ is likely needed to signal the end of pseudopod extension and the appropriate actin remodeling necessary for successful internalization. Indeed, sequestration of $PI(4,5)P_2$ molecules is associated with weakening of membrane integrity (Terebiznit *et al.* 2002), possibly owing to modifications in actin structure. DAG formation is conjointly required for phagocytosis (Botelho *et al.* 2000), although its precise function is unclear. Previous data have suggested that exogenous addition of DAG or DAG mimetics promotes phagocytosis by activating PKC (Larsen *et al.* 2000), however, there is much debate regarding the involvement of PKC during this process (see below).

Phospholipase D

Phospholipase D catalyzes the breakdown of phosphatidylcholine, one of the major membrane phospholipids, to choline and phosphatidic acid. PKC, ARF, RhoA, and tyrosine kinases regulate PLD activation. Interestingly, $PI(4,5)P_2$ was shown to activate PLD, which then stimulates PI4P5K, creating a positive feedback loop during phagocytosis. PLD activation is stimulated during phagocytosis of IgG-coated particles in leukocytes, as evinced by an increase in phosphatidic acid and, in the presence of ethanol, by the formation of phosphatidylethanol (Kusner *et al.* 1999). Relatively specific chemical inhibitors of PLD diminish phagocytic activity (Kusner *et al.* 1999), suggesting that FcγR-mediated phagocytosis requires the activity of the lipase. Ceramide, generated during the sphingomyelin cycle, mirrors the effect of chemical PLD inhibitors and negatively regulates PLD activation and phagocytosis (Suchard *et al.* 1997; Hinkovska-Galcheva *et al.* 2003). In leukocytes, ceramide production at the plasma membrane is coordinated with phagocytosis, further intimating that ceramide and PLD are regulators of this process (Hinkovska-Galcheva *et al.* 1998).

Until recently, the identity of PLD isozymes involved in particle engulfment was not known. The activity of both known isozymes, PLD1 and PLD2, appears to be compulsory for phagocytosis. However, only PLD1 is localized to the phagocytic cup (Iyer *et al.* 2004). The effect of PLD2 is likely to be indirect, influencing the function of effectors that may participate in phagocytosis.

The mechanisms engaged by PLD to regulate phagocytosis are unclear. Phosphatidic acid, which can be easily converted to DAG, could serve as an additional source for PKC activation. In addition, phosphatidic acid has been documented to activate other lipid-modifying enzymes involved in phagocytosis, including PLA_2 and PLCγ.

Phospholipase A_2 (PLA_2)

PLA_2 catalyzes the release of arachidonic acid from phosphatidylcholine or phosphatidylethanolamine. The released fatty acid can be further modified by cyclooxygenases and lipoxygenases to release proinflammatory mediators commonly associated with the allergic response. Mammalian cells express up to three isoforms of PLA_2: a secreted Ca^{2+}-dependent PLA_2, a cytosolic Ca^{2+}-dependent PLA ($cPLA_2$), and a cytosolic Ca^{2+}-independent PLA_2 ($iPLA_2$). These isoforms are differentially regulated among phagocyte cell types. In monocytes and macrophages, IgG-stimulated phagocytosis activates $iPLA_2$ (Lennartz *et al.* 1993), a reaction that is seemingly mandatory for internalization to proceed (Lennartz & Brown 1991). Regulation of the other forms is less well defined.

Inhibition of PLA_2 ablates phagocytosis; this effect can be reverted by exogenous addition of arachidonic acid (Lennartz *et al.* 1997). As mentioned above, arachidonic acid can be metabolized into inflammatory mediators, yet these active metabolites do not appear to function in phagocytosis (Zheng *et al.* 1999). Instead, arachidonic acid itself appears to be the critical agent in the regulation of phagocytosis, although the precise mechanism is still obscure. Arachidonic acid has been implicated in exocytosis of membranes and may contribute to the delivery of endomembranes to the forming phagosome during particle internalization (Lennartz *et al.* 1997).

Protein kinase C (PKC)

Isozymes of PKC are members of a large group of serine threonine kinases. These enzymes are classified into four subfamilies based on structure and co-factor requirements. They families comprise the conventional or classical (α, βI, βII, γ), novel (δ, ϵ, η, ϕ), atypical (ζ, λ/ι), and the recently identified (μ, ν) PKCs. Classical PKCs (cPKCs) are activated by DAG and Ca^{2+}; novel PKCs (nPKCs) respond only to Ca^{2+}; atypical PKCs (aPKCs) do not respond to DAG or Ca^{2+} stimulation. Whereas the involvement of PKC in CR3-mediated ingestion is well established, its function in FcγR-stimulated phagocytosis is much more complex. Ingestion of IgG-opsonized targets in monocytes is accompanied by PKC activation and translocation of one or more isoforms to the phagosomal membrane (Zheleznyak & Brown 1992). Studies employing chemical PKC inhibitors or PKC dominant-negative constructs demonstrate that active PKC is required for maximum phagocytosis (Allen & Aderem 1995; Zheng *et al.* 1995; Karimi *et al.* 1999), although conflicting data have been reported (Greenberg *et al.* 1993). These discrepancies may reflect the different isozymes studied or the particle and cell type of choice. To date, PKCα (Larsen *et al.* 2000; Allen & Aderem 1995; Breton & Descoteaux 2000), PKCβ (Dekker *et al.* 2000), PKCγ (Melendez *et al.* 1999), PKCδ (Larsen *et al.* 2000), and PKCϵ (Larsen *et al.* 2000) have been reported to be recruited to the plasma membrane during phagocytosis. The variety of isoforms recruited may be required to coordinate different aspects of phagocytosis or may instead reflect the specific class of FcγR triggered.

Recent studies suggest that cPKCs are required for initiation of the respiratory burst during phagocytosis (Larsen *et al.* 2000), whereas nPKCs signal earlier events related to phagocytosis itself (Larsen *et al.* 2000, 2002). PKCε overexpression was able to increase the rate of phagocytosis in addition to enhancing actin polymerization (Larsen *et al.* 2002). Careful analysis of the downstream effectors of PKC could also provide important clues to its function during phagocytosis. PKCs may modulate phagocytosis indirectly through the regulation of PLA_2 and PLD isoforms. In addition, MARCKS, a known substrate of PKCα, is rapidly phosphorylated and recruited to the forming phagosome along with F-actin, thus linking PKC activation to actin assembly during internalization (Garcia-Garcia & Rosales 2002). Despite the obvious connection of PKCs to phagocytosis, the function of PKC isozymes in this process is incompletely defined and requires further study.

Small GTPases

Several small GTPases of the Rho family, Rho, Rac, and Cdc42, are fundamental in the reorganization of the actin cytoskeleton to form stress fibres, filopodia, and lamellipodia. During phagocytosis, actin structures underlying the participating membrane are actively rearranged to accommodate the particle to be internalized. All three Rho family members have been implicated in IgG-evoked phagocytosis. Once activated, these GTPases recruit effector molecules that aid in the reorganization of actin structures.

Inactivation of Rho by treatment with C3 transferase was reported to abrogate phagocytosis of IgG-coated particles (Hackam *et al.* 1997). On the other hand, reports based on dominant negative constructs have discounted the participation of RhoA in FcγR-mediated phagocytosis (Caron & Hall 1998; Olazabal *et al.* 2002). By contrast, Rac and Cdc42 are found at the phagocytic cup (Figure 3.4A, B) and are firmly established as being critical links in FcγR signaling (Figure 3.2). Inhibition of either molecule by expression of its dominant-negative counterpart inhibits F-actin accumulation at the cup, thereby inhibiting phagocytosis (Caron & Hall 1998; Cox *et al.* 1997). Analogously, phagocytosis of IgE-coated particles via the FcεR is similarly blocked by Cdc42 and Rac inhibition (Massol *et al.* 1998). Although the inhibitory effects of both GTPases appear to be similar, detailed studies have provided insights into their differential regulatory roles. Recruitment of constitutively active Cdc42 to clustered membrane receptors induces the formation of filopodium-like structures around particles, yet fails to support *bona fide* phagocytosis (Castellano *et al.* 1999). On the other hand,

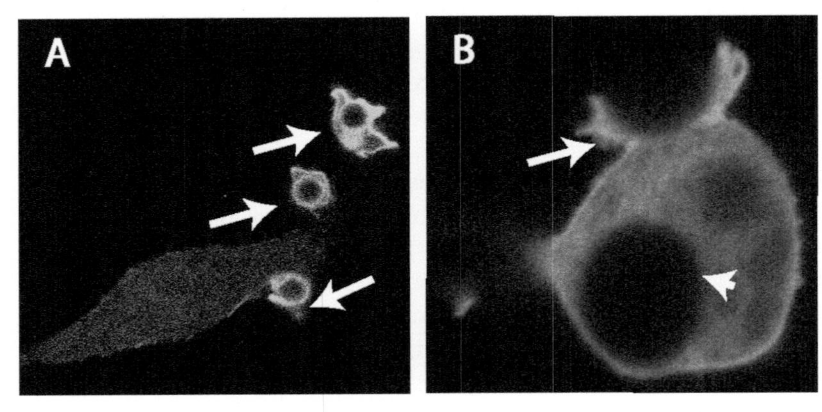

Figure 3.4 Visualization of Rac and Cdc42 during phagocytosis. (A) Macrophages expressing PBD-YFP, a probe for active Rac or Cdc42. During phagocytosis, fluorescence is concentrated at pseudopodial extensions and persists briefly on newly formed phagosomes. (B) Macrophages expressing Rac1-GFP. Rac1 accumulates at the phagocytic cup during internalization. Phagocytic cups are indicated by arrows and sealed phagosomes by arrowheads.

concentrated activation of Rac1 at the plasma membrane suffices to trigger phagocytosis, but these nascent phagosomes are without the same filopodial architecture (Castellano *et al.* 2000). Furthermore, during phagocytosis, the activation of the two GTPases is different. Using ratiometric fluorescence imaging and fluorescence resonance energy transfer techniques, Swanson and colleagues revealed the spatial and temporal distribution of both Rac and Cdc42 during phagocytosis. Two phases of G-protein stimulation were observed: activated Cdc42 is recruited to the tips of extending pseudopods, whereas Rac1 is present equally throughout the phagocytic cup and persisted therein through phagosome closure. Rac2 was also found to be activated at a late stage of phagosome closure (Hoppe & Swanson 2004). Together these data point to a strict assignment of functions for each GTPase during phagocytosis, even though they eventually converge at the point of actin assembly.

Vav, a guanine nucleotide exchange factor, appears at the phagocytic cup during phagocytosis, where it acts to promote GTP-loading of Rac and not Cdc42 (Patel *et al.* 2002). Experiments using dominant-negative constructs suggested that Vav was responsible for Rac activation during Fc-mediated phagocytosis (Patel *et al.* 2002). However, more recent experiments with Vav-deficient mice found no impairment of Fc-dependent particle uptake (Brugge, JS, personal communication).

GTPases of the ADP-ribosylation factor (ARF) family are known regulators of intracellular membrane dynamics, but also share a role in phagocytosis. Only some ARF family members are sensitive to brefeldin A inhibition; the activity of these members does not appear to be essential for phagocytosis. On the other hand, active ARF6, a brefeldin-A-insensitive GTPase, mediates F-actin accumulation underneath sites of IgG-stimulated FcγR (Q. Zhang *et al.* 1998). Activated ARF6 transiently localizes to the phagocytic cup and is at least partly responsible for delivery of VAMP3-positive endomembranes to support extending pseudopods (Niedergang *et al.* 2003). Moreover, PAG3, a GTPase activating protein of ARF6, is localized coordinately with ARF6 and F-actin at sites of IgG-stimulated phagocytosis (Uchida *et al.* 2001). Mutant forms or overexpression of PAG3, which disrupt the balance between ARF6-GTP and ARF-GDP, preclude F-actin accumulation and attenuate phagocytosis (Uchida *et al.* 2001). These data support the need for turnover of ARF6 during phagocytosis.

Phagocytosis: a size-dependent process

The preceding sections discussed the myriad of signals produced during phagocytosis as a generic event that occurs consistently after receptors are engaged. However, it is becoming increasingly apparent that the signals initiating phagocytosis may differ depending on the ligand and the receptor engaged. One parameter that is often overlooked, but equally important, is the size of the particle in question. If the particle is large, the number of receptors engaged is likely to differ from that activated by smaller particles. Thus, the intensity and geometric disposition of the signals generated will change, whether or not each receptor signals independently of the other or even if they interact. In addition, the demand for membrane delivery and the requirement for actin rearrangement are lessened during uptake of smaller particles. Therefore, the coordination of FcγR signaling is subject to size thresholds, despite the fact that identical ligands and receptors are involved in all cases.

Several observations indicate that the molecular machinery engaged during the uptake of large and small particles is not identical, as follows. (1) As mentioned earlier, PI3K is compulsory for ingestion of large particles, possibly because PI3K regulates pseudopodial projections needed to encircle large targets (Vieira *et al.* 2001; Cox *et al.* 1999). Consistently, SHIP-null macrophages or overexpression of inactive SHIP, a phosphatase that converts PIP_3 into $PI(3,4)P_2$, enhances the phagocytosis of large particles but has no effect on smaller ones (Cox *et al.* 2001). (2) Actin-dependent uptake has a size threshold of greater than 2–3 μm, whereas particles of diameter less than

1 μm are thought to be endocytosed via clathrin-based mechanisms (Koval *et al.* 1998). (3) New data suggest that the adaptor complex that is associated with the budding of clathrin-vesicles from endosomes and trans-Golgi network, AP-1, is localized at the phagocytic cup and is necessary for efficient uptake. *Dictyostelium* mutant cells (apm1(-) cells) disrupted for AP-1 medium chain have abnormal phagocytosis, an inhibitory effect that is more pronounced with large particles, implicating a role for this adaptor complex in actin-based pseudopod extension (Lefkir *et al.* 2004). (4) Soluble immune particles and IgG are both recognized by identical receptors, namely FcγR. However, endocytosis of soluble immune complexes requires ubiquitylation, whereas phagocytosis of IgG-opsonized particles occurs independently of this modification (Booth *et al.* 2002). Collectively, these data suggest that phagocytosis discriminates between small and large ligands and that, by a mechanism that has not been elucidated, receptors are able to decipher size differences to coordinate appropriate signals for their subsequent uptake.

CONCLUDING REMARKS

Phagocytosis is an extremely complex process; it is not a single phenomenon, but a collection of related yet different intricate molecular processes. This complexity can be attributed in part to the vast selection of receptors capable of initiating internalization, and to the permutations in the types, numbers and relative density of engaged receptors and in the geometric disposition and abundance of the ligands coating the particles. In this chapter, we have attempted to review signaling pathways that are activated in one type of phagocytosis, that mediated by FcγR, while recognizing that this paradigm is not likely to apply precisely to other types of receptor–ligand pair. Moreover, as discussed above, even for the IgG–FcγR cognate pair, signaling events can vary depending on the size and shape of the target particle, the complement of FcγR isoforms engaged, and the cell type under analysis.

The admittedly simplistic picture derived from the study of a single receptor–ligand pair *in vitro*, though useful as an experimental model, most likely fails to reflect the global response that ensues when phagocytes encounter pathogens in their native environment. A single pathogen will contain several different endogenous ligands and will be coated with multiple opsonins, which could interact with a varied combination of receptors, greatly complicating the outcome. Furthermore, receptors can engage in cross-talk and synergy, and the resultant signals will depend on the type of phagocytic cell and on its state of activation. Therefore, ongoing studies in model systems should be considered necessary baby steps towards a more comprehensive understanding of the biology of phagocytosis.

ACKNOWLEDGEMENTS

Original work in the authors' laboratory is supported by the Canadian Institutes of Health Research (CIHR) and the NIH. KKH is a recipient of a graduate studentship from the Canadian Cystic Fibrosis Foundation. SG is the current holder of the Pitblado Chair in Cell Biology at the Hospital for Sick Children.

REFERENCES

Adachi, Y., *et al.*, Characterization of beta-glucan recognition site on C-Type lectin, Dectin 1. *Infect Immun*, 2004. **72**(7): p. 4159–71.

Aderem, A. and D. M. Underhill, Mechanisms of phagocytosis in macrophages. *Annu Rev Immunol*, 1999. **17**: p. 593–623.

Aderem, A. A., *et al.*, Ligated complement receptors do not activate the arachidonic acid cascade in resident peritoneal macrophages. *J Exp Med*, 1985. **161**(3): p. 617–22.

Akira, S. and K. Takeda, Toll-like receptor signalling. *Nat Rev Immunol*, 2004. 4(7): p. 499–511.

Allen, L. H. and A. Aderem, A role for MARCKS, the alpha isozyme of protein kinase C and myosin I in zymosan phagocytosis by macrophages. *J Exp Med*, 1995. **182**(3): p. 829–40.

Antal-Szalmas, P., Evaluation of CD14 in host defence. *Eur J Clin Invest*, 2000. **30**(2): p. 167–79.

Araki, N., M. T. Johnson, and J. A. Swanson, A role for phosphoinositide 3-kinase in the completion of macropinocytosis and phagocytosis by macrophages. *J Cell Biol*, 1996. **135**(5): p. 1249–60.

Ariizumi, K., *et al.*, Identification of a novel, dendritic cell-associated molecule, dectin-1, by subtractive cDNA cloning. *J Biol Chem*, 2000. **275**(26): p. 20157–67.

Azzoni, L., *et al.*, Stimulation of Fc gamma RIIIA results in phospholipase C-gamma 1 tyrosine phosphorylation and p56lck activation. *J Exp Med*, 1992. **176**(6): p. 1745–50.

Barabe, F., *et al.*, Early events in the activation of Fc gamma RIIA in human neutrophils: stimulated insolubilization, translocation to detergent-resistant domains, and degradation of Fc gamma RIIA. *J Immunol*, 2002. **168**(8): p. 4042–9.

Beutler, B., Inferences, questions and possibilities in Toll-like receptor signalling. *Nature,* 2004. **430**(6996): p. 257–63.

Blander, J. M. and R. Medzhitov, Regulation of phagosome maturation by signals from toll-like receptors. *Science,* 2004. **304**(5673): p. 1014–18.

Bodin, S., *et al.*, A critical role of lipid rafts in the organization of a key FcgammaRIIa-mediated signaling pathway in human platelets. *Thromb Haemost,* 2003. **89**(2): p. 318–30.

Bonilla, F. A., *et al.*, Adapter proteins SLP-76 and BLNK both are expressed by murine macrophages and are linked to signaling via Fcgamma receptors I and II/III. *Proc Natl Acad Sci USA,* 2000. **97**(4): p. 1725–30.

Booth, J. W., *et al.*, Contrasting requirements for ubiquitylation during Fc receptor-mediated endocytosis and phagocytosis. *EMBO J,* 2002. **21**(3): p. 251–8.

Botelho, R. J., *et al.*, Localized biphasic changes in phosphatidylinositol-4,5-bisphosphate at sites of phagocytosis. *J Cell Biol,* 2000. **151**(7): p. 1353–68.

Breton, A. and A. Descoteaux, Protein kinase C-alpha participates in FcgammaR-mediated phagocytosis in macrophages. *Biochem Biophys Res Commun,* 2000. **276**(2): p. 472–6.

Brown, F. D., *et al.*, Phosphatidylinositol 4,5-bisphosphate and Arf6-regulated membrane traffic. *J Cell Biol,* 2001. **154**(5): p. 1007–17.

Brown, G. D. and S. Gordon, Immune recognition. A new receptor for beta-glucans. *Nature,* 2001. **413**(6851): p. 36–7.

Brown, G. D. and S. Gordon, Fungal beta-glucans and mammalian immunity. *Immunity,* 2003. **19**(3): p. 311–15.

Brown, G. D., *et al.*, Dectin-1 mediates the biological effects of beta-glucans. *J Exp Med,* 2003. **197**(9): p. 1119–24.

Cao, X., *et al.*, The inositol 3-phosphatase PTEN negatively regulates Fc gamma receptor signaling, but supports Toll-like receptor 4 signaling in murine peritoneal macrophages. *J Immunol,* 2004. **172**(8): p. 4851–7.

Caron, E. and A. Hall, Identification of two distinct mechanisms of phagocytosis controlled by different Rho GTPases. *Science,* 1998. **282**(5394): p. 1717–21.

Caron, E., A. J. Self, and A. Hall, The GTPase Rap1 controls functional activation of macrophage integrin alphaMbeta2 by LPS and other inflammatory mediators. *Curr Biol,* 2000. **10**(16): p. 974–8.

Castellano, F., *et al.*, Inducible recruitment of Cdc42 or WASP to a cell-surface receptor triggers actin polymerization and filopodium formation. *Curr Biol,* 1999. **9**(7): p. 351–60.

Castellano, F., P. Montcourrier, and P. Chavrier, Membrane recruitment of Rac1 triggers phagocytosis. *J Cell Sci,* 2000. **113**(17): p. 2955–61.

Cooney, D. S., *et al.*, Signal transduction by human-restricted Fc gamma RIIa involves three distinct cytoplasmic kinase families leading to phagocytosis. *J Immunol*, 2001. **167**(2): p. 844–54.

Coppolino, M. G., *et al.*, Evidence for a molecular complex consisting of Fyb/SLAP, SLP-76, Nck, VASP and WASP that links the actin cytoskeleton to Fcgamma receptor signalling during phagocytosis. *J Cell Sci*, 2001. **114** (23): p. 4307–18.

Coppolino, M. G., *et al.*, Inhibition of phosphatidylinositol-4-phosphate 5-kinase Ialpha impairs localized actin remodeling and suppresses phagocytosis. *J Biol Chem*, 2002. **277**(46): p. 43849–57.

Cox, D., *et al.*, Syk tyrosine kinase is required for immunoreceptor tyrosine activation motif-dependent actin assembly. *J Biol Chem*, 1996. **271**(28): p. 16597–602.

Cox, D., *et al.*, Requirements for both Rac1 and Cdc42 in membrane ruffling and phagocytosis in leukocytes. *J Exp Med*, 1997. **186**(9): p. 1487–94.

Cox, D., *et al.*, A requirement for phosphatidylinositol 3-kinase in pseudopod extension. *J Biol Chem*, 1999. **274**(3): p. 1240–7.

Cox, D., *et al.*, A regulatory role for Src homology 2 domain-containing inositol 5′-phosphatase (SHIP) in phagocytosis mediated by Fc gamma receptors and complement receptor 3 (alpha(M)beta(2); CD11b/CD18). *J Exp Med*, 2001. **193**(1): p. 61–71.

Cox, D., *et al.*, Myosin X is a downstream effector of PI(3)K during phagocytosis. *Nat Cell Biol*, 2002. **4**(7): p. 469–77.

Crowley, M. T., *et al.*, A critical role for Syk in signal transduction and phagocytosis mediated by Fcgamma receptors on macrophages. *J Exp Med*, 1997. **186**(7): p. 1027–39.

Dekker, L. V., *et al.*, Protein kinase C-beta contributes to NADPH oxidase activation in neutrophils. *Biochem J*, 2000. **347**(1): p. 285–9.

Della Bianca, V., M. Grzeskowiak, and F. Rossi, Studies on molecular regulation of phagocytosis and activation of the NADPH oxidase in neutrophils. IgG- and C3b-mediated ingestion and associated respiratory burst independent of phospholipid turnover and Ca^{2+} transients. *J Immunol*, 1990. **144**(4): p. 1411–17.

Della Bianca, V., *et al.*, Source and role of diacylglycerol formed during phagocytosis of opsonized yeast particles and associated respiratory burst in human neutrophils. *Biochem Biophys Res Commun*, 1991. **177**(3): p. 948–55.

Dobrovolskaia, M. A. and S. N. Vogel, Toll receptors, CD14, and macrophage activation and deactivation by LPS. *Microbes Infect*, 2002. **4**(9): p. 903–14.

Dombrosky-Ferlan, P., *et al.*, Felic (CIP4b), a novel binding partner with the Src kinase Lyn and Cdc42, localizes to the phagocytic cup. *Blood*, 2003. **101**(7): p. 2804–9.

Doyle, S. E., *et al.*, Toll-like receptors induce a phagocytic gene program through p38. *J Exp Med*, 2004. **199**(1): p. 81–90.

Dunne, D. W., *et al.*, The type I macrophage scavenger receptor binds to gram-positive bacteria and recognizes lipoteichoic acid. *Proc Natl Acad Sci USA*, 1994. **91**(5): p. 1863–7.

Dunzendorfer, S., *et al.*, TLR4 is the signaling but not the lipopolysaccharide uptake receptor. *J Immunol*, 2004. **173**(2): p. 1166–70.

Dusi, S., *et al.*, Tyrosine phosphorylation of phospholipase C-gamma 2 is involved in the activation of phosphoinositide hydrolysis by Fc receptors in human neutrophils. *Biochem Biophys Res Commun*, 1994. **201**(3): p. 1100–8.

East, L. and C. M. Isacke, The mannose receptor family. *Biochim Biophys Acta*, 2002. **1572**(2–3): p. 364–86.

Egan, B. S., R. Abdolrasulnia, and V. L. Shepherd, IL-4 modulates transcriptional control of the mannose receptor in mouse FSDC dendritic cells. *Arch Biochem Biophys*, 2004. **428**(2): p. 119–30.

Elomaa, O., *et al.*, Cloning of a novel bacteria-binding receptor structurally related to scavenger receptors and expressed in a subset of macrophages. *Cell*, 1995. **80**(4): p. 603–9.

Elshourbagy, N. A., *et al.*, Molecular characterization of a human scavenger receptor, human MARCO. *Eur J Biochem*, 2000. **267**(3): p. 919–26.

Ezekowitz, R. A., *et al.*, Molecular characterization of the human macrophage mannose receptor: demonstration of multiple carbohydrate recognition-like domains and phagocytosis of yeasts in Cos-1 cells. *J Exp Med*, 1990. **172**(6): p. 1785–94.

Fallman, M., *et al.*, Complement receptor-mediated phagocytosis is associated with accumulation of phosphatidylcholine-derived diglyceride in human neutrophils. Involvement of phospholipase D and direct evidence for a positive feedback signal of protein kinase. *J Biol Chem*, 1992. **267**(4): p. 2656–63.

Fernandez, N., *et al.*, Release of arachidonic acid by stimulation of opsonic receptors in human monocytes: the FcgammaR and the complement receptor 3 pathways. *J Biol Chem*, 2003. **278**(52): p. 52179–87.

Fitzer-Attas, C. J., *et al.*, Fcgamma receptor-mediated phagocytosis in macrophages lacking the Src family tyrosine kinases Hck, Fgr, and Lyn. *J Exp Med*, 2000. **191**(4): p. 669–82.

Fong, D. C., *et al.*, Selective in vivo recruitment of the phosphatidylinositol phosphatase SHIP by phosphorylated Fc gammaRIIB during negative regulation of IgE-dependent mouse mast cell activation. *Immunol Lett,* 1996. **54**(2–3): p. 83–91.

Forsberg, M., *et al.*, Activation of Rac2 and Cdc42 on Fc and complement receptor ligation in human neutrophils. *J Leukoc Biol,* 2003. **74**(4): p. 611–19.

Fujihara, M., *et al.*, Molecular mechanisms of macrophage activation and deactivation by lipopolysaccharide: roles of the receptor complex. *Pharmacol Ther,* 2003. **100**(2): p. 171–94.

Ganesan, L. P., *et al.*, The protein-tyrosine phosphatase SHP-1 associates with the phosphorylated immunoreceptor tyrosine-based activation motif of Fc gamma RIIa to modulate signaling events in myeloid cells. *J Biol Chem,* 2003. **278**(37): p. 35710–17.

Gantner, B. N., *et al.*, Collaborative induction of inflammatory responses by dectin-1 and Toll-like receptor 2. *J Exp Med,* 2003. **197**(9): p. 1107–17.

Garcia-Garcia, E. and C. Rosales, Signal transduction during Fc receptor-mediated phagocytosis. *J Leukoc Biol,* 2002. **72**(6): p. 1092–108.

Ghiran, I., *et al.*, Complement receptor 1/CD35 is a receptor for mannan-binding lectin. *J Exp Med,* 2000. **192**(12): p. 1797–808.

Gough, P. J. and S. Gordon, The role of scavenger receptors in the innate immune system. *Microbes Infect,* 2000. **2**(3): p. 305–11.

Greenberg, S., *et al.*, Clustered syk tyrosine kinase domains trigger phagocytosis. *Proc Natl Acad Sci USA,* 1996. **93**(3): p. 1103–7.

Greenberg, S., P. Chang, and S. C. Silverstein, Tyrosine phosphorylation is required for Fc receptor-mediated phagocytosis in mouse macrophages. *J Exp Med,* 1993. **177**(2): p. 529–34.

Grunwald, U., *et al.*, Monocytes can phagocytose Gram-negative bacteria by a CD14-dependent mechanism. *J Immunol,* 1996. **157**(9): p. 4119–25.

Gu, H., *et al.*, Critical role for scaffolding adapter Gab2 in Fc gamma R-mediated phagocytosis. *J Cell Biol,* 2003. **161**(6): p. 1151–61.

Hackam, D. J., *et al.*, Rho is required for the initiation of calcium signaling and phagocytosis by Fcgamma receptors in macrophages. *J Exp Med,* 1997. **186**(6): p. 955–66.

Hampton, R. Y., *et al.*, Recognition and plasma clearance of endotoxin by scavenger receptors. *Nature,* 1991. **352**(6333): p. 342–4.

Haseki, K., *et al.*, Role of Syk in Fc gamma receptor-coupled tyrosine phosphorylation of Cbl in a manner susceptible to inhibition by protein kinase C. *Eur J Immunol,* 1999. **29**(10): p. 3302–12.

Heale, J. P., *et al.*, Two distinct receptors mediate nonopsonic phagocytosis of different strains of Pseudomonas aeruginosa. *J Infect Dis,* 2001. **183**(8): p. 1214–20.

Herre, J., *et al.*, Dectin-1 utilizes novel mechanisms for yeast phagocytosis in macrophages. *Blood*, 2004. **104** (13): p. 4038–45.

Hilpela, P., M. K. Vartiainen, and P. Lappalainen, Regulation of the actin cytoskeleton by PI(4,5)P_2 and PI(3,4,5)P3. *Curr Top Microbiol Immunol*, 2004. **282**: p. 117–63.

Hinkovska-Galcheva, V., *et al.*, Activation of a plasma membrane-associated neutral sphingomyelinase and concomitant ceramide accumulation during IgG-dependent phagocytosis in human polymorphonuclear leukocytes. *Blood*, 1998. **91**(12): p. 4761–9.

Hinkovska-Galcheva, V., *et al.*, Enhanced phagocytosis through inhibition of de novo ceramide synthesis. *J Biol Chem*, 2003. **278**(2): p. 974–82.

Hoppe, A. D. and J. A. Swanson, Cdc42, Rac1, and Rac2 display distinct patterns of activation during phagocytosis. *Mol Biol Cell*, 2004. **15**(8): p. 3509–19.

Hoshino, K., *et al.*, Cutting edge: Toll-like receptor 4 (TLR4)-deficient mice are hyporesponsive to lipopolysaccharide: evidence for TLR4 as the Lps gene product. *J Immunol*, 1999. **162**(7): p. 3749–52.

Huang, Z. Y., *et al.*, The effect of phosphatases SHP-1 and SHIP-1 on signaling by the ITIM- and ITAM-containing Fcgamma receptors FcgammaRIIB and FcgammaRIIA. *J Leukoc Biol*, 2003. **73**(6): p. 823–9.

Hunter, S., *et al.*, Structural requirements of Syk kinase for Fcgamma receptor-mediated phagocytosis. *Exp Hematol*, 1999. **27**(5): p. 875–84.

Ibarrola, I., *et al.*, Influence of tyrosine phosphorylation on protein interaction with FcgammaRIIa. *Biochim Biophys Acta*, 1997. **1357**(3): p. 348–58.

Iyer, S. S., *et al.*, Phospholipases D1 and D2 coordinately regulate macrophage phagocytosis. *J Immunol*, 2004. **173**(4): p. 2615–23.

Izadi, K. D., *et al.*, Characterization of Cbl-Nck and Nck-Pak1 interactions in myeloid FcgammaRII signaling. *Exp Cell Res*, 1998. **245**(2): p. 330–42.

Kang, B. K. and L. S. Schlesinger, Characterization of mannose receptor-dependent phagocytosis mediated by *Mycobacterium tuberculosis* lipoarabinomannan. *Infect Immun*, 1998. **66**(6): p. 2769–77.

Kant, A. M., *et al.*, SHP-1 regulates Fcgamma receptor-mediated phagocytosis and the activation of RAC. *Blood*, 2002. **100**(5): p. 1852–9.

Karimi, K., T. R. Gemmill, and M. R. Lennartz, Protein kinase C and a calcium-independent phospholipase are required for IgG-mediated phagocytosis by Mono-Mac-6 cells. *J Leukoc Biol*, 1999. **65**(6): p. 854–62.

Katsumata, O., *et al.*, Association of FcgammaRII with low-density detergent-resistant membranes is important for cross-linking-dependent initiation of the tyrosine phosphorylation pathway and superoxide generation. *J Immunol*, 2001. **167**(10): p. 5814–23.

Kiefer, F., *et al.*, The Syk protein tyrosine kinase is essential for Fcgamma receptor signaling in macrophages and neutrophils. *Mol Cell Biol*, 1998. **18**(7): p. 4209–20.

Kim, J. S., *et al.*, PTEN controls immunoreceptor (immunoreceptor tyrosine-based activation motif) signaling and the activation of Rac. *Blood*, 2002. **99**(2): p. 694–7.

Kono, H., *et al.*, Spatial raft coalescence represents an initial step in Fc gamma R signaling. *J Immunol*, 2002. **169**(1): p. 193–203.

Koval, M., *et al.*, Size of IgG-opsonized particles determines macrophage response during internalization. *Exp Cell Res*, 1998. **242**(1): p. 265–73.

Koziel, H., *et al.*, Reduced binding and phagocytosis of *Pneumocystis carinii* by alveolar macrophages from persons infected with HIV-1 correlates with mannose receptor downregulation. *J Clin Invest*, 1998. **102**(7): p. 1332–44.

Kusner, D. J., C. F. Hall, and S. Jackson, Fc gamma receptor-mediated activation of phospholipase D regulates macrophage phagocytosis of IgG-opsonized particles. *J Immunol*, 1999. **162**(4): p. 2266–74.

Kwiatkowska, K. and A. Sobota, The clustered Fcgamma receptor II is recruited to Lyn-containing membrane domains and undergoes phosphorylation in a cholesterol-dependent manner. *Eur J Immunol*, 2001. **31**(4): p. 989–98.

Kwiatkowska, K., J. Frey, and A. Sobota, Phosphorylation of FcgammaRIIA is required for the receptor-induced actin rearrangement and capping: the role of membrane rafts. *J Cell Sci*, 2003. **116**(3): p. 537–50.

Kyono, W. T., *et al.*, Differential interaction of Crkl with Cbl or C3G, Hef-1, and gamma subunit immunoreceptor tyrosine-based activation motif in signaling of myeloid high affinity Fc receptor for IgG (Fc gamma RI). *J Immunol*, 1998. **161**(10): p. 5555–63.

Laan, L. J. van der *et al.*, Regulation and functional involvement of macrophage scavenger receptor MARCO in clearance of bacteria in vivo. *J Immunol*, 1999. **162**(2): p. 939–47.

Larsen, E. C., *et al.*, Differential requirement for classic and novel PKC isoforms in respiratory burst and phagocytosis in RAW 264.7 cells. *J Immunol*, 2000. **165**(5): p. 2809–17.

Larsen, E. C., *et al.*, A role for PKC-epsilon in Fc gammaR-mediated phagocytosis by RAW 264.7 cells. *J Cell Biol*, 2002. **159**(6): p. 939–44.

Lee, S. J., *et al.*, Mannose receptor-mediated regulation of serum glycoprotein homeostasis. *Science*, 2002. **295**(5561): p. 1898–901.

Lefkir, Y., *et al.*, Involvement of the AP-1 adaptor complex in early steps of phagocytosis and macropinocytosis. *Mol Biol Cell*, 2004. **15**(2): p. 861–9.

Lennartz, M. R. and E. J. Brown, Arachidonic acid is essential for IgG Fc receptor-mediated phagocytosis by human monocytes. *J Immunol*, 1991. **147**(2): p. 621–6.

Lennartz, M. R., *et al.*, Immunoglobulin G-mediated phagocytosis activates a calcium-independent, phosphatidylethanolamine-specific phospholipase. *J Leukoc Biol*, 1993. **54**(5): p. 389–98.

Lennartz, M. R., *et al.*, Phospholipase A2 inhibition results in sequestration of plasma membrane into electronlucent vesicles during IgG-mediated phagocytosis. *J Cell Sci*, 1997. **110** (Pt 17): p. 2041–52.

Liao, F., H. S. Shin, and S. G. Rhee, Tyrosine phosphorylation of phospholipase C-gamma 1 induced by cross-linking of the high-affinity or low-affinity Fc receptor for IgG in U937 cells. *Proc Natl Acad Sci USA*, 1992. **89**(8): p. 3659–63.

Linehan, S. A., L. Martinez-Pomares, and S. Gordon, Macrophage lectins in host defence. *Microbes Infect*, 2000. **2**(3): p. 279–88.

Lowry, M. B., *et al.*, Chimeric receptors composed of phosphoinositide 3-kinase domains and FCgamma receptor ligand-binding domains mediate phagocytosis in COS fibroblasts. *J Biol Chem*, 1998. **273**(38): p. 24513–20.

Lutz, M. A. and P. H. Correll, Activation of CR3-mediated phagocytosis by MSP requires the RON receptor, tyrosine kinase activity, phosphatidylinositol 3-kinase, and protein kinase C zeta. *J Leukoc Biol*, 2003. **73**(6): p. 802–14.

Makranz, C., *et al.*, Phosphatidylinositol 3-kinase, phosphoinositide-specific phospholipase-Cgamma and protein kinase-C signal myelin phagocytosis mediated by complement receptor-3 alone and combined with scavenger receptor-AI/II in macrophages. *Neurobiol Dis*, 2004. **15**(2): p. 279–86.

Mancek, M., P. Pristovsek, and R. Jerala, Identification of LPS-binding peptide fragment of MD-2, a toll-receptor accessory protein. *Biochem Biophys Res Commun*, 2002. **292**(4): p. 880–5.

Maresco, D. L., *et al.*, The SH2-containing 5′-inositol phosphatase (SHIP) is tyrosine phosphorylated after Fc gamma receptor clustering in monocytes. *J Immunol*, 1999. **162**(11): p. 6458–65.

Marshall, J. G., *et al.*, Restricted accumulation of phosphatidylinositol 3-kinase products in a plasmalemmal subdomain during Fc gamma receptor-mediated phagocytosis. *J Cell Biol*, 2001. **153**(7): p. 1369–80.

Martinez-Pomares, L., *et al.*, Analysis of mannose receptor regulation by IL-4, IL-10, and proteolytic processing using novel monoclonal antibodies. *J Leukoc Biol*, 2003. **73**(5): p. 604–13.

Massol, P., *et al.*, Fc receptor-mediated phagocytosis requires CDC42 and Rac1. *EMBO J*, 1998. **17**(21): p. 6219–29.

Matsuo, T., *et al.*, Specific association of phosphatidylinositol 3-kinase with the protooncogene product Cbl in Fc gamma receptor signaling. *FEBS Lett*, 1996. **382**(1–2): p. 11–14.

Melendez, A. J., M. M. Harnett, and J. M. Allen, FcgammaRI activation of phospholipase Cgamma1 and protein kinase C in dibutyryl cAMP-differentiated U937 cells is dependent solely on the tyrosine-kinase activated form of phosphatidylinositol-3-kinase. *Immunology*, 1999. **98**(1): p. 1–8.

Momose, H., *et al.*, Dual phosphorylation of phosphoinositide 3-kinase adaptor Grb2-associated binder 2 is responsible for superoxide formation synergistically stimulated by Fc gamma and formyl-methionyl-leucyl-phenylalanine receptors in differentiated THP-1 cells. *J Immunol*, 2003. **171**(8): p. 4227–34.

Nichols, K. E., *et al.*, Macrophage activation and Fcgamma receptor-mediated signaling do not require expression of the SLP-76 and SLP-65 adaptors. *J Leukoc Biol*, 2004. **75**(3): p. 541–52.

Niedergang, F., *et al.*, ADP ribosylation factor 6 is activated and controls membrane delivery during phagocytosis in macrophages. *J Cell Biol*, 2003. **161**(6): p. 1143–50.

Ninomiya, N., *et al.*, Involvement of phosphatidylinositol 3-kinase in Fc gamma receptor signaling. *J Biol Chem*, 1994. **269**(36): p. 22732–7.

O'Neill, L. A., Toll-like receptor signal transduction and the tailoring of innate immunity: a role for Mal? *Trends Immunol*, 2002. **23**(6): p. 296–300.

O'Shea, J. J., *et al.*, Activation of the C3b receptor: effect of diacylglycerols and calcium mobilization. *J Immunol*, 1985. **135**(5): p. 3381–7.

Olazabal, I. M., *et al.*, Rho-kinase and myosin-II control phagocytic cup formation during CR, but not FcgammaR, phagocytosis. *Curr Biol*, 2002. **12**(16): p. 1413–18.

Onozuka, K., *et al.*, Participation of CD14 in the phagocytosis of smooth-type *Salmonella typhimurium* by the macrophage-like cell line, J774.1. *Microbiol Immunol*, 1997. **41**(10): p. 765–72.

Ozinsky, A., *et al.*, The repertoire for pattern recognition of pathogens by the innate immune system is defined by cooperation between toll-like receptors. *Proc Natl Acad Sci USA*, 2000. **97**(25): p. 13766–71.

Palecanda, A., *et al.*, Role of the scavenger receptor MARCO in alveolar macrophage binding of unopsonized environmental particles. *J Exp Med*, 1999. **189**(9): p. 1497–506.

Park, R. K., *et al.*, CBL-GRB2 interaction in myeloid immunoreceptor tyrosine activation motif signaling. *J Immunol*, 1998. **160**(10): p. 5018–27.

Patel, J. C., A. Hall, and E. Caron, Vav regulates activation of Rac but not Cdc42 during FcgammaR-mediated phagocytosis. *Mol Biol Cell*, 2002. **13**(4): p. 1215–26.

Peiser, L., *et al.*, Macrophage class A scavenger receptor-mediated phagocytosis of *Escherichia coli*: role of cell heterogeneity, microbial strain, and culture conditions *in vitro*. *Infect Immun*, 2000. **68**(4): p. 1953–63.

Peiser, L., *et al.*, The class A macrophage scavenger receptor is a major pattern recognition receptor for *Neisseria meningitidis* which is independent of lipopolysaccharide and not required for secretory responses. *Infect Immun*, 2002. **70**(10): p. 5346–54.

Platt, N. and S. Gordon, Is the class A macrophage scavenger receptor (SR-A) multifunctional? – The mouse's tale. *J Clin Invest*, 2001. **108**(5): p. 649–54.

Platt, N., *et al.*, Role for the class A macrophage scavenger receptor in the phagocytosis of apoptotic thymocytes in vitro. *Proc Natl Acad Sci USA*, 1996. **93**(22): p. 12456–60.

Platt, N., R. P. da Silva, and S. Gordon, Recognizing death: the phagocytosis of apoptotic cells. *Trends Cell Biol*, 1998. **8**(9): p. 365–72.

Pommier, C. G., *et al.*, Plasma fibronectin enhances phagocytosis of opsonized particles by human peripheral blood monocytes. *J Exp Med*, 1983. **157**(6): p. 1844–54.

Raeder, E. M., *et al.*, Syk activation initiates downstream signaling events during human polymorphonuclear leukocyte phagocytosis. *J Immunol*, 1999. **163**(12): p. 6785–93.

Raveh, D., *et al.*, Th1 and Th2 cytokines cooperate to stimulate mannose-receptor-mediated phagocytosis. *J Leukoc Biol*, 1998. **64**(1): p. 108–13.

Rhee, S. G., Regulation of phosphoinositide-specific phospholipase C. *Annu Rev Biochem*, 2001. **70**: p. 281–312.

Sato, N., M. K. Kim, and A. D. Schreiber, Enhancement of fcgamma receptor-mediated phagocytosis by transforming mutants of Cbl. *J Immunol*, 1999. **163**(11): p. 6123–31.

Schmitz, G. and E. Orso, CD14 signalling in lipid rafts: new ligands and co-receptors. *Curr Opin Lipidol*, 2002. **13**(5): p. 513–21.

Schreiber, S., *et al.*, Prostaglandin E specifically upregulates the expression of the mannose-receptor on mouse bone marrow-derived macrophages. *Cell Regul*, 1990. **1**(5): p. 403–13.

Serrander, L., M. Fallman, and O. Stendahl, Activation of phospholipase D is an early event in integrin-mediated signalling leading to phagocytosis in human neutrophils. *Inflammation*, 1996. **20**(4): p. 439–50.

Shimazu, R., *et al.*, MD-2, a molecule that confers lipopolysaccharide responsiveness on Toll-like receptor 4. *J Exp Med*, 1999. **189**(11): p. 1777–82.

Shiratsuchi, A., *et al.*, Inhibitory effect of Toll-like receptor 4 on fusion between phagosomes and endosomes/lysosomes in macrophages. *J Immunol*, 2004. **172**(4): p. 2039–47.

Skippen, A., *et al.*, Mechanism of ADP ribosylation factor-stimulated phosphatidylinositol 4,5-bisphosphate synthesis in HL60 cells. *J Biol Chem*, 2002. **277**(8): p. 5823–31.

Strzelecka-Kiliszek, A. and A. Sobota, Sequential translocation of tyrosine kinases Lyn and Syk to the activated Fcgamma receptors during phagocytosis. *Folia Histochem Cytobiol*, 2002. **40**(2): p. 131–2.

Strzelecka-Kiliszek, A., K. Kwiatkowska, and A. Sobota, Lyn and Syk kinases are sequentially engaged in phagocytosis mediated by Fc gamma R. *J Immunol*, 2002. **169**(12): p. 6787–94.

Suchard, S. J., *et al.*, Ceramide inhibits IgG-dependent phagocytosis in human polymorphonuclear leukocytes. *Blood*, 1997. **89**(6): p. 2139–47.

Suzuki, H., *et al.*, A role for macrophage scavenger receptors in atherosclerosis and susceptibility to infection. *Nature*, 1997. **386**(6622): p. 292–6.

Takeda, K., T. Kaisho, and S. Akira, Toll-like receptors. *Annu Rev Immunol*, 2003. **21**: p. 335–76.

Terebiznik, M. R., *et al.*, Elimination of host cell PtdIns(4,5)P(2) by bacterial SigD promotes membrane fission during invasion by Salmonella. *Nat Cell Biol*, 2002. **4**(10): p. 766–73.

Thomas, C. A., *et al.*, Protection from lethal gram-positive infection by macrophage scavenger receptor-dependent phagocytosis. *J Exp Med*, 2000. **191**(1): p. 147–56.

Traynor, J. R. and K. S. Authi, Phospholipase A2 activity of lysosomal origin secreted by polymorphonuclear leucocytes during phagocytosis or on treatment with calcium. *Biochim Biophys Acta*, 1981. **665**(3): p. 571–7.

Tridandapani, S., *et al.*, Recruitment and phosphorylation of SH2-containing inositol phosphatase and Shc to the B-cell Fc gamma immunoreceptor tyrosine-based inhibition motif peptide motif. *Mol Cell Biol*, 1997. **17**(8): p. 4305–11.

Tridandapani, S., *et al.*, The adapter protein LAT enhances fcgamma receptor-mediated signal transduction in myeloid cells. *J Biol Chem*, 2000. **275**(27): p. 20480–7.

Turner, M., *et al.*, Tyrosine kinase SYK: essential functions for immunoreceptor signalling. *Immunol Today*, 2000. **21**(3): p. 148–54.

Uchida, H., *et al.*, PAG3/Papalpha/KIAA0400, a GTPase-activating protein for ADP-ribosylation factor (ARF), regulates ARF6 in Fcgamma receptor-mediated phagocytosis of macrophages. *J Exp Med*, 2001. **193**(8): p. 955–66.

Underhill, D. M. and A. Ozinsky, Phagocytosis of microbes: complexity in action. *Annu Rev Immunol*, 2002. **20**: p. 825–52.

Underhill, D. M., *et al.*, The Toll-like receptor 2 is recruited to macrophage phagosomes and discriminates between pathogens. *Nature*, 1999. **401**(6755): p. 811–15.

Vieira, O. V., *et al.*, Distinct roles of class I and class III phosphatidylinositol 3-kinases in phagosome formation and maturation. *J Cell Biol*, 2001. **155**(1): p. 19–25.

Vita, N., *et al.*, Detection and biochemical characteristics of the receptor for complexes of soluble CD14 and bacterial lipopolysaccharide. *J Immunol*, 1997. **158**(7): p. 3457–62.

Weernink, P. A., *et al.*, Activation of type I phosphatidylinositol 4-phosphate 5-kinase isoforms by the Rho GTPases, RhoA, Rac1, and Cdc42. *J Biol Chem*, 2004. **279**(9): p. 7840–9.

West, M. A., *et al.*, Enhanced dendritic cell antigen capture via toll-like receptor-induced actin remodeling. *Science*, 2004. **305**(5687): p. 1153–7.

Wright, S. D., L. S. Craigmyle, and S. C. Silverstein, Fibronectin and serum amyloid P component stimulate C3b- and C3bi-mediated phagocytosis in cultured human monocytes. *J Exp Med*, 1983. **158**(4): p. 1338–43.

Zhang, Q., *et al.*, A requirement for ARF6 in Fcgamma receptor-mediated phagocytosis in macrophages. *J Biol Chem*, 1998. **273**(32): p. 19977–81.

Zhang, W., *et al.*, LAT: the ZAP-70 tyrosine kinase substrate that links T cell receptor to cellular activation. *Cell*, 1998. **92**(1): p. 83–92.

Zheleznyak, A. and E. J. Brown, Immunoglobulin-mediated phagocytosis by human monocytes requires protein kinase C activation. Evidence for protein kinase C translocation to phagosomes. *J Biol Chem*, 1992. **267**(17): p. 12042–8.

Zheng, L., *et al.*, Arachidonic acid, but not its metabolites, is essential for FcgammaR-stimulated intracellular killing of *Staphylococcus aureus* by human monocytes. *Immunology*, 1999. **96**(1): p. 90–7.

Zheng, L., *et al.*, Role of protein kinase C isozymes in Fc gamma receptor-mediated intracellular killing of *Staphylococcus aureus* by human monocytes. *J Immunol*, 1995. **155**(2): p. 776–84.

CHAPTER 4

Life, death, and inflammation: manipulation of phagocyte function by *Helicobacter pylori*

Lee-Ann H. Allen

Helicobacter pylori is a spiral-shaped, flagellated, microaerophilic Gram-negative bacterium that colonizes the gastric epithelium of *c.*50% of humans (Blaser & Berg 2001). These organisms elicit a strong immune response that does not resolve the infection; once acquired, bacteria persist for a lifetime in the absence of antibiotic treatment. All persons infected with *H. pylori* have gastritis and 20%–30% will develop severe disease that includes gastric and duodenal ulcers, gastric adenocarcinoma or MALT lymphoma (Covacci *et al.* 1999). A characteristic feature of *H. pylori* infection is the massive recruitment of phagocytes (particularly neutrophils) to the gastric mucosa (Allen 2000). Of interest here is the fact that *H. pylori* survives for years in a phagocyte-rich environment, and a growing body of data demonstrates that these organisms modulate the host inflammatory response and phagocyte function. By this mechanism *H. pylori* evades phagocytic killing and promotes host tissue damage.

COLONIZATION OF THE GASTRIC EPITHELIUM

Helicobacter pylori is the only microbe that survives in the hostile environment of the human stomach (Blaser & Berg 2001, Montecucco & Rappuoli 2001). Urease is a nickel-containing enzyme that is essential for colonization; ammonia generated by this enzyme buffers *H. pylori* as it passes through the highly acidic gastric lumen. Bacteria establish residence in the mucus layer over the epithelium; motility in this milieu is enhanced by the spiral shape

Phagocytosis and Bacterial Pathogenicity, ed. J. D. Ernst and O. Stendahl. Published by Cambridge University Press.

of the organism and multiple polar flagella. Organisms in the mucus layer secrete a "vacuolating cytotoxin" called VacA. Released VacA oligomerizes, is activated by exposure to low pH, and binds to the epithelium. Intoxicated cells accumulate large vacuoles derived from late endosomal compartments. In addition, VacA and ammonia disrupt epithelial barrier function and a fraction of VacA reaches the host cell cytosol where it associates with mitochondria and promotes apoptosis via release of cytochrome *c*. Although all strains of *H. pylori* contain the *vacA* gene, not all organisms produce an active toxin. The s1m1 *vacA* allele encodes a potent cytotoxin that causes ulceration and necrosis in mice. In contrast, s2m2 VacA is synthesized and secreted but is inactive *in vivo* and *in vitro*. A subset of *H. pylori* strains also contain the *cag* pathogenicity island (PAI), which encodes a type IV secretion apparatus. Tight binding of *cag*-positive *H. pylori* to the epithelium stimulates delivery of CagA into the cytosol. Simultaneously, a signaling pathway is activated that results in the synthesis and secretion of IL-8. In addition to IL-8, urease, a cecropin-like peptide called Hp(2-20), and the neutrophil-activating protein (HP-NAP) recruit phagocytes to the stomach (Allen 2000; Montecucco & Rappuoli 2001; Bylund *et al.* 2001). Neutrophils traverse the epithelium and enter the mucus layer in large numbers (Allen 2000). In contrast, macrophages encounter bacteria in regions of tissue damage and ulceration (Allen 2000). These interactions are summarized in Figure 4.1.

Both VacA and the *cag* PAI are central to the ability of *H. pylori* to cause severe disease (Blaser & Berg 2000; Montecucco & Rappuoli 2001). Strains lacking s1m1 *vacA* or the *cag* PAI rarely induce ulcers. The *cag* PAI is also essential for tumor development. Consequently, strains of *H. pylori* have been divided into two groups. Type I (*cag*+/s1m1 *vacA*) strains cause ulceration and cancer, and type II (*cag*−/s2m2 *vacA*) organisms are common in persons with asymptomatic gastritis. Importantly, type II *H. pylori* induce less inflammation and tissue damage than type I bacteria and it has been suggested that these organisms occupy distinct niches in the gastric mucosa (Blaser & Berg 2001).

INFLAMMATION

A distinguishing feature of *H. pylori* infection is the chronic, neutrophil-dominant inflammation of the gastric mucosa. How this state is established and maintained is only partly understood, but the data available suggest key roles for cyclooxygenase-2 (COX-2), prostaglandin E_2 (PGE_2) and bacterial virulence factors that modulate levels of cytokines such as IL-8 and IL-6. The net result is a unique cytokine milieu that stimulates neutrophil recruitment to, and retention in, the gastric mucosa.

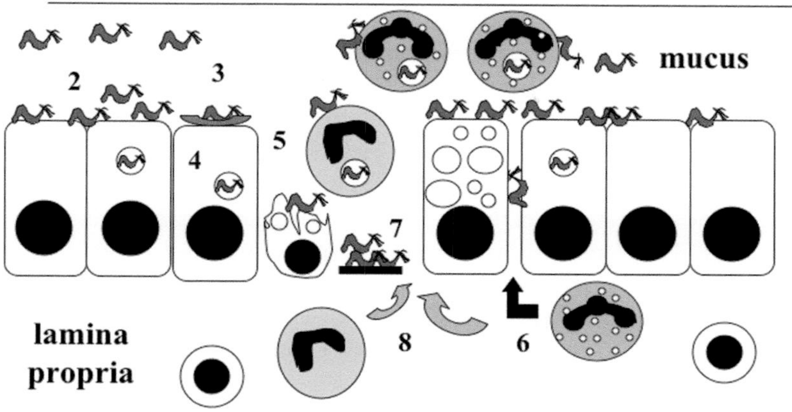

Figure 4.1 Schematic view of the *Helicobacter pylori*-infected stomach. (1) Urease generates an ammonia cloud that buffers *H. pylori* transiting through the gastric lumen. (2) *H. pylori* grows in the mucus layer and binds to the epithelium near tight junctions. (3) Actin-rich adhesion pedestals form beneath attached organisms. (4) Bacteria invade the epithelium at low frequency. (5) The secreted VacA toxin causes cell vacuolation and death. Ammonia and VacA also disrupt intercellular junctions. (6) In response to IL-8, neutrophils cross the epithelium in large numbers and phagocytose bacteria in the mucus layer. (7) At sites of cell death and ulceration *H. pylori* forms microcolonies on the extracellular matrix. (8) Macrophages traverse pores in exposed basement membranes to phagocytose bacteria.

Signaling through Toll-like receptors

Recognition of microbial non-self is central to innate immune function. In this regard, Toll-like receptors (TLRs) allow mammals to detect invading microbes via their ability to bind conserved microbial components. At least four of the ten human TLRs modulate responses to Gram-negative bacteria: TLR2 binds bacterial lipoproteins, TLR4 binds LPS in the context of MD-2, TLR5 binds flagellin, and TLR9 binds CpG motifs in bacterial DNA (Underhill & Ozinsky 2002a; Beutler 2004; Gioannini *et al.* 2004). Signaling downstream of each receptor–ligand complex triggers the synthesis and secretion of proinflammatory cytokines and the inflammatory response is modulated, in part, by the repertoire of pattern-recognition receptors engaged by each microbe. The available data indicate that TLR-dependent signaling pathways can be manipulated by pathogens such *Mycobacterium tuberculosis* and human immunodeficiency virus (van Kooyk & Geijtenbeek 2003) and this mechanism may also be used by *H. pylori* to survive in the gastric mucosa.

Recent studies have only begun to dissect TLR expression in the human stomach (Backhed *et al.* 2003; Schmauber *et al.* 2004; Wen *et al.* 2004; Ishihara *et al.* 2004). Gastric biopsy samples from healthy *H. pylori*-negative individuals contain cells that express TLR4, TLR5, and TLR9. Specific analysis of antral epithelial cells within these samples demonstrates apical and basolateral TLR5 and TLR9. TLR2 is not present and TLR4 has been detected in some studies but not in others. *H. pylori* infection upregulates TLR4 and MD-2 at the apical surface of the epithelium and triggers a concomitant redistribution of TLR5 and TLR9 to the basolateral membrane. Gastric neutrophils express TLR4 and TLR9 and macrophages express TLR4 and TLR5. Although the effects of whole *H. pylori* and bacterial components on TLR-dependent signaling pathways are largely undefined, it is clear that mononuclear phagocytes respond to LPS in a CD14- and TLR4-dependent manner with release of IL-8, TNFα and IL-1β (Maeda *et al.* 2001; Bhattacharyya *et al.* 2002; Bliss *et al.* 1998; Innocenti *et al.* 2001). In contrast, *H. pylori* does not activate cells through TLR5 (Lee *et al.* 2003; Gewirtz *et al.* 2004). TLR5 is present on macrophages and at the basolateral suface of the infected epithelium but is not detected in gastric PMNs. *H. pylori* does not release release flagellins; purified FlaA and FlaB are at least 1,000-fold less potent agonists of TLR5 than flagellins of other organisms, as judged by activation of NFκB. Clearly, more studies are needed to define the impact of *H. pylori* on individual TLRs and TLR-dependent signaling pathways. Nevertheless, the low bioactivity of *H. pylori* LPS (Muotiala *et al.* 1992; Birkholz *et al.* 1993; Pece *et al.* 1995) suggests that the canonical TLR4-dependent signaling pathway may be only a minor player in the gastric mucosa and that the *H. pylori*-specific inflammatory response is driven largely by bacterial virulence factors that act via TLR-independent mechanisms.

Role of virulence factors and COX-2

Evolution of the inflammatory response is controlled by the concerted actions of cytokines, chemokines and arachidonic acid metabolites (Lawrence *et al.* 2002; Aliberti *et al.* 2002; Hurst *et al.* 2001; Hachicha *et al.* 1999; Levy *et al.* 2001; Gronert *et al.* 1998). COX-2 drives the onset of inflammation by catalyzing the production of PGE_2, which, in turn, recruits neutrophils to the site of infection. The effects of PGE_2 are enhanced by release of IL-8 and TNFα and upregulation of adhesion molecules such as ICAM-1 and VCAM-1. Subsequent production of leukotriene B_4 (LTB_4) enhances PMN chemotaxis and promotes cell activation. Later, the effects of PGE_2 and LTB_4

are antagonized directly by IL-6 and lipoxin-A_4 (LXA_4). Both IL-6 and LXA_4 are potent agonists of monocyte adhesion and chemotaxis and simultaneously function as stop signals that halt neutrophil influx. Ultimately, LXA_4 is also essential for resolution of inflammation and restoration of normal tissue homeostasis. During *H. pylori* infection the inflammatory response is arrested at the early, PMN-dominant phase as a direct result of the combined actions of COX-2, PGE_2 and bacterial virulence determinants. The net result is an unusual inflammatory milieu distinguished by the presence of INFγ and IL-12 along with IL-10, IL-8, IL-6 and TNFα (Zevering *et al.* 1999).

Unlike the gastric mucosa of healthy individuals or persons with *H. pylori*-negative gastritis, COX-2 is dramatically upregulated during *H. pylori* infection and this results in a 25–90-fold increase in PGE_2 (Meyer *et al.* 2003; Juttner *et al.* 2003; Pomorski *et al.* 2001; Tatsuguchi *et al.* 2000; Gambero *et al.* 2003; Fu *et al.* 1999). Use of a rat air pouch model of inflammation suggests that COX-2 and PGE_2 rise rapidly after instillation of bacteria; a robust influx of PMNs occurs in less than six hours (Gambero *et al.* 2003). In humans, COX-2 and PGE_2 are generated *in situ* by epithelial and parietal cells (Pomorski *et al.* 2001; Jackson *et al.* 2000) and by neutrophils and mononuclear phagocytes migrating into the gastric mucosa (Meyer *et al.* 2003; Fu *et al.* 1999; Jackson *et al.* 2000; Kim *et al.* 2003; Wu *et al.* 1999). COX-2 expression is elevated further in persons with gastric ulcers and COX-2 is particularly high in macrophages at ulcer margins (Tatsuguchi *et al.* 2000; Jackson *et al.* 2000). Epithelial cell production of COX-2 and PGE_2 is inhibited markedly by disruption of the *cag* PAI but is independent of *vacA* (Pomorski *et al.* 2001; Gambero *et al.* 2003; Sakai *et al.* 2003). Whether the *cag* PAI is essential for COX-2 synthesis by phagocytes has not been tested directly, but monocytes synthesize COX-2 in response to whole *H. pylori*, bacterial lysates or purifed urease, suggesting that the ability of individual virulence factors to stimulate production of PGE_2 may be cell-type-specific (Meyer *et al.* 2003).

Recent data indicate that COX-2 and PGE_2 regulate cytokine balance during *H. pylori* infection via their ability to limit accumulation of IFNγ and IL-12 while enhancing production of IL-10 (Meyer *et al.* 2003). In the absence of COX-2, *H. pylori*-infected monocytes generate large amounts of IL-12 that stimulates lymphocyte production of IFNγ; IL-10 decreases markedly, tissue damage increases, and bacterial survival is compromised (Zevering *et al.* 1999; Meyer *et al.* 2003). Conversely, elimination of IFNγ reduces tissue damage and enhances bacterial proliferation. That the *H. pylori*-induced cytokine profile is tightly controlled is reinforced by the fact that IL-4 is not present in the *H. pylori*-infected mucosa, and neither IL-10 nor IL-4 accumulate in persons with other forms of gastritis (Covacci *et al.* 1999; Zevering *et al.* 1999; Ceponis *et al.* 2003).

Although it is clear that COX-2 and PGE_2 promote neutrophil accumulation at sites of infection, it is also apparent that the acute inflammatory state in the gastric mucosa is amplified and sustained by *H. pylori* virulence determinants that act on endothelial cells, epithelial cells and macrophages. Tight binding of type I *H. pylori* to the gastric epithelium activates NFκB and results in the synthesis and secretion of the potent neutrophil attractant IL-8 as well as TNFα and IL-6 (Backhed *et al.* 2003). Chemokine and cytokine production is TLR-independent and is triggered only by bacteria with an intact type IV secretion apparatus (Backhed *et al.* 2003; Maeda *et al.* 2001). However, IL-8 release does not require the secreted effector CagA, and as such may be triggered by a signaling cascade that is activated when the secretion apparatus interacts with specific plasma membrane proteins of the epithelium (Crabtree *et al.* 1995; Backert *et al.* 2000; Fischer *et al.* 2001). Secreted IL-8 imprints the extracellular matrix and in conjunction with ICAM-1 promotes neutrophil influx and migration into the mucus layer (Hofman *et al.* 2000; Mori *et al.* 2000). Macrophages also stimulate neutrophil influx via synthesis and release of IL-1β, IL-8 and TNFα (Maeda *et al.* 2001; Bhattacharyya *et al.* 2002; Bliss *et al.* 1998; Innocenti *et al.* 2001; Moese *et al.* 2002). IL-1β and IL-8 act sequentially such that IL-1β promotes neutrophil attachment to the endothelium whereas IL-8 favors chemotaxis (Suzuki *et al.* 2004). Analysis of biopsy organ cultures validates this model and also demonstrates that whole bacteria or organ culture supernatants activate endothelial cells *in vitro* as judged by upregulation of P- and E-selectin, ICAM-1, VCAM-1 and synthesis and release of IL-6, IL-8 and GRO-α (Suzuki *et al.* 2004; Innocenti *et al.* 2002; Byrne *et al.* 2002). At the same time, bacterial Hsp60 stimulates macrophage production of IL-6 and IL-10 (Zevering *et al.* 1999; Gobert *et al.* 2004). In this case, macrophage cytokine release is independent of both TLRs and the *cag* PAI.

In sum, host inflammatory mediators and bacterial virulence factors collaborate to create a distinct cytokine milieu in the infected gastric mucosa. Inasmuch as COX-2 and PGE_2 establish conditions that favor PMN recruitment to the gastric mucosa, the available data suggest that bacterial virulence factors amplify and sustain PMN influx and, in this manner mimic LTB_4. In the future it will be of interest to determine whether LTB_4 is produced during *H. pylori* infection, whether *H. pylori* actively inhibits production of LXA_4, and whether inflammation is modulated by members of the peroxisome proliferator-activated receptor family (Daynes & Jones 2002). It is also of interest that COX-2 has also been linked to gastric cancer. This enzyme is expressed in premalignant lesions and adenocarcinomas; very high concentrations of COX-2 are associated with tissue invasion and metastasis, and tumor growth is inhibited by COX-2 blockade (Oshima *et al.* 2004; Hu *et al.* 2004).

PHAGOCYTOSIS

A large body of data demonstrates that *H. pylori* is not efficiently killed by macrophages or neutrophils either *in vivo* or *in vitro* (Allen 2000; Marshall 1991; Marwali *et al.* 2003; Chmiela *et al.* 1994; Kist *et al.* 1993; Pruul 1987; Andersen *et al.* 1993; Hori & Tsutsumi 1996; Allen *et al.* 2000; Allen 2001; Allen & Allgood 2002; Zu *et al.* 2000; Ramarao & Meyer 2001; Ramarao *et al.* 2000a); substantive killing appears to occur only when phagocytes are present in vast excess (Andersen *et al.* 1993; Zu *et al.* 2000). Although we are only beginning to understand bacterium–phagocyte interactions at the molecular level, the available data suggest that *H. pylori* uses its virulence factors to delay its entry into phagocytes, to disrupt membrane trafficking and phagosome–lysosome fusion in macrophages and to manipulate the neutrophil respiratory burst. Collectively, these alterations of phagocyte function promote host tissue damage and are essential for *H. pylori* survival in the gastric mucosa.

Attachment of *H. pylori* to phagocytes

Since the seminal studies of Silverstein and colleagues in 1975 it has been clear that phagocytosis is driven by specific receptor–ligand interactions that allow monocytes, macrophages and neutrophils to extend their plasma membranes around attached particles, usually in a "zipper-like" manner (Aderem & Underhill 1999; Underhill & Ozinsky 2002b). In most cases particle uptake commences rapidly after receptor engagement and a large number of signaling pathways coordinate local actin polymerization, membrane expansion, and phagosome closure (Aderem & Underhill 1999; Underhill & Ozinsky 2002b). Although all types of phagocytosis are defined by their requirement for actin filaments, it is now clear that multiple mechanisms of phagocytosis exist and that mechanism of entry impacts significantly the fate of the ingested microbe (Aderem & Underhill 1999; Underhill & Ozinsky 2002b; Allen & Aderem 1996a,b). With regard to *H. pylori,* neither the bacterial adhesins nor the phagocyte receptors that mediate bacterial binding and engulfment have been defined. Nevertheless, a rapidly expanding body of data suggests that although most *H. pylori* isolates can bind to macrophages and neutrophils in the absence of opsonins, the efficiency of this interaction varies, particularly with regard to PMNs. *H. pylori* adhesion is complex, and surface carbohydrates and lipids as well as proteins have been implicated. At the same time, elucidation of interactions that are essential for phagocytosis and bacterial survival remain elusive. None the less, the diversity of interactions identified thus far suggests that *H. pylori* internalization may occur by more than one mechanism, each of which could employ more than one receptor–ligand pair.

Some of the first studies of *H. pylori*–phagocyte interactions demonstrated that bacterial binding to mononuclear phagocytes is conferred by a combination of sialic-acid-dependent and sialic-acid-independent hemagglutinins (Chmiela *et al.* 1995a,b, 1997; Hirmo *et al.* 1996). Sialic-acid-dependent interactions are blocked by treatment of macrophages with neuraminidase or incubation of bacteria with fetuin; both types of interaction are reduced markedly in the presence of heparin. In contrast, sialic-acid-dependent interactions may be essential for *H. pylori* binding to PMNs (Chimela *et al.* 1994, 1996; Rautelin *et al.* 1994b; Miller-Podraza *et al.* 1999). Whole bacteria bind to several gangliosides present in isolated membrane and granule fractions of PMNs and to glycosphingolipids separated on thin layer chromatograms (Miller-Podraza *et al.* 1999; Teneberg *et al.* 2000; Karlsson *et al.* 2001). *H. pylori* attachment to (and activation of) PMNs is reduced up to 60% in the presence of 3′-sialyllactose or 3′-sialyllactosamine and somewhat less by sialyl-Lewis^x (Teneberg *et al.* 2000). Taken together, the data suggest that *H. pylori* binding to phagocytes is multifactorial. In this regard it is of interest that nearly all *H. pylori* strains bind well to human monocytes and monocyte-derived macrophages (MDM), to primary murine macrophages (peritoneal and bone-marrow-derived) and to the murine macrophage-like cell lines J774 and RAW 264.7 (Allen 2000; Allen & Allgood 2002; Ramarao & Meyer 2001; Ramarao *et al.* 2000a; Chmiela *et al.* 1995b, 1997; Zheng & Jones 2003). In contrast, only a subset of *H. pylori* isolates bind well to human neutrophils (Chimela *et al.* 1994, 1996; Allen 2001; Rautelin *et al.* 1994b; Miller-Podraza *et al.* 1999; Teneberg *et al.* 2000; Ramarao *et al.* 2000a) and infection of the transformed human cell lines THP-1 and U937 is often poor regardless of agents used to promote cell differentiation (Zheng & Jones 2003; Rittig *et al.* 2003). The efficiency of *H. pylori*–phagocyte interactions may also be influenced by bacterial growth conditions (agar vs. broth), and binding to sulfatides is specifically enhanced by exposure to low pH (Miller-Podraza *et al.* 1999; Huesca *et al.* 1996).

Although most studies have focused on the interactions of *H. pylori* with the gastric epithelium, a few adhesins that modulate *H. pylori*–phagocyte interactions are beginning to emerge. First, HpaA is a 29 kDa *N*-acetylneuraminyl-(α-2,3)-lactose-binding hemagglutinin homolog that confers attachment of *H. pylori* strains (such as NCTC 11637) to sialic acid (Evans *et al.* 1993, 1995; O'Toole *et al.* 1995; Valkonen *et al.* 1997; Blom *et al.* 2001). The *hpaA* gene is found in all isolates but, as is true for many *H. pylori* proteins, the amount of HpaA that is present on the bacterial surface can vary markedly and may be affected by growth phase and/or medium composition (Blom *et al.* 2001, Lundstrom *et al.* 2001). At the same time, certain *H. pylori* strains (such as NCTC 11638) express little if any HpaA and have been classified as sialic-acid-independent binders (Chmiela *et al.* 1995b, 1997; Hirmo *et al.* 1995, 1996, 1998). The fact that both 11637 and 11638 are type I strains of *H. pylori* suggests that there is considerable diversity in mechanisms of attachment even among ulcerogenic organisms (Chmiela *et al.* 1997; Hirmo *et al.* 1995, 1996, 1998; Teneberg *et al.* 2000). HpaA-deficient *H. pylori* may utilize preferentially polyglucosylceramide-binding adhesins (Evans & Evans 2000).

Second, *H. pylori*–phagocyte interactions may be modulated by LPS in a strain-specific manner. The structure of *Helicobacter* LPS is phase-variable, owing to slipped-strand mispairing within polyC tracts that regulate expression of two α3-fucosyltransferase genes (Appelmelk *et al.* 1999). When the fucosyltransferases are active, LPS O-antigen is modified by the addition of multiple Lewis^x determinants $(Le^x)_n$ that occasionally terminate with Le^y motifs ($(Le^x)_n$ Le^y) (Appelmelk *et al.* 1998, 1999). Theoretically, phase variation is a random process that occurs at a rate of *c*.0.5%. However, distinct phenotypes may be favored under certain conditions. $(Le^x)_n$ Le^y motifs are retained by bacteria exposed to low pH *in vitro*, suggesting that expression of these epitopes may be favored in the infected gastric mucosa (Moran *et al.* 2002). Accordingly, the LPS of 82% of type I *H. pylori* isolates retains Le^x/Le^y for long periods of time, yet the LPS of type II organisms does not (Wirth *et al.* 1996). Targeted disruption of *cagA* impairs incorporation of Le^y into LPS, suggesting a direct role for CagA in modulation of endotoxin structure or, more likely, coordinate regulation of virulence factor expression by an as yet unidentified mechanism (Wirth *et al.* 1996; Barnard *et al.* 2004). In some cases, LPS structure is modified further by the addition of Le^a or Le^b or sialylation of Le^x (Appelmelk *et al.* 2000). Sialyl-Le^x epitopes on *H. pylori* endotoxin may allow bacteria to bind members of the scavenger receptor family (Peiser *et al.* 2002; Peiser & Gordon 2001). Specifically, *H. pylori* binds to salivary gp-340 in a sialyl-Le^x-dependent manner and this may explain the presence of these organisms in the oral cavity (Prakobphol *et al.* 2000). Whether *H. pylori* interacts with macrophage scavenger receptors is unknown.

Third, sialyl-Le^x (sCD15) is also present on the surface of neutrophils and is induced on macrophages and epithelial cells during inflammation (Mahdavi *et al.* 2002). *H. pylori* expresses a sialic-acid-specific adhesin called SabA, which is a 70 kDa protein that binds to sialyl-Le^x dimers (Mahdavi *et al.* 2002; Roche *et al.* 2004). SabA allows bacteria to attach tightly to the inflamed gastric epithelium and it has been suggested that phase variation of SabA could modulate bacterial binding to leukocytes (Mahdavi *et al.* 2002). Thus, it is of interest that engagement of sCD15 on phagocytes triggers signaling that upregulates and/or activates the β2 integrin receptor CD11b/CD18. Consequently, it will be important to determine whether sCD15 and β2 integrins function as co-receptors during *H. pylori* phagocytosis. To date, a role for SabA in bacterial attachment to phagocytes has not been demonstrated directly. Nevertheless, there is precedent for this type of interaction. The parasite *Schistosoma mansoni* expresses two surface receptors that modulate its interaction with macrophages. One receptor carries Le^x and the other binds to sCD15 on macrophages; concomitant engagement of these receptors along with CD11b/CD18 activates macrophage signaling pathways that culminate in antibody-dependent cytotoxicity (Trottein *et al.* 1997).

Finally, *H. pylori* also binds to extracellular matrix components including laminin, vitronectin, type IV collagen, heparan sulfate, and hyaluronic acid (Valkonen *et al.* 1994, 1997; Ringner *et al.* 1994; Trust *et al.* 1991). Damage to the gastric epithelium allows *H. pylori* to form microcolonies on exposed basement membranes and promotes contact with macrophages (Allen 2000; Marshall 1991; Valkonen *et al.* 1994). Both LPS and a 25 kDa adhesin (perhaps HpaA) have been implicated in *H. pylori* attachment to laminin (Valkonen *et al.* 1994, 1997). How the organism interacts with other extracellular matrix components is unknown, but incubation of *H. pylori* with vitronectin in the presence of serum impairs attachment to phagocytes (Chmiela *et al.* 1997) whereas, heparan sulfate-sensitive adhesins enhance binding of some strains to macrophages (Chmiela *et al.* 1996).

In sum, *H. pylori* binding to macrophages and neutrophils is a complex, multifactorial process. How the candidate adhesins function alone or in concert with other bacterial surface components remains to be determined. Ultimately, it will be important to define distinct binding patterns and to dissect how different mechanisms of attachment modulate bacterial engulfment and the subsequent ability of bacteria to resist intracellular killing. Given the apparent complexity of the bacterium–phagocyte interface, it will may be necessary to create mutant organisms in which genes for several adhesins and accessory enzymes (such as fucosyltransferases) have been disrupted.

Delayed phagocytosis

Although the receptors that mediate *H. pylori* internalization have not been defined, there is strong evidence that two distinct mechanisms are used by these bacteria to gain entry into macrophages. Less virulent type II organisms are rapidly and efficiently ingested, and in this manner resemble most phagocytic stimuli including those that engage Fcγ receptors, mannose receptors, and β1 or β2 integrins (Allen 2000; Allen *et al.* 2000; Allen & Aderem 1995, 1996b). In marked contrast, ulcerogenic type I *H. pylori* actively retard their own uptake, and there is a lag of several minutes between bacterial binding, the onset of cytoskeletal rearrangements, and membrane protrusion (Allen 2000; Allen *et al.* 2000; Allen & Allgood 2002). Delayed phagocytosis has not been described for other particulate stimuli and is observed only with live, metabolically active type I *H. pylori*. Indeed, slow phagocytosis is ablated by opsonization with specific IgG, blockade of protein synthesis or heat-denaturation of bacterial proteins (Allen *et al.* 2000; Ramarao & Meyer 2001; Ramarao *et al.* 2000a). Mutagenesis studies indicate that deletion of the entire *cag* PAI or disruption of genes that encode key elements of the type IV secretion apparatus (*virB7* and *virB11*) enhances bacterial engulfment but disruption of the secreted effector CagA or other virulence proteins such as VacA and urease are without effect (Ramarao & Meyer 2001; Ramarao *et al.* 2000a). Taken together, the data suggest that additional effectors of the type IV secretion system await discovery or that the secretion apparatus itself might act as an adhesin. In this regard it is of interest that effectors of the type III secretion systems of *Salmonella* and *Shigella* regulate the actin cytoskeleton, and enteropathogenic *Escherichia coli* uses its type III secretion apparatus to deliver a receptor for intimin into epithelial cells (Celli *et al.* 2000; Cornelis & Van Gijsegem 2000; Nhieu *et al.* 2000).

Recent data also indicate that slow phagocytosis of type I *H. pylori* defines a novel phagocytic pathway in macrophages that is regulated by atypical protein kinase C-ζ (PKCζ) (Allen & Allgood 2002). PKCζ and PKCε are activated by type I *H. pylori* and accumulate on forming phagosomes. Pharmacologic inhibitors and antisense oligodeoxynucleotides demonstrate that PKCζ is essential for *H. pylori* engulfment but other PKC isoforms are not. In this regard it is significant that atypical PKC isoforms lack structural motifs that confer regulation by diacyglycerol or phorbol esters (Ways *et al.* 1992). Rather, PKCζ is activated by the lipid products of class IA phosphoinositide 3-kinases (PI3K) (Herrera-Velit *et al.* 1997; Standaert *et al.* 1997). Sequential activation of PI3K and PKCζ regulates local actin polymerization; *H. pylori* engulfment and uptake by this mechanism is essential for bacterial survival. PI3Ks are also essential for phagocytosis of large IgG-coated particles (Araki *et al.* 1996; Cox *et al.* 1999), but in this case signaling downstream of PI3K regulates local exocytic events required for pseudopod extension and phagosome closure without affecting PKC or the actin cytoskeleton (Araki *et al.* 1996; Cox *et al.* 1999; Raeder *et al.* 1999). PKCζ is neither activated by IgG-coated particles nor required for Fcγ receptor-mediated phagocytosis (Allen & Allgood 2002). Whether *H. pylori* and IgG-coated particles utilize distinct PI3K isoforms is unknown.

Collectively, the data suggest that internalization of virulent *H. pylori* may be a multi-step process wherein initial contact is followed by more intimate adhesion that triggers signaling pathways that lead to activation of PI3K and PKCζ and phagosome formation. As *H. pylori* uptake is slow, at least two scenarios can be envisioned. First, the uptake pathway utilized by *H. pylori* may be novel but rather inefficient. Alternatively, signaling pathways that favor rapid ingestion and phagosome maturation may not be engaged (or could be actively inhibited). These scenarios are not mutually exclusive and many additional studies are needed. Nevertheless, it is clear that mode of entry has a dramatic effect on bacterial fate. Delayed phagocytosis is linked to survival of type I organisms but rapid uptake of type II *H. pylori* or IgG-opsonized bacteria is not (Allen *et al.* 2000; Allen & Allgood 2002).

Megasome formation

Another distinguishing feature of type I *H. pylori* infection of macrophages is the ability of these organisms to stimulate phagosome–phagosome fusion (Allen 2000; Allen *et al.* 2000). Ultrastructural analysis of *H. pylori*-infected cells demonstrates that immediately following phagosome closure bacteria reside in conventional phagosomes each containing a single organism. Shortly thereafter, these organelles cluster in the cytoplasm and undergo homotypic fusion. The resulting compartments are called "megasomes" and *H. pylori* survive within these structures for at least 24 hours. Megasome formation occurs similarly in primary human and murine macrophages as well as J774 and RAW 264.7 macrophage-like cell lines (Allen *et al.* 2000; Zheng & Jones 2003). Phagosome clustering and fusion requires intact microtubules in the macrophage cytoplasm; these structures are observed only in cells containing live, metabolically active type I *H. pylori* (Allen *et al.* 2000).

Moreover, the results of a recent study demonstrate that type I *H. pylori* inhibits phagosome maturation (Zheng & Jones 2003). Megasomes acquire and retain early phagosome markers such as coronin and early endosome antigen 1, but phagosome acidification and acquisition of late endosomal markers are impaired. Use of isogenic *H. pylori* mutants indicates a specific role for VacA in megasome formation. Notably, type II *H. pylori* produce an inactive (s2m2) form of VacA (Allen 2000; Xiang *et al.* 1995; Atherton *et al.* 1995). Phagosomes containing these organisms do not undergo clustering and fusion; bacteria are killed inside compartments with lysosomal characteristics (Allen 2000; Allen *et al.* 2000; Zheng & Jones 2003). It will be of interest to determine whether delayed uptake of *H. pylori* is coupled to VacA secretion and whether VacA delivery is altered by opsonization. Most studies of VacA function have examined the effects of the pure cytotoxin on epithelial cells. In this case, VacA binds to one or more surface receptors (epidermal growth factor receptor or receptor protein tyrosine phosphatase-α or -β) (Seto *et al.* 1998; Yahiro *et al.* 1999, 2003), is endocytosed, translocates to the cytosol, and over the course of several hours induces the formation of moderately acidic vacuoles that accumulate rab7 and lamp-1, but not mature lysosomal hydrolases (Montecucco & de Bernard 2003; Satin *et al.* 1997; Molinari *et al.* 1997; Papini *et al.* 1997). However, macrophages would be exposed to VacA that is secreted by intraphagosomal bacteria. Where VacA accumulates in macrophages is unknown and whether its activity is altered by other bacterial components or the intraphagosomal environment remains to be determined.

Manipulation of PKC signaling may also be essential for *H. pylori* survival. During phagocytosis of most particles and microbes PKCα is activated. This enzyme has been linked to intracellular killing because of its roles in activation of the respiratory burst and phagosome–lysosome fusion (Allen & Aderem 1995, 1996b; Larsen *et al.* 2000; St-Denis *et al.* 1999). Indeed, the ability of pathogens such as group B streptococci, *Legionella pneumophila*, and *Leishmania donovani* to inhibit or downregulate PKCα is essential for their survival in mononuclear phagocytes (St-Denis *et al.* 1999; Jacob *et al.* 1994; Cornacchione *et al.* 1998; Giorgione *et al.* 1996; Holm *et al.* 2001), and overexpression of dominant-negative PKCα in RAW 264.7 cells impairs maturation of latex bead-containing phagosomes as judged by acquisition of *lamp-1, flotillin-1, rab7,* and the acid hydrolases cathepsin S and cathepsin D (Hing *et al.* 2004). In this same vein, slow uptake of type I *H. pylori* occurs without activation of PKCα, and bacteria ingested by this mechanism persist inside megasomes (Allen *et al.* 2000; Allen & Allgood 2000). As uptake of *H. pylori* is PKCζ-dependent, it is also of interest that active PKCα inhibits PKCζ in skeletal muscle cells (Condorelli *et al.* 2001). Whether this is also true in macrophages is unknown. Nevertheless, it is tempting to speculate that PKCα and PKCζ may be activated in a mutually exclusive manner in phagocytes and it will be of interest to determine whether PKCζ regulates engulfment of other pathogens that resist phagocytic killing.

Role of CagA

In the past few years it has become apparent that type IV secretion systems are central to the ability of many Gram-negative bacteria to cause disease including *Legionella, Brucella, Coxiella* and *H. pylori* (Covacci *et al.* 1999; Cascales & Christie 2003). Like type III secretion systems, type IV apparati are complex protein assemblies that span the inner and outer membrane of the organism and, in most cases, function as contact-activated "molecular syringes" that translocate effector proteins from the bacterial cytosol directly into host cells. Variations on this theme include *Agrobacterium tumefaciens,* which uses type IV secretion to transport DNA rather than proteins, and *Bordetella pertussis,* which uses type IV secretion to export pertussis toxin in a contact-independent manner. Although it is clear that type IV secretion systems allow *Legionella* and *Brucella* to disrupt phagosome maturation in macrophages, only a few secreted effector proteins have been identified.

With regard to *H. pylori,* a type IV secretion apparatus is encoded by the *cag* PAI and as such is one of the distinguishing features of virulent type I organisms. CagA is the only known effector of this secretion system; tight binding of bacteria to epithelial cells triggers export of CagA into the host cell cytosol (Segal *et al.* 1999; Odenbreit *et al.* 2000; Asahi *et al.* 2000; Stein *et al.* 2000). Thereafter, CagA associates with the plasma membrane beneath attached bacteria and over the course of several hours a fraction of the protein is phosphorylated on one or more tyrosine residues located in the C-terminal third of the protein. After 24–48 hours, infected cells scatter and become elongated. At the membrane CagA associates with the adapter protein Grb2 and stimulates sustained activation of MEK that disrupts tight junctions and epithelial barrier function via delocalization of ZO-1 (Mimuro *et al.* 2002; Amieva *et al.* 2003). At the same time, phosphorylated CagA triggers another signaling cascade that modulates the actin cytoskeleton via inactivation of Src family kinases and dephosphorylation of cortactin (Selbach *et al.* 2003). These data indicate that CagA promotes cell motility while phosphorylated CagA modulates changes in cell shape. In this regard it is of interest that CagA secretion and phosphorylation are defective in both of the *H. pylori* genome strains (Odenbreit *et al.* 2000). J99 CagA lacks functional tyrosine phosphorylation motifs, and a frame-shift in the *cag* PAI of strain 26695 impairs the ability of the type IV secretion apparatus to deliver CagA into host cells.

H. pylori also injects CagA into macrophage-like cell lines and PMNs; however, phosphorylated CagA has not been detected in infected phagocytes (Odenbreit *et al.* 2001; Moese *et al.* 2001). Surprisingly, these cells contain a tyrosine-phosphorylated 35 kDa protein fragment that is derived from the C-terminus of full length CagA. This p35 contains all of the EPIYA motifs present in the holoprotein, and phosphorylated p35 accumulates after several hours. Nevertheless, the mechanism of CagA processing remains elusive and the function of p35 is unclear. In view of these data, it is tempting to speculate that other effectors exist and than these proteins may affect preferentially *H. pylori*–phagocyte interactions.

NEUTROPHIL ACTIVATION

Phagocyte density correlates directly with tissue damage in the *H. pylori*-infected gastric mucosa. A large body of data now indicates that a cadre of virulence factors associated with, and released from, the infecting organism orchestrate phagocyte recruitment, priming, and activation during infection. Of particular interest is the ability of *H. pylori* to modulate both phagocyte oxidative metabolism and neutrophil degranulation.

Endotoxin and priming

It has been known for some time that phagocytes at sites of infection are "primed" and, as a result, generate more reactive oxygen species (ROS) following stimulation than their naïve counterparts. *In vitro*, the primed state can be induced by adhesion of cells to extracellular matrix proteins or by exposure to LPS; cell activation ensues following subsequent exposure to TNFα, immune complexes, whole microbes or microbial products (Aderem *et al.* 1986; Guthrie *et al.* 1984; Karlsson *et al.* 1998; Nathan 1987; Berger *et al.* 2002). Most commonly, neutrophils or monocytes are primed with low doses of LPS and then stimulated with fMLP. Although the molecular basis of priming is incompletely understood, LPS-primed neutrophils exhibit increased surface expression of formyl peptide receptors (FPRs), β2 integrins and cytochrome b_{558} (Williams & Solomkin 1999; Condliffe *et al.* 1998; Downey *et al.* 1995; De Leo *et al.* 1998). Therefore, one consequence of endotoxin priming appears to be mobilization of gelatinase and/or specific granules. Importantly, however, cells exposed to LPS are not fully activated and do not synthesize ROS.

In accordance with this model, *H. pylori* extracts or purified *Helicobacter* endotoxin also prime neutrophils for enhanced production of ROS (and IL-8 release) after stimulation with fMLP (Nielsen *et al.* 1994; Shimoyama *et al.* 2003; Norgaard *et al.* 1995; Nielsen & Andersen 1992a). Nevertheless, the bioactivity of *H. pylori* LPS is much lower than that of endotoxins of enteric organisms such as *E. coli* or *Salmonella*. Specifically, *H. pylori* LPS is

500- to 1,000-fold less potent as judged by its mitogenicity, pyrogenicity, toxicity, and reactivity in the classical *Limulus* amebocyte lysate assay (Muotiala *et al.* 1992; Birkholz *et al.* 1993; Pece *et al.* 1995). Moreover, *Salmonella* LPS primes neutrophils for mobilization of azurophilic granules in response to fMLP or opsonized zymosan whereas *H. pylori* LPS does not (Nielsen *et al.* 1994; Norgaard *et al.* 1995). The unique bioactivity of *H. pylori* LPS is directly related to its unusual structure (Pece *et al.* 1995; Moran *et al.* 1992; Suda *et al.* 1997). The lipid A core of LPS is underacylated, the fatty acids that are incorporated are unusually long (16:0 and 18:0), and the 4′-position is only weakly phosphorylated. Some of these structural features are shared by other endotoxins with low bioactivity, including the LPS of *Francisella tularensis* (Vinogradov *et al.* 2002). At the molecular level, cell activation is reduced owing to the failure of *H. pylori* LPS to interact efficiently with LPS-binding protein and CD14 (Cunningham *et al.* 1996, 2000).

NADPH oxidase activity and targeting

The multi-component NADPH oxidase is an essential element of innate defense that catalyzes the conversion of molecular oxygen into superoxide anions (DeLeo & Quinn 1996). Because ROS are toxic to host tissue as well as microbes, NADPH oxidase activity is tightly controlled. In resting phagocytes the enzyme is disassembled and inactive with subunits segregated in the membrane and cytosol. Following cell activation, p47phox, p67phox, p40phox, and Rac translocate to the membrane where they bind tightly to flavocytochrome b_{558} (gp91phox/p22phox heterodimers). Importantly, the site of oxidant generation depends on the nature of the stimulus. In general, soluble agonists promote oxidase assembly at the plasma membrane and superoxide is generated in the extracellular milieu. On the other hand, particulate stimuli (such as opsonized zymosan particles or *Neisseria meningitidis* serogroup B) target the NADPH oxidase to forming phagosomes, and superoxide accumulates in the phagosome lumen (Allen *et al.* 1999; DeLeo *et al.* 1999; Dahlgren & Karlsson 1999). Concentrating ROS inside phagosomes enhances killing of ingested microbes and protects host cells from oxidative damage. Notably, superoxide anions generated by the NADPH oxidase can be converted into more toxic ROS. Superoxide rapidly and spontaneously dismutates into hydrogen peroxide; in the presence of myeloperoxidase (released from neutrophil azurophilic granules or monocyte lysosomes) hydrogen peroxide is converted into highly toxic hypochlorous acid. Pathogens that resist oxygen-dependent killing mechanisms of phagocytes must either inhibit oxidant generation or evade or withstand toxic ROS.

A characteristic feature of *H. pylori* infection is a chronic, neutrophil-dominant inflammation. It has been suggested that bacterial and host factors collaborate to damage the gastric mucosa, and a large body of data now demonstrates that a subset of *H. pylori* virulence factors recruit phagocytes to the stomach and promote oxidative tissue damage. Sonicates or extracts of all *H. pylori* strains examined to date can directly activate neutrophils and monocytes *in vitro* as judged by stimulation of chemotaxis, upregulation of β2 integrins, and synthesis and release of ROS (Allen 2000; Nielsen & Andersen 1992a,b; Mai *et al.* 1991, 1992; Craig *et al.* 1992). Interestingly, *H. felis* (which does not infect humans) is a more potent chemotactic agent than *H. pylori* but cannot trigger a respiratory burst (T. K. Hansen *et al.* 2001). These data suggest that *H. pylori* contains virulence factors that activate specifically distinct phagocyte responses. In agreement with this notion, three proteins with phagocyte-activating activity have been described to date: urease, the cecropin-like peptide Hp(2-20), and the *H. pylori* neutrophil-activating protein (HP-NAP) (Bylund *et al.* 2001; Mai *et al.* 1991, 1992; Gobert *et al.* 2002b; Evans *et al.* 1995b; Satin *et al.* 2000; Putsep *et al.* 1999). Importantly, all of these proteins are found in the bacterial cytosol and are not substrates for any known secretion pathway. Thus, it is generally believed that they are released from dying bacteria via "altruistic autolysis" (Allen 2000; Schraw *et al.* 1999; Dunn *et al.* 1997; Phadnis *et al.* 1996; Krishnamurthy *et al.* 1998).

HP-NAP is a 15 kDa iron-binding protein that forms dodecamers in solution (Allen 2000; Evans *et al.* 1995a,b; Tonello *et al.* 1999). Like urease, pure HP-NAP is a chemotactic agent that promotes upregulation of β2 integrins on monocytes and PMNs (Evans *et al.* 1995b; Satin *et al.* 2000). HP-NAP also activates the phagocyte NADPH oxidase as judged by nitroblue tetrazolium staining; the amount of H_2O_2 released from monocytes or PMNs is modest and comparable to that of cells treated with fMLP (Evans *et al.* 1995b; Satin *et al.* 2000). Cell activation occurs via a signaling pathway that includes Src family tyrosine kinases, PI3K and p38 MAP kinase, and is sensitive to inhibition by pertussis toxin (Satin *et al.* 2000; Nishioka *et al.* 2003). Specific cell-surface carbohydrates bind HP-NAP (Teneberg *et al.* 1997), but cell-association does not reach saturation (Nishioka *et al.* 2002). This may explain the unusual time course of the HP-NAP-triggered oxidative burst wherein H_2O_2 release increases gradually for up to 60 minutes (Satin *et al.* 2000). *H. pylori* also produces a cecropin-like peptide called Hp(2-20) that is highly toxic to both *E. coli* and *Candida albicans* and, like HP-NAP, activates phagocytes in a pertussis-toxin-sensitive manner (Bylund *et al.* 2001; Putsep *et al.* 1999; Lee *et al.* 2002). Hp(2-20) binds specifically to the FPR-like receptors FPRL1 and FPRL2 on neutrophils and mononuclear phagocytes (Bylund *et al.* 2001; Betten *et al.* 2001). Neutrophils treated with Hp(2-20) upregulate CD11b/CD18 and undergo a rapid and transient respiratory burst that is comparable to that seen in cells treated with fMLP (Bylund *et al.* 2001). Interestingly, responses to both HP-NAP and Hp(2-20) can be primed (Satin *et al.* 2000; Bylund *et al.* 2002). TNFα or INFγ enhance cell activation by HP-NAP; *E. coli* LPS primes cells for activation by Hp(2-20). Whether *H. pylori* LPS has a similar effect has not been explored. Urease is a potent chemoattractant for monocytes and to a lesser extent PMNs; although its mechanism of action is largely undefined, it is clearly independent of FPRs (Mai *et al.* 1991, 1992). Unlike HP-NAP and Hp(2-20), urease does not activate the NADPH oxidase. However, ammonia generated by urease can react with hypochlorous acid to generated long-lived, highly toxic monochloramines (Ishihara *et al.* 2002; Suzuki *et al.* 1999; Murakami *et al.* 1995); as such, the effects of urease on oxidative tissue damage are significant but indirect. In accordance with this model, monochloramine scavengers impair monocyte influx and neutrophil activation in the gastric mucosa (Ishihara *et al.* 2002). Taken together, the data demonstrate that *H. pylori* releases virulence factors that recruit phagocytes to the gastric mucosa and promote cell activation. Whether these same factors modulate macrophage and neutrophil function during phagocytosis of whole bacteria remains to be determined.

In marked contrast to the ability of all *H. pylori* extracts to activate phagocytes, two distinct phenotypes are observed when bacteria obtained from biopsy samples are incubated with PMNs *in vitro* (Rautelin *et al.* 1993, 1994b; Teneberg *et al.* 1997; Das *et al.* 1988; Bernatowska *et al.* 1989; Danielsson & Jurstrand 1998). Some *H. pylori* isolates strongly activate neutrophils, yet others exhibit little or no activity in the absence of serum opsonins. The molecular basis for these two phenotypes is unknown, but the work of Danielsson and colleagues suggests that neutrophil-activating capacity is a distinct virulence determinant that is often associated with, yet independent of, *vacA* and the *cag* PAI (Teneberg *et al.* 1997; Rautelin *et al.* 1993; Danielsson & Jurstrand 1998). One important caveat is the fact that nearly all of these studies used luminol chemiluminescence to measure PMN activation. Because the luminol assay detects ROS generated by myeloperoxidase (and not superoxide or H_2O_2) it is unclear whether non-activating *H. pylori* strains lack the ability to activate the NADPH oxidase or fail to stimulate mobilization of azurophilic granules (Allen 2001). Alternatively, non-activating strains may contain neutrophil-activating factors but lack adhesins that promote efficient cell binding. Notably absent is a molecular characterization of the *H. pylori* phagosome in PMNs and there is no direct evidence that any *H. pylori* strain alters NADPH oxidase targeting or degranulation. However, inefficient phagocytosis of *H. pylori* may favor release of ROS into the extracellular milieu, and bacterial superoxide dismutase and catalase protect these organisms from the toxic effects of ROS *in vivo* and *in vitro* (Seyler *et al.* 2001; Spiegelhaldes *et al.* 1993; Odenbreit *et al.* 1996; Ramarao *et al.* 2000b; Harris *et al.* 2002, 2003; Basu *et al.* 2004). The importance of the NADPH oxidase to host defense is underscored by the fact that persons with inherited mutations in this enzyme have chronic granulomatous disease (CGD) and suffer from repeated, life-threatening infections with catalase-positive microbes (Forehand *et al.* 1995). As *H. pylori* is catalase-positive, one might predict that *H. pylori* infection would be more common in persons with CGD. Although this question has not been addressed in humans, the results of a recent study indicate that the mouse-adapted *H. pylori* strain SS1 is not able to colonize *gp91phox*-null mice (Blanchard *et al.* 2003). Consequently, it is tempting to speculate that oxidant-induced tissue damage is essential for *H. pylori* survival.

Several lines of evidence suggest that neutrophil activation by whole bacteria differs significantly from the effects of *H. pylori* extracts or purified proteins. First, urease, HP-NAP and Hp(2-20) are found in all *H. pylori* strains, yet only a subset of intact organisms activate PMNs *in vitro* (Allen 2001, 2003; Rautelin *et al.* 1994a, b; Teneberg *et al.* 2000; Das *et al.* 1988; Bernatowska *et al.* 1989; Danielsson & Jurstrand 1998; Suzuki *et al.* 1994; Abe *et al.* 2000). Second, in contrast to virulence factors such as VacA, the sequences of UreA, UreB and HP-NAP are conserved and the amount of protein synthesized by different strains is relatively constant (Allen 2001; Montecucco & de Bernard 2003; Leakey *et al.* 2000). Third, responses to intact *H. pylori* are robust and comparable to those of cells stimulated with phorbol esters, but cell activation by HP-NAP and Hp(2-20) is modest and comparable in magnitude to that seen in cells treated with fMLP (Bylund *et al.* 2001; Satin *et al.* 2000; Allen 2003). Fourth, whole *H. pylori* activate both the NADPH oxidase and degranulation as judged by luminol-dependent chemiluminescence, but bacterial sonicates and HP-NAP do not (Allen 2001; Teneberg *et al.* 2000). Fifth, phagocytes can be activated potently by mechanisms that are independent of pertussis-toxin-dependent signaling pathways (P. S. Hansen *et al.* 2001). Sixth, there is evidence that urease and HP-NAP released from dying bacteria can be adsorbed by live organisms, but the extent of this interaction is a matter of debate (Blom *et al.* 2001; Thorensen *et al.* 2000; Sabarth *et al.* 2002) and whether Hp(2-20) binds to intact bacteria is unknown. An important consideration is the fact that *H. pylori* is very fragile and tends to become leaky when manipulated *in vitro*; bacterial composition may also vary with culture conditions (Krishnamurthy *et al.* 1998; Blom *et al.* 2002). A more detailed analysis of intact organisms, perhaps by immunoelectron microscopy, is needed to resolve this important question.

Taken together, the data indicate that phagocyte responses to *H. pylori* may be coupled to receptor–ligand interactions that occur during phagocytosis. This idea is supported by the fact that opsonins modulate neutrophil responses to whole organisms, enhancing cell activation in some cases and inhibiting it in others (Pruul *et al.* 1987; Teneberg *et al.* 2000; Rautelin *et al.* 1993, 1994a; Das *et al.* 1988; Bernatowska *et al.* 1989; Danielsson & Jurstrand 1998). Additional studies will be needed to define bacterial factor(s) that promote mobilization of azurophilic granules. Also of interest is the fact that HP-NAP and Hp(2-20) act similarly on monocytes and PMNs *in vitro*, yet PMNs are preferentially recruited to, and activated in, the *H. pylori*-infected stomach. Neutrophil recruitment is enhanced by IL-8 released from epithelial cells infected with type I organisms, but PMNs also outnumber macrophages in persons infected with less virulent II strains (although total phagocyte numbers are lower) (Allen 2000). Thus, the inflammatory state achieved *in vivo* requires the cooperation of many bacterial factors and host inflammatory mediators. Regardless of the mechanism, it is clear that PMN influx and activation is central to pathogenesis and the ability of *H. pylori* to cause disease.

Resistance to reactive nitrogen intermediates

Activated murine macrophages upregulate inducible nitric oxide synthase (iNOS, also called NOS2) in response to IFNγ, TNFα, and/or LPS; this enzyme catalyzes the conversion of L-arginine into NO. Reactive nitrogen intermediates such as peroxynitrite ($ONOO^-$) are generated by the reaction of NO with superoxide anions. $ONOO^-$ is a strong oxidant and nitrating agent that is highly toxic to several intracellular pathogens, including *Mycobacterium tuberculosis* and *Salmonella enterica* serovar *typhimurium* (Bryk *et al.* 2000; Kuwahara *et al.* 2000). In contrast, reactive nitrogen intermediates are relatively ineffective against *H. pylori*. This is not due to a failure of activated macrophages to synthesize iNOS (Pignatelli *et al.* 1998, 2001; Gobert *et al.* 2001, 2002b; Wilson *et al.* 1996) this enzyme is elevated in the gastric mucosa of infected patients. However, the results of several studies suggest that NO fails to accumulate in infected macrophages owing to the synergistic actions of *H. pylori* arginase, AhpC, and urease (Bryk *et al.* 2000; Kuwahara *et al.* 2000; Gobert *et al.* 2001). L-Arginine is an essential nutrient for *H. pylori* that is rapidly and specifically acquired from the environment *in vivo*, or growth medium *in vitro* (De Reuse & Skouloubris 2001). Thereafter, a constitutively active arginase (encoded by *rocF*) rapidly converts arginine into urea. Subsequent consumption of urea by *H. pylori* urease drives arginine uptake and catabolism (Gobert *et al.* 2001; De Reuse & Skouloubris 2001). Because L-arginine is also the substrate of macrophage iNOS, consumption of this amino acid by *H. pylori* limits NO production by macrophages and thereby promotes bacterial survival. In support of this model, *rocF* mutants of *H. pylori* are unable to restrict macrophage NO synthesis and are killed in an NO-dependent manner (Gobert *et al.* 2001). Conversely, null mutations in iNOS restore the viability of *rocF* mutants in macrophages *in vitro* (Gobert *et al.* 2001). Along these same lines, *H. pylori* infection enhances macrophage arginase II activity eight-fold further, limiting phagocyte NO synthesis and perhaps favoring macrophage apoptotic death (Gobert *et al.* 2001; and see below).

In addition to its ability to retard NO synthesis, *H. pylori* can also detoxify peroxynitrite that is generated by activated macrophages. All *H. pylori* strains tested to date contain an *ahpC* gene, which encodes an alkylhydroperoxide reductase (AhpC) (Bryk *et al.* 2000). These 26 kDa enzymes are found in many Gram-negative bacteria and catalyze the detoxification of peroxynitrite, thereby limiting oxidative DNA damage and protein tyrosine nitrosylation (Bryk *et al.* 2000). *H. pylori* AhpC exhibits broad substrate specificity, is regulated by pH, and may be essential for *H. pylori* survival since mutants are difficult to isolate (Baker *et al.* 2001; Lundstrom & Bolin 2000). Indeed, it appears that viability of *ahpC* mutants may require concomitant upregulation of genes for other virulence factors that also confer protection against oxidative stress (Barnard *et al.* 2004; Olczak *et al.* 2002).

EVASION OF COMPLEMENT-MEDIATED LYSIS

Both antibodies and complement proteins (such as C3b) can be detected in the *et al*-infected gastric mucosa, but in this locale they do not control bacterial infection (Darwin *et al.* 1996; Berstad *et al.* 1997). This is likely due to the fact that immunoglobulin levels are relatively low as well as the ability of *H. pylori* to cloak itself in proteins that prevent complement-mediated lysis (Rautemaa *et al.* 2001). *H. pylori* infection in general, and neutrophil influx in particular, are associated with an increase in expression of two inhibitors of complement: CD55 (also called decay-accelerating factor or DAF) and CD59 (protectin) (Sasaki *et al.* 1998). In contrast, CD46 (membrane cofactor protein or MCP) is not affected (Sasaki *et al.* 1998). In the uninfected stomach, protectin and DAF are localized to the apical membrane of the epithelium and MCP is found at the basolateral surface (Sasaki *et al.* 1998). In the infected mucosa, protectin also associates with the surface of *H. pylori* and is distributed in a granular pattern throughout the mucus layer (Rautemaa *et al.* 2001). Consequently, *H. pylori* do not acquire C5b-C9, membrane attack complexes to do not form, and bacteria are protected from complement-mediated destruction.

Laboratory-grown *H. pylori* is not coated with CD59, and most organisms bind C3 *in vitro* (Gonzalez-Valencia *et al.* 1996). Nevertheless, the majority of bacteria resist lysis by low concentrations of non-immune serum and, under these conditions, complement opsonization does not increase killing of *H. pylori* by phagocytes (Chmiela *et al.* 1997; Rautelin *et al.* 1993; Berstad *et al.* 1997; Gonzalez-Valencia *et al.* 1996; Rokita *et al.* 1998). On the other hand, immune serum improves significantly the ability of macrophages and neutrophils to kill *H. pylori in vitro* (Kist *et al.* 1993; Andersen *et al.* 1993) and opsonization with specific IgG in the absence of serum can enhance killing by macrophages (Allen & Allgood 2002; Peppoloni *et al.* 2002). How *H. pylori* is protected from complement *in vitro* is an area of active exploration. Interestingly, *ureB* mutants derived from the type I strain N6 do not express UreB or UreA; deposition of C3c and C3b on these organisms is enhanced approximately 40-fold relative to the urease-positive organisms (Rokita *et al.* 1998). How cytoplasmic urease allows *H. pylori* to evade complement deposition is unknown (Allen 2000). In this regard, sialylation of *Neisseria gonorrhoeae* endotoxin retards complement activation, and it has been suggested that Le^x and Le^y motifs of *H. pylori* LPS may serve a similar function (Moran *et al.* 1996). Lastly, complement deposition may also be modulated by association of *H. pylori* with components of the extracellular matrix such as heparan sulfate, vitronectin, and hyaluronic acid (Chmiela *et al.* 1997).

APOPTOSIS

In the past few years it has become apparent that bacterial pathogens can manipulate host cell death pathways to their advantage. This was first documented for *Shigella flexneri*: and uptake of *Shigella* by macrophages induces rapid cell death via a novel pathway that involves activation of caspase-1 (Hilbi *et al.* 1997, 1998). A similar pathway is triggered in macrophages containing *Salmonella enterica* serovar *typhimurium* (Brennan & Cookson 2000; Hersh *et al.* 1999). Caspase-1-mediated macrophage death exhibits features of apoptosis and necrosis, and it is generally believed that release of viable bacteria from dying cells is essential for bacterial persistence (Hilbi *et al.* 1997, 1998; Brennan & Cookson 2000; Hersh *et al.* 1999). On the other hand, apoptotic death is one mechanism by which activated macrophages control

Mycobacterium tuberculosis (Oddo *et al.* 1998; Saunders *et al.* 2003). In this case, apoptosis is triggered by the extrinsic pathway via the concerted action of Fas, TNFα, and ATP. Sequestration of *Mycobacteria* inside apoptotic bodies deprives these organisms of their intracellular niche and limits the spread of infection to neighboring cells.

Unlike tissue macrophages, PMNs are short-lived cells that die by spontaneous apoptosis. Interestingly, DeLeo and colleagues recently described an "apoptosis differentiation program" in neutrophils that is triggered by phagocytosis of a variety of pathogenic bacteria including *Staphylococcus aureus, Borrelia, Burkholderia,* and *Listeria* (Kobayashi *et al.* 2003). In all cases, bacteria are efficiently internalized (within 10 min) and the vast majority of ingested organisms are killed within 90 min. At the same time, phagocytosis induces a global change in gene expression in PMNs such that pro-apoptotic effectors are upregulated and receptors and other elements of innate immunity are suppressed. Consequently, the rate of apoptosis of infected cells is accelerated relative to naïve PMNs and this may provide a mechanism to control inflammation and tissue damage during the immune response. That certain pathogens can manipulate this program is suggested by the fact that cell death is accelerated further in PMNs infected with *Streptococcus pyogenes* (Kobayashi *et al.* 2003). Whether similar changes in gene expression occur in infected macrophages is unknown. Collectively, the data suggest that phagocytosis of pathogenic bacteria often promotes apoptotic death of macrophages and/or neutrophils and that this may impair or enhance bacterial killing in an organism-specific manner.

In this regard, the available data suggest that *H. pylori* infection enhances neutrophil survival (Kim *et al.* 2001a,b) yet promotes macrophage death (Zheng & Jones 2003; Gobert *et al.* 2002a; Menaker *et al.* 2004). Specifically, macrophage death ensues *c.*24 hours after *H. pylori* engulfment (Zheng & Jones 2003; Gobert *et al.* 2002; Menaker *et al.* 2004). Dying cells exhibit morphological features of apoptosis including cytoplasmic vacuolation, membrane blebbing, and chromatin condensation; death occurs, at least in part, via the intrinsic pathway as judged by activation of caspase-8, enhanced cleavage of Bid, and release of cytochrome *c* from mitrochondria (Menaker *et al.* 2004). To date, the role of *H. pylori* virulence factors in macrophage death is only beginning to be explored. Null mutations in *vacA* or *cagA* enhance macrophage survival by 40%–55% relative to cells infected with the parental type I strain 60190 (Menaker *et al.* 2004), but whether this indicates a direct effect of these virulence factors on apoptotic signaling pathways or reflects decreased intracellular survival of mutant bacteria remains to be determined.

The results of another study demonstrate a key role for polyamines in *H. pylori*-induced apoptosis (Gobert *et al.* 2002). Infection with *H. pylori* triggers a rapid and specific induction of macrophage arginase II *in vitro* and in the human gastric mucosa. Arginase II transcription is regulated by NFκB, and enzyme synthesis and activity increase eight- to ten-fold. Upregulation of arginase II is accompanied by activation of ornithine decarboxylase (ODCase). Together, these two enzymes catalyze the conversion of L-arginine into polyamines at the expense of iNOS and NO. Death of *H. pylori*-infected macrophages is ablated by inhibition of arginase II or ODCase, and direct exposure of macrophages to the polyamines spermine or spermidine restores apoptosis in a dose-dependent manner. These data demonstrate that viability of macrophages, like that of intestinal epithelial cells (Li *et al.* 2001), is modulated by polyamines, and indicate a central role for arginase and ODCase in this process. Notably, arginase II is a mitochondrial enzyme. Although it is tempting to speculate that polyamines trigger macrophage death by modulating the intrinsic apoptotic pathway, perhaps in conjunction with VacA and/or CagA, this has not yet been demonstrated. In future studies it will also be of interest to determine whether bacteria released from dying cells are viable and whether their ability to interact with naïve phagocytes has been altered.

In marked contrast to the effects of *H. pylori* on macrophages, the results of two studies show that infection of neutrophils with *H. pylori*, or exposure of PMNs to *H. pylori* extracts, significantly prolongs cell survival relative to naïve PMNs (Kim *et al.* 2001a,b). The effects of *H. pylori* are most pronounced between 12 and 48 hours and may be specific for this organism since preparations of *E. coli*, *Campylobacter jejuni*, and *C. fetus* were without effect. In support of this notion, each of the pathogens studied by DeLeo and colleagues accelerated neutrophil death (Kobayashi *et al.* 2003), yet *H. pylori* appears to enhance cell survival. At the molecular level, *H. pylori* manipulates neutrophil viability by increasing expression of the anti-apoptotic protein Bcl-X_L, suppressing expression of Fas ligand and TNF receptor 1, and impairing activation of caspase-3 and caspase-8. A role for *H. pylori* virulence factors in this process has not yet been demonstrated, but PMN survival may be directly related to the high concentrations of PGE_2 in the gastric mucosa (Kim *et al.* 2003).

MANIPULATION OF ADAPTIVE IMMUNITY

In addition to its effects on phagocytes, *H. pylori* also modulates the adaptive immune response. T cells that have been incubated with live bacteria are unable to respond normally to phytohemagglutinin; mutagenesis studies suggests a role for both *cagA* and *vacA* in this process (Paziak-Domanska *et al.* 2000). Other studies using purified VacA demonstrate that this toxin is sufficient to block T cell activation and suppress T cell proliferation (Boncristiano *et al.* 2003; Gebert *et al.* 2003; Sundrud *et al.* 2004). Moreover, T cell apoptosis is induced directly by unidentified components of the *cag* PAI and indirectly via ROS released by Hp(2-20)-activated mononuclear phagocytes (Betten *et al.* 2001; Wang *et al.* 2001). With regard to B cells, VacA disrupts selectively a subset of MHC Class II-dependent antigen presentation events (Molinari *et al.* 1998). At the same time, the fact that the dendritic cell-specific ICAM-3-grabbing non-integrin (DC-SIGN) binds with high affinity to fucose residues of Le^x motifs suggests that *H. pylori* may also interact directly with DCs (Appelmelk *et al.* 2003; Van Die *et al.* 2003). Indeed, a recent study indicates that *et al* promotes DC maturation and cytokine synthesis in a dose-dependent manner, but whether this response requires Le^x or DC-SIGN is unknown (Kranzer *et al.* 2004). This finding has important implications since alterations in DC function may be particularly important in the context of MALT lymphoma (Taki *et al.* 2002). These data, together with the known effects of *H. pylori* on cytokine production, affect our understanding of *H. pylori* pathogenesis and the development of novel therapies and effective vaccines.

CONCLUSIONS

In the past ten years we have learned a great deal about *H. pylori* pathogenesis and bacterium–phagocyte interactions. *H. pylori* uses its virulence factors to induce a chronic neutrophil-rich inflammatory response. Phagocytes are recruited to the stomach in large numbers and encounter bacteria in the mucus layer and in regions of ulceration. Bacterial binding to phagocytes is complex and multifactorial. Virulent organisms modulate their entry into phagocytes and engulfment occurs by a unique mechanism. Nevertheless, the receptor–ligand interactions required for delayed phagocytosis have not been defined and whether the type IV secretion apparatus acts as an adhesin is unknown. VacA disrupts phagosome maturation in macrophages and inhibits lymphocyte function. A cadre of virulence factors act in concert to stimulate the neutrophil respiratory burst and detoxify reactive nitrogen intermediates generated by activated macrophages. Infection prolongs PMN survival and enhances macrophage death; the tissue damage that ensues allows *H. pylori* to survive in the hostile environment of the human stomach.

REFERENCES

Abe, T., T. Shimoyama, S. Fukuda *et al.* 2000. Effects of *Helicobacter pylori* in the stomach on neutrophil chemiluminescence in patients with gastric cancer. *Luminescence* **15**: 267–71.

Aderem, A., and D. M. Underhill. 1999. Mechanisms of phagocytosis in macrophages. *A. Rev. Immunol.* **17**: 593–623.

Aderem, A. A., D. S. Cohen, S. D. Wright, and Z. A. Cohn. 1986. Bacterial lipopolysaccharides prime macrophages for enhanced release of arachidonic acid metabolites. *J. Exp. Med.* **164**: 165–79.

Aliberti, J., S. Hieny, C. R. E. Sousa, C. N. Serhan, and A. Sher. 2002. Lipoxin-mediated inhibition of IL-12 production by DCs: a mechanism for regulation of microbial immunity. *Nat. Immunol.* **3**: 76–82.

Allen, L.-A. H. 2000. Modulating phagocyte activation: the pros and cons of *Helicobacter pylori* virulence factors. *J. Exp. Med.* **191**: 1451–4.

Allen, L.-A. H. 2001. The role of the neutrophil and phagocytosis in infection caused by *Helicobacter pylori*. *Curr. Opin. Infect. Dis.* **14**: 273–7.

Allen, L. A. H. 2003. Mechanisms of pathogenesis: evasion of killing by polymorphonuclear leukocytes. *Microbes Infect.* **5**: 1329–35.

Allen, L.-A. H., and A. Aderem. 1995. A role for MARCKS, the α isozyme of protein kinase C and myosin I in zymosan phagocytosis by macrophages. *J. Exp. Med.* **182**: 829–40.

Allen, L.-A. H., and A. Aderem. 1996a. Mechanisms of phagocytosis. *Curr. Opin. Immunol.* **8**: 36–40.

Allen, L.-A. H., and A. Aderem. 1996b. Molecular definition of distinct cytoskeletal structures involved in complement- and Fc receptor-mediated phagocytosis in macrophages. *J. Exp. Med.* **184**: 627–37.

Allen, L.-A. H., and J. A. Allgood. 2002. Atypical protein kinase C-ζ is essential for delayed phagocytosis of *Helicobacter pylori*. *Curr. Biol.* **13**: 1762–6.

Allen, L.-A. H., F. R. DeLeo, A. Gallois *et al.* 1999. Transient association of the nicotinamide adenine dinucleotide phosphate oxidase subunits p47phox and p67phox with phagosomes in neutrophils from patients with X-linked chronic granulomatous disease. *Blood* **93**: 3521–30.

Allen, L.-A. H., L. S. Schlesinger, and B. Kang. 2000. Virulent strains of *Helicobacter pylori* demonstrate delayed phagocytosis and stimulate homotypic phagosome fusion in macrophages. *J. Exp. Med.* **191**: 115–27.

Amieva, M. R., R. Vogelmann, A. Covacci *et al.* 2003. Disruption of the epithelial apical-junctional complex by *Helicobacter pylori* CagA. *Science* **300**: 1430–4.

Andersen, L. P., J. Blom, and H. Nielsen. 1993. Survival and ultrastructural changes of *Helicobacter pylori* after phagocytosis by human polymorphonuclear phagocytes and monocytes. *APMIS* **101**: 61–72.

Appelmelk, B. J., B. Shiberu, C. Trinks *et al.* 1998. Phase variation in *Helicobacter pylori* lipopolysaccharide. *Infect. Immun.* **66**: 70–6.

Appelmelk, B. J., S. L. Martin, M. A. Monteiro *et al.* 1999. Phase variation in *Helicobacter pylori* lipopolysaccharide due to changes in the lengths of poly(C) tracts in alpha 3-fucosyltransferase genes. *Infect. Immun.* **67**: 5361–6.

Appelmelk, B. J., M. C. Martino, E. Veenhof *et al.* 2000. Phase variation in H type I and Lewis a epitopes of *Helicobacter pylori* lipopolysaccharide. *Infect. Immun.* **68**: 5928–32.

Appelmelk, B. J., I. van Die, S. J. van Vliet *et al.* 2003. Cutting edge: Carbohydrate profiling identifies new pathogens that interact with dendritic cell-specific ICAM-3-grabbing nonintegrin on dendritic cells. *J. Immunol.* **170**: 1635–9.

Araki, N., M. T. Johnson, and J. A. Swanson. 1996. A role for phosphoinositide 3-kinase in the completion of macropinocytosis and phagocytosis by macrophages. *J. Cell Biol.* **135**: 1249–60.

Asahi, M., T. Azuma, S. Ito *et al.* 2000. *Helicobacter pylori* CagA protein can be tyrosine phosphorylated in gastric epithelial cells. *J. Exp. Med.* **191**: 593–602.

Atherton, J. C., P. Cao, R. M. J. Peek *et al.* 1995. Mosaicism in vacuolating cytotoxin alleles of *Helicobacter pylori*. *J. Biol. Chem.* **270**: 17771–7.

Backert, S., E. Ziska, V. Brinkmann *et al.* 2000. Translocation of the *Helicobacter pylori* CagA protein in gastric epithelial cells by a type IV secretion apparatus. *Cell. Microbiol.* **2**: 155–64.

Backhed, F., B. Rokbi, E. Torstensson *et al.* 2003. Gastric mucosal recognition of *Helicobacter pylori* is independent of Toll-like receptor 4. *J. Infect. Dis.* **187**: 829–36.

Baker, L. M. S., A. Raudonikiene, P. S. Hoffman, and L. B. Poole. 2001. Essential thioredoxin-dependent peroxiredoxin system from *Helicobacter pylori*: Genetic and kinetic characterization. *J. Bacteriol.* **183**: 1961–73.

Barnard, F. M., M. F. Loughlin, H. P. Fainberg *et al.* 2004. Global regulation of virulence and the stress response by CsrA in the highly adapted human gastric pathogen *Helicobacter pylori*. *Mol. Microbiol.* **51**: 15–32.

Basu, M., S. J. Czinn, and T. G. Blanchard. 2004. Absence of catalase reduces long-term survival of *Helicobacter pylori* in macrophage phagosomes. *Helicobacter* **9**: 211–6.

Berger, M., S. Budhu, E. Lu *et al.* 2002. Different G(i)-coupled chemoattractant receptors signal qualitatively different functions in human neutrophils. *J. Leukoc. Biol.* **71**: 798–806.

Bernatowska, E., P. Jose, H. Davies, M. Stephenson, and D. Webster. 1989. Interaction of *Campylobacter* species with antibody, complement and phagocytosis. *Gut* **30**: 906–11.

Berstad, A. E., P. Brandtzaeg, R. Stave, and T. S. Halstensen. 1997. Epithelium related deposition of activated complement in *Helicobacter pylori* associated gastritis. *Gut* **40**: 196–203.

Betten, A., J. Bylund, T. Cristophe *et al.* 2001. A proinflammatory peptide from *Helicobacter pylori* activates monocytes to induce lymphocyte dysfunction and apoptosis. *J. Clin. Invest.* **108**: 1221–8.

Beutler, B. 2004. Inferences, questions and possibilities in toll-like receptor signalling. *Nature* **430**: 257–63.

Bhattacharyya, A., S. Pathak, S. Datta *et al.* 2002. Mitogen-activated protein kinases and nuclear factor-kappa B regulate *Helicobacter pylori*-mediated interleukin-8 release from macrophages. *Biochem. J.* **368**: 121–9.

Birkholz, S., U. Knipp, C. Nietzki, R. J. Adamek, and W. Opferkuch. 1993. Immunological activity of lipopolysaccharide of *Helicobacter pylori* on human peripheral mononuclear blood cells in comparison to lipopolysaccharides of other intestinal bacteria. *FEMS Immunol. Med. Microbiol.* **6**: 317–24.

Blanchard, T. G., F. W. Yu, C. L. Hsieh, and R. W. Redline. 2003. Severe inflammation and reduced bacteria load in murine *Helicobacter* infection caused by lack of phagocyte oxidase activity. *J. Infect. Dis.* **187**: 1609–15.

Blaser, M. J., and D. E. Berg. 2001. *Helicobacter pylori* genetic diversity and risk of human disease. *J. Clin. Invest.* **107**: 767–73.

Bliss, C. M., D. T. Golenbock, S. Keates, J. K. Linevsky, and C. P. Kelly. 1998. *Helicobacter pylori* lipopolysaccharide binds to CD14 and stimulates release of interleukin-8, epithelial neutrophil-activating peptide 78, and monocyte chemotactic protein 1 by human monocytes. *Infect. Immun.* **66**: 5357–63.

Blom, K., B. S. Lundin, I. Bolin, and A. M. Svennerholm. 2001. Flow cytometric analysis of the localization of *Helicobacter pylori* antigens during different growth phases. *FEMS Immunol. Med. Microbiol.* **30**: 173–9.

Blom, K., A. M. Svennerholm, and I. Bolin. 2002. The expression of the *Helicobacter pylori* genes ureA and nap is higher in vivo than in vitro as measured by quantitative competitive reverse transcriptase-PCR. *FEMS Immunol. Med. Microbiol.* **32**: 219–26.

Boncristiano, M., S. R. Paccani, S. Barone *et al.* 2003. The *Helicobacter pylori* vacuolating toxin inhibits T cell activation by two independent mechanisms. *J. Exp. Med.* **198**: 1887–97.

Brennan, M. A., and B. T. Cookson. 2000. *Salmonella* induces macrophage death by caspase-1-dependent necrosis. *Mol. Microbiol.* **38**: 31–40.

Bryk, R., P. Griffin, and C. Nathan. 2000. Peroxynitrite reductase activity of bacterial peroxiredoxins. *Nature* **407**: 211–15.

Bylund, J., T. Christophe, F. Boulay *et al.* 2001. Proinflammatory activity of a cecropin-like antibacterial peptide from *Helicobacter pylori*. *Antimicrob. Agents Chemother.* **45**: 1700–4.

Bylund, J., A. Karlsson, F. Boulay, and C. Dahlgren. 2002. Lipopolysaccharide-induced granule mobilization and priming of the neutrophil response to *Helicobacter pylori* peptide Hp(2–20), which activates formyl peptide receptor-like 1. *Infect. Immun.* **70**: 2908–14.

Byrne, M. F., P. A. Corcoran, J. C. Atherton *et al.* 2002. Stimulation of adhesion molecule expression by *Helicobacter pylori* and increased neutrophil adhesion to human umbilical vein endothelial cells. *FEBS Lett.* **532**: 411–4.

Cascales, E., and P. J. Christie. 2003. The versatile bacterial type IV secretion systems. *Nat. Rev. Microbiol.* **1**: 137–49.

Celli, J., W. Y. Deng, and B. B. Finlay. 2000. Enteropathogenic Escherichia coli (EPEC) attachment to epithelial cells: exploiting the host cell cytoskeleton from the outside. *Cell. Microbiol.* **2**: 1–9.

Ceponis, P. J. M., D. M. McKay, R. J. Menaker, E. Galindo-Mata, and N. L. Jones. 2003. *Helicobacter pylori* infection interferes with epithelial Stat6-mediated interleukin-4 signal transduction independent of cagA, cagE, or VacA. *J. Immunol.* **171**: 2035–41.

Chmiela, M., J. Lelwala-Guruge, and T. Wadstrom. 1994. Interaction of cells of *Helicobacter pylori* with human polymorphonuclear leukocytes: possible role of hemagglutinins. *FEMS Immunol. Med. Microbiol.* **9**: 41–8.

Chmiela, M., B. Paziak-Domanska, and T. Wadstrom. 1995a. Attachment, ingestion, and intracellular killing of *Helicobacter pylori* by human peripheral blood mononuclear leukocytes and mouse peritoneal inflammatory macrophages. *FEMS Immunol. Med. Microbiol.* **10**: 307–16.

Chmiela, M., B. Paziak-Domanska, W. Rudnicka, and T. Wadstrom. 1995b. The role of heparan sulfate-binding activity of *Helicobacter pylori* bacteria in their adhesion to murine macrophages. *APMIS* **103**: 469–74.

Chmiela, M., A. Ljungh, W. Rudnicka, and T. Wadstrom. 1996. Phagocytosis of *Helicobacter pylori* bacteria differing in the heparan sulfate binding by human polymorphonuclear leukocytes. *Zentralbl. Bakteriol.* **283**: 346–50.

Chmiela, M., E. Czkwianianc, T. Wadstrom, and W. Rudnicka. 1997. Role of *Helicobacter pylori* surface structures in bacterial interaction with macrophages. *Gut* **40**: 20–4.

Condliffe, A. M., E. Kitchen, and E. R. Chilvers. 1998. Neutrophil priming – pathophysiological consequences and underlying mechanisms. *Clin. Sci.* **94**: 461–71.

Condorelli, G., G. Vigliotta, A. Trencia *et al.* 2001. Protein kinase C (PLC)-α activation inhibits PKC-ζ and mediates the action of PED/PEA-15 on glucose transport in the L6 skeletal muscle cells. *Diabetes* **50**: 1244–52.

Cornacchione, P., L. Scaringi, K. Fettucciari *et al.* 1998. Group B streptococci persist inside macrophages. *Immunology* **93**: 86–95.

Cornelis, G. R., and F. Van Gijsegem. 2000. Assembly and function of type III secretory systems. *A. Rev. Microbiol.* **54**: 735–74.

Covacci, A., J. L. Telford, G. Del Giudice, J. Parsonnet, and R. Rappuoli. 1999. *Helicobacter pylori* virulence and genetic geography. *Science* **284**: 1328–33.

Cox, D., C.-C. Tseng, G. Bjekic, and S. Greenberg. 1999. A requirement for phosphatidylinositol 3-kinase in pseudopod extension. *J. Biol. Chem.* **274**: 1240–7.

Crabtree, J. E., Z. Xiang, I. J. D. Lindley *et al.* 1995. Induction of interleukin-8 secretion from gastric epithelial cells by a caga negative isogenic mutant of *Helicobacter pylori*. *J. Clin. Pathol.* **48**: 967–9.

Craig, P. M., M. C. Territo, W. E. Karnes, and J. H. Walsh. 1992. *Helicobacter pylori* secretes a chemotactic factor for monocytes and neutrophils. *Gut* **33**: 1020–3.

Cunningham, M. D., C. Seachord, K. Ratcliffe *et al.* 1996. *Helicobacter pylori* and *Porphyromonas gingivalis* lipopolysaccharides are poorly transferred to recombinant soluble CD14. *Infect. Immun.* **64**: 3601–8.

Cunningham, M. D., R. A. Shapiro, C. Seachord *et al.* 2000. CD14 employs hydrophilic regions to "capture" lipopolysaccharides. *J. Immunol.* **164**: 3255–63.

Dahlgren, C., and A. Karlsson. 1999. Respiratory burst in human neutrophils. *J. Immmunol. Methods* **232**: 3–14.

Danielsson, D., and M. Jurstrand. 1998. Nonopsonic activation of neutrophils by *Helicobacter pylori* is inhibited by rebamipide. *Dig. Dis. Sci.* **43**: S167–73.

Darwin, P. E., M. B. Sztein, Q. X. Zheng, S. P. James, and G. T. Fantry. 1996. Immune evasion by *Helicobacter pylori*: gastric spiral bacteria lack surface immunoglobulin deposition and reactivity with homologous antibodies. *Helicobacter* **1**: 20–7.

Das, S. S., Q. N. Karim, and C. S. F. Easmon. 1988. Opsonophagocytosis of *Campylobacter pylori*. *J. Med. Microbiol.* **27**: 125–30.

Daynes, R. A., and D. C. Jones. 2002. Emerging roles of PPARs in inflammation and immunity. *Nat. Rev. Immunol.* **2**: 748–59.

De Reuse, H., and S. Skouloubris. 2001. Nitrogen metabolism. In *Helicobacter pylori Physiology and Genetics*, H. L. T. Mobley, G. L. Mendz, and S. L. Hazell, editors, pp. 125–33. Washington, DC: ASM Press.

DeLeo, F. R., and M. T. Quinn. 1996. Assembly of the phagocyte NADPH oxidase: molecular interaction of oxidase proteins. *J. Leukoc. Biol.* **60**: 677–91.

DeLeo, F. R., J. Renee, S. McCormick *et al.* 1998. Neutrophils exposed to bacterial lipopolysaccharide upregulate NADPH oxidase assembly. *J. Clin. Invest.* **101**: 455–63.

DeLeo, F. R., L.-A. H. Allen, M. Apicella, and W. M. Nauseef. 1999. NADPH oxidase activation and assembly during phagocytosis. *J. Immunol.* **163**: 6732–40.

Die, I. van, S. J. van Vliet, A. K. Nyame *et al.* 2003. The dendritic cell-specific C-type lectin DC-SIGN is a receptor for *Schistosoma mansoni* egg antigens and recognizes the glycan antigen Lewis x. *Glycobiology* **13**: 471–8.

Downey, G. P., T. Fukushima, L. Fialkow, and T. K. Waddell. 1995. Intracellular signaling in neutrophil priming and activation. *Semin. Cell Biol.* **6**: 345–56.

Dunn, B. E., N. B. Vakil, B. G. Schneider *et al.* 1997. Localization of *Helicobacter pylori* urease and heat shock protein in human gastric biopsies. *Infect. Immun.* **65**: 1181–8.

Evans, D. G., T. K. Karjalainen, D. J. Evans, Jr., D. Y. Graham, and C. H. Lee. 1993. Cloning, nucleotide sequence, and expression of a gene encoding an adhesin subunit protein of *Helicobacter pylori*. *J. Bacteriol.* **175**: 674–83.

Evans, D. G., D. J. Evans, Jr., H. C. Lampert, and D. Y. Graham. 1995. Restriction fragment length polymorphism in the adhesin gene hpaA of *Helicobacter pylori*. *Am. J. Gastroenterol.* **90**: 1282–8.

Evans, D. J., and D. G. Evans. 2000. *Helicobacter pylori* adhesins: Review and perspectives. *Helicobacter* **5**: 183–95.

Evans, D. J., Jr., D. G. Evans, H. C. Lampert, and H. Nakano. 1995a. Identification of four new prokaryotic bacterioferritins, from *Helicobacter pylori, Anabaena variabilis, Bacillus subtilis,* and *Treponema pallidum,* by analysis of gene sequences. *Gene* **153**: 123–7.

Evans, D. J., Jr., D. G. Evans, T. Takemura *et al.* 1995b. Characterization of a *Helicobacter pylori* neutrophil-activating protein. *Infect. Immun.* **63**: 2213–20.

Fischer, W., J. Puls, R. Buhrdorf *et al.* 2001. Systematic mutagenesis of the *Helicobacter pylori* cag pathogenicity island: essential genes for CagA translocation in host cells and induction of interleukin-8. *Mol. Microbiol.* **42**: 1337–48.

Forehand, J. R., W. M. Nauseef, J. T. Curnutte, and R. B. J. Johnston. 1995. Inherited disorders of phagocyte killing. In *The Metabolic and Molecular Bases of Inherited Disease,* C. R. Scriver, A. L. Beaudet, W. S. Sly, and D. Valle, editors, pp. 3995–4026. New York: McGraw-Hill.

Fu, S. D., K. S. Ramanujam, A. Wong *et al.* 1999. Increased expression and cellular localization of inducible nitric oxide synthase and cyclooxygenase 2 in *Helicobacter pylori* gastritis. *Gastroenterology* **116**: 1319–29.

Gambero, A., T. L. Becker, S. A. Gurgueira *et al.* 2003. Acute inflammatory response induced by *Helicobacter pylori* in the rat air pouch. *FEMS Immunol. Med. Microbiol.* **38**: 193–8.

Gebert, B., W. Fischer, E. Weiss, R. Hoffmann, and R. Haas. 2003. *Helicobacter pylori* vacuolating cytotoxin inhibits T lymphocyte activation. *Science* **301**: 1099–102.

Gewirtz, A. T., Y. M. Yu, U. S. Krishna *et al.* 2004. *Helicobacter pylori* flagellin evades toll-like receptor 5-mediated innate immunity. *J. Infect. Dis.* **189**: 1914–20.

Gioannini, T. L., A. Teghanemt, D. S. Zhang *et al.* 2004. Isolation of an endotoxin-MD-2 complex that produces Toll-like receptor 4-dependent cell activation at picomolar concentrations. *Proc. Natl. Acad. Sci. USA* **101**: 4186–91.

Giorgione, J. R., S. T. Turco, and R. M. Epand. 1996. Transbilayer inhibition of protein kinase C by the lipophophoglycan from *Leishmania donovani. Proc. Natl. Acad. Sci. USA* **93**: 11634–9.

Gobert, A. P., D. J. McGee, M. Akhtar *et al.* 2001. *Helicobacter pylori* arginase inhibits nitric oxide production by eukaryotic cells: A strategy for bacterial survival. *Proc. Natl. Acad. Sci. USA* **98**: 13844–9.

Gobert, A. P., Y. L. Cheng, J. Y. Wang *et al.* 2002a. *Helicobacter pylori* induces macrophage apoptosis by activation of arginase II. *J. Immunol.* **168**: 4692–700.

Gobert, A. P., B. D. Mersey, Y. L. Cheng *et al.* 2002b. Cutting edge: Urease release by *Helicobacter pylori* stimulates macrophage inducible nitric oxide synthase. *J. Immunol.* **168**: 6002–6.

Gobert, A. P., J. C. Bambou, C. Werts *et al.* 2004. *Helicobacter pylori* heat shock protein 60 mediates interleukin-6 production by macrophages via a Toll-like receptor (TLR)-2-, TLR-4-, and myeloid differentiation factor 88-independent mechanism. *J. Biol. Chem.* **279**: 245–50.

Gonzalez-Valencia, G., G. I. Perez-Perez, R. G. Washburn, and M. J. Blaser. 1996. Susceptibility of *Helicobacter pylori* to the bacteriocidal activity of human serum. *Helicobacter* **1**: 28–33.

Gronert, K., A. Gewirtz, J. L. Madara, and C. N. Serhan. 1998. Identification of a human enterocyte lipoxin a(4) receptor that is regulated by interleukin (il)-13 and interferon gamma and inhibits tumor necrosis factor alpha-induced il-8 release. *J. Exp. Med.* **187**: 1285–94.

Guthrie, L. A., L. C. McPhail, P. M. Henson, and R. B. J. Johnston. 1984. Priming of neutrophils for enhanced release of oxygen metabolites by bacterial lipopolysaccharide. *J. Exp. Med.* **160**: 1656–71.

Hachicha, M., M. Pouliot, N. A. Petasis, and C. N. Serhan. 1999. Lipoxin (LX)A(4) and aspirin-triggered 15-epi-LXA(4) inhibit tumor necrosis factor 1 alpha-initiated neutrophil responses and trafficking: Regulators of a cytokine-chemokine axis. *J. Exp. Med.* **189**: 1923–9.

Hansen, P. S., P. H. Madsen, S. B. Petersen, and H. Nielsen. 2001. Inflammatory activation of neutrophils by *Helicobacter pylori*; a mechanism insensitive to pertussis toxin. *Clin. Exp. Immunol.* **123**: 73–80.

Hansen, T. K., P. S. Hansen, A. Norgaard *et al.* 2001. *Helicobacter felis* does not stimulate human neutrophil oxidative burst in contrast to '*Gastrospirillum hominis*' and *Helicobacter pylori*. *FEMS Immunol. Med. Microbiol.* **30**: 187–95.

Harris, A. G., F. E. Hinds, A. G. Beckhouse, T. Kolesnikow, and S. L. Hazell. 2002. Resistance to hydrogen peroxide in *Helicobacter pylori*: role of catalase (KatA) and Fur, and functional analysis of a novel gene product designated 'KatA-associated protein', KapA (HP0874). *Microbiology* **148**: 3813–25.

Harris, A. G., J. E. Wilson, S. J. Danon *et al.* 2003. Catalase (KatA) and KatA-associated protein (KapA) are essential to persistent colonization in the *Helicobacter pylori* SS1 mouse model. *Microbiology* **149**: 665–72.

Herrera-Velit, P., K. L. Knutson, and N. E. Reiner. 1997. Phosphatidylinositol 3-kinase-dependent activation of protein kinase C-zeta in bacterial lipopolysaccharide-treated human monocytes. *J. Biol. Chem.* **272**: 16445–52.

Hersh, D., D. M. Monack, M. R. Smith *et al.* 1999. The *Salmonella* invasin SipB induces macrophage apoptosis by binding to caspase-1. *Proc. Natl. Acad. Sci. USA* **96**: 2396–401.

Hilbi, H., Y. J. Chen, K. Thirumalai, and A. Zychlinsky. 1997. The interleukin 1-beta-converting enzyme, caspase 1, is activated during *Shigella flexneri*-induced apoptosis in human monocyte-derived macrophages. *Infect. Immun.* **65**: 5165–70.

Hilbi, H., J. E. Moss, D. Hersh *et al.* 1998. *Shigella*-induced apoptosis is dependent on Caspase-1 which binds to IpaB. *J. Biol. Chem.* **273**: 32895–900.

Hing, J. D. N. Y., M. Desjardins, and A. Descoteaux. 2004. Proteomic analysis reveals a role for protein kinase C-alpha in phagosome maturation. *Biochem. Biophys. Res. Commun.* **319**: 810–16.

Hirmo, S., M. Utt, M. Ringner, and T. Wadstrom. 1995. Inhibition of heparan sulphate and other glycosaminoglycans binding to *Helicobacter pylori* by various polysulphated carbohydrates. *FEMS Immunol. Med. Microbiol.* **10**: 301–6.

Hirmo, S., S. Kelm, R. Schauer, B. Nilsson, and T. Wadstrom. 1996. Adhesion of *Helicobacter pylori* strains to alpha-2,3-linked sialic acids. *Glycoconjugate J.* **13**: 1005–11.

Hirmo, S., E. Artursson, G. Puu, T. Wadstrom, and B. Nilsson. 1998. Characterization of *Helicobacter pylori* interactions with sialylglycoconjugates using a resonant mirror biosensor. *Analyt. Biochem.* **257**: 63–6.

Hofman, V., V. Ricci, A. Galmiche *et al.* 2000. Effect of *Helicobacter pylori* on polymorphonuclear leukocyte migration across polarized T84 epithelial cell monolayers: role of vacuolating cytotoxin VacA and cag pathogenicity island. *Infect. Immun.* **68**: 5225–33.

Holm, A., K. Tejle, K.-E. Magnusson, A. Descoteaux, and B. Rasmusson. 2001. *Leishmania donovani* lipophosphoglycan causes periphagosomal actin accumulation: correlation with impaired translocation of PKCα and defective phagosome maturation. *Cell. Microbiol.* **3**: 439–47.

Hori, S., and Y. Tsutsumi. 1996. *Helicobacter pylori* infection in gastric xanthomas: immunohistochemical analysis of 145 lesions. *Pathol. Int.* **46**: 589–93.

Hu, P. J., J. Yu, Z. R. Zeng *et al.* 2004. Chemoprevention of gastric cancer by celecoxib in rats. *Gut* **53**: 195–200.

Huesca, M., S. Borgia, P. Hoffman, and C. A. Lingwood. 1996. Acidic pH changes receptor binding specificity of *Helicobacter pylori*: a binary adhesion model in which surface heat shock (stress) proteins mediate sulfatide recognition in gastric colonization. *Infect. Immun.* **64**: 2643–8.

Hurst, S. M., T. S. Wilkinson, R. M. McLoughlin *et al.* 2001. IL-6 and its soluble receptor orchestrate a temporal switch in the pattern of leukocyte recruitment seen during acute inflammation. *Immunity* **14**: 705–14.

Innocenti, M., A. M. Svennerholm, and M. Quiding-Jarbrink. 2001. *Helicobacter pylori* lipopolysaccharides preferentially induce CXC chemokine production in human monocytes. *Infect. Immun.* **69**: 3800–8.

Innocenti, M., A. C. Thoreson, R. L. Ferrero *et al.* 2002. *Helicobacter pylori*-induced activation of human endothelial cells. *Infect. Immun.* **70**: 4581–90.

Ishihara, R., H. Iishi, N. Sakai *et al.* 2002. Polaprezinc attenuates *Helicobacter pylori*-associated gastritis in Mongolian gerbils. *Helicobacter* **7**: 384–9.

Ishihara, S., M. A. K. Rumi, Y. Kadowaki *et al.* 2004. Essential Role of MD-2 in TLR4-dependent signaling during *Helicobacter pylori*-associated gastritis. *J. Immunol.* **173**: 1406–16.

Jackson, L. M., K. C. Wu, Y. R. Mahida, D. Jenkins, and C. J. Hawkey. 2000. Cyclooxygenase (COX) 1 and 2 in normal, inflamed, and ulcerated human gastric mucosa. *Gut* **47**: 762–70.

Jacob, T., J. C. Escallier, M. V. Sanguedolce *et al.* 1994. *Legionella pneumophila* inhibits superoxide generation in human monocytes via the down-modulation of α and β protein kinase C isotypes. *J. Leukoc. Biol.* **55**: 310–12.

Juttner, S., T. Cramer, S. Wessler *et al.* 2003. *Helicobacter pylori* stimulates host cyclooxygenase-2 gene transcription: critical importance of MEK/ERK-dependent activation of USF1/-2 and CREB transcription factors. *Cell. Microbiol.* **5**: 821–34.

Karlsson, A., P. Follin, H. Leffler, and C. Dahlgren. 1998. Galectin-3 activates the NADPH oxidase in exudated but not peripheral blood neutrophils. *Blood* **91**: 3430–8.

Karlsson, A., H. Miller-Podraza, P. Johansson *et al.* 2001. Different glycosphingolipid composition in human neutrophil subcellular compartments. *Glycoconjugate. J.* **18**: 231–43.

Kim, J. S., J. M. Kim, H. C. Jung, and I. S. Song. 2001a. Caspase-3 activity and expression of Bcl-2 family in human neutrophils by *Helicobacter pylori* water-soluble proteins. *Helicobacter* **6**: 207–15.

Kim, J. S., J. M. Kim, H. C. Jung, I. S. Song, and C. Y. Kim. 2001b. Inhibition of apoptosis in human neutrophils by *Helicobacter pylori* water-soluble surface proteins. *Scand. J. Gastroenterol.* **36**: 589–600.

Kim, J. S., J. M. Kim, H. C. Jung, and I. S. Song. 2003. The effect of rebamipide on the expression of proinflammatory mediators and apoptosis in human neutrophils by *Helicobacter pylori* water-soluble surface proteins. *Aliment. Pharmacol. Ther.* **18**: 45–54.

Kist, M., C. Spiegelhalder, T. Moriki, and H. E. Schaefer. 1993. Interaction of *Helicobacter pylori* (strain 151) and *Campylobacter coli* with human peripheral polymorphonuclear granulocytes. *Zentralbl. Bakteriol.* **280**: 58–72.

Kobayashi, S. D., K. R. Braughton, A. R. Whitney *et al.* 2003. Bacterial pathogens modulate an apoptosis differentiation program in human neutrophils. *Proc. Natl. Acad. Sci. USA* **100**: 10948–53.

Kranzer, K., A. Eckhardt, M. Aigner *et al.* 2004. Induction of maturation and cytokine release of human dendritic cells by *Helicobacter pylori*. *Infect. Immun.* **72**: 4416–23.

Krishnamurthy, P., M. Parlow, N. B. Vakil *et al.* 1998. *Helicobacter pylori* containing only cytoplasmic urease is susceptible to acid. *Infect. Immun.* **66**: 5060–6.

Kuwahara, H., Y. Miyamoto, T. Akaike *et al.* 2000. *Helicobacter pylori* urease suppresses bactericidal activity of peroxynitrite via carbon dioxide production. *Infect. Immun.* **68**: 4378–83.

Larsen, E. C., J. A. DiGennaro, N. Saito *et al.* 2000. Differential requirement for classic and novel PKC isoforms in respiratory burst and phagocytosis in RAW 264.7 cells. *J. Immunol.* **165**: 2809–17.

Lawrence, T., D. A. Willoughby, and D. W. Gilroy. 2002. Anti-inflammatory lipid mediators and insights into the resolution of inflammation. *Nat. Rev. Immunol.* **2**: 787–95.

Leakey, A., J. La Brooy, and R. Hirst. 2000. The ability of *Helicobacter pylori* to activate neutrophils is determined by factors other than *H. pylori* neutrophil-activating protein. *J. Infect. Dis.* **182**: 1749–55.

Lee, D. G., Y. Park, H. N. Kim, H. K. Kim *et al.* 2002. Antifungal mechanism of an antimicrobial peptide, HP(2–20), derived from N-terminus of *Helicobacter pylori* ribosomal protein L1 against *Candida albicans*. *Biochem. Biophys. Res. Commun.* **291**: 1006–13.

Lee, S. K., A. Stack, E. Katzowitsch *et al.* 2003. *Helicobacter pylori* flagellins have very low intrinsic activity to stimulate human gastric epithelial cells via TLR5. *Microbes Infect.* **5**: 1345–56.

Levy, B. D., C. B. Clish, B. Schmidt, K. Gronert, and C. N. Serhan. 2001. Lipid mediator class switching during acute inflammation: signals in resolution. *Nat. Immunol.* **2**: 612–19.

Li, L., J. N. Rao, B. L. Bass, and J. Y. Wang. 2001. NF-kappa B activation and susceptibility to apoptosis after polyamine depletion in intestinal epithelial cells. *Am. J. Physiol. Gastrointest. Liver Physiol.* **280**: G992–1004.

Lundstrom, A. M., and I. Bolin. 2000. A 26 kDa protein of *Helicobacter pylori* shows alkyl hydroperoxide reductase (AhpC) activity and the mono-cistronic transcription of the gene is affected by pH. *Microb. Pathog.* **29**: 257–66.

Lundstrom, A. M., K. Blom, V. Sundaeus, and I. Bolin. 2001. HpaA shows variable surface localization but the gene expression is similar in different *Helicobacter pylori* strains. *Microb. Pathog.* **31**: 243–53.

Maeda, S., M. Akanuma, Y. Mitsuno *et al.* 2001. Distinct mechanism of *Helicobacter pylori*-mediated NF-kappa B activation between gastric cancer cells and monocytic cells. *J. Biol. Chem.* **276**: 44856–64.

Mahdavi, J., B. Sonden, M. Hurtig *et al.* 2002. *Helicobacter pylori* SabA adhesin in persistent infection and chronic inflammation. *Science* **297**: 573–8.

Mai, U. E., G. I. Perez-Perez, L. M. Wahl *et al.* 1991. Soluble surface proteins from *Helicobacter pylori* activate monocytes/macrophages by lipopolysaccharide-independent mechanism. *J. Clin. Invest.* **87**: 894–900.

Mai, U. E., G. I. Perez-Perez, J. B. Allen *et al.* 1992. Surface proteins from *Helicobacter pylori* exhibit chemotactic activity for human leukocytes and are present in gastric mucosa. *J. Exp. Med.* **175**: 517–25.

Marshall, B. J. 1991. III. Virulence and pathogenicity of *Helicobacter pylori*. *J. Gastroenterol. Hepatol.* **6**: 121–4.

Marwali, M. R., J. Rey-Ladino, L. Dreolini, D. Shaw, and F. Takei. 2003. Membrane cholesterol regulates LFA-1 function and lipid raft heterogeneity. *Blood* **102**: 215–22.

Menaker, R. J., P. J. M. Ceponis, and N. L. Jones. 2004. *Helicobacter pylori* induces apoptosis of macrophages in association with alterations in the mitochondrial pathway. *Infect. Immun.* **72**: 2889–98.

Meyer, F., K. S. Ramanujam, A. P. Gobert, S. P. James, and K. T. Wilson. 2003. Cutting edge: Cyclooxygenase-2 activation suppresses Th1 polarization in response to *Helicobacter pylori*. *J. Immunol.* **171**: 3913–17.

Miller-Podraza, H., J. Bergstrom, S. Teneberg *et al.* 1999. *Helicobacter pylori* and neutrophils: sialic acid-dependent binding to various isolated glycoconjugates. *Infect. Immun.* **67**: 6309–13.

Mimuro, H., T. Suzuki, J. Tanaka *et al.* 2002. Grb2 is a key mediator of *Helicobacter pylori* CagA protein activities. *Mol. Cell* **10**: 745–55.

Moese, S., M. Selbach, U. Zimny-Arndt *et al.* 2001. Identification of a tyrosine-phosphorylated 35 kDa carboxy-terminal fragment (p35(CagA)) of the *Helicobacter pylori* CagA protein in phagocytic cells: Processing or breakage? *Proteomics* **1**: 618–29.

Moese, S., M. Selbach, T. F. Meyer, and S. Backert. 2002. cag(+) *Helicobacter pylori* induces homotypic aggregation of macrophage-like cells by up-regulation and recruitment of intracellular adhesion molecule 1 to the cell surface. *Infect. Immun.* **70**: 4687–91.

Molinari, M., C. Galli, N. Norais *et al.* 1997. Vacuoles induced by *Helicobacter pylori* toxin contain both late endosomal and lysosomal markers. *J. Biol. Chem.* **272**: 25339–44.

Molinari, M., M. Salio, C. Galli *et al.* 1998. Selective inhibition of Ii-dependent antigen presentation by *Helicobacter pylori* toxin VacA. *J. Exp. Med.* **187**: 135–40.

Montecucco, C., and M. de Bernard. 2003. Molecular and cellular mechanisms of action of the vacuolating cytotoxin (VacA) and neutrophil-activating protein (HP-NAP) virulence factors of *Helicobacter pylori*. *Microbes Infect.* **5**: 715–21.

Montecucco, C., and R. Rappuoli. 2001. Living dangerously: How *Helicobacter pylori* survives in the human stomach. *Nat. Rev. Mol. Cell Biol.* **2**: 457–66.

Moran, A. P., I. M. Helander, and T. U. Kosunen. 1992. Compositional analysis of *Helicobacter pylori* rough-form lipopolysaccharides. *J. Bacteriol.* **174**: 1370–77.

Moran, A. P., M. M. Prendergast, and B. J. Appelmelk. 1996. Molecular mimicry of host structures by bacterial lipopolysaccharides and its contribution to disease. *FEMS Immunol. Med. Microbiol.* **16**: 105–15.

Moran, A. P., Y. A. Knirel, S. N. Senchenkova *et al.* 2002. Phenotypic variation in molecular mimicry between *Helicobacter pylori* lipopolysaccharides and human gastric epithelial cell surface glycoforms – Acid-induced phase variation in Lewis(X) and Lewis(Y) expression by *H. pylori* lipopolysaccharides. *J. Biol. Chem.* **277**: 5785–95.

Mori, M., A. Wada, T. Hirayama *et al.* 2000. Activation of intercellular adhesion molecule 1 expression by *Helicobacter pylori* is regulated by NF-κB in gastric epithelial cancer cells. *Infect. Immun.* **68**: 1806–14.

Muotiala, A., I. M. Helander, L. Pyhala, T. U. Kosunen, and A. P. Moran. 1992. Low biological activity of *Helicobacter pylori* lipopolysaccharide. *Infect. Immun.* **60**: 1714–16.

Murakami, M., K. Asagoe, H. Dekigai *et al.* 1995. Products of neutrophil metabolism increase ammonia-induced gastric mucosal damage. *Dig. Dis. Sci.* **40**: 268–73.

Nathan, C. F. 1987. Neutrophil activation on biological surfaces. *J. Clin. Invest.* **80**: 1550–60.

Nhieu, G. T. V., R. Bourdet-Sicard, G. Dumenil, A. Blocker, and P. J. Sansonetti. 2000. Bacterial signals and cell responses during *Shigella* entry into epithelial cells. *Cell. Microbiol.* **2**: 187–93.

Nielsen, H., and L. P. Andersen. 1992a. Activation of human phagocyte oxidative metabolism by *Helicobacter pylori. Gastroenterology.* **103**: 1747–53.

Nielsen, H., and L. P. Andersen. 1992b. Chemotactic activity of *Helicobacter pylori* sonicate for human polymorphonuclear leucocytes and monocytes. *Gut* **33**: 738–42.

Nielsen, H., S. Birkholz, L. P. Andersen, and A. P. Moran. 1994. Neutrophil activation by *Helicobacter pylori* lipopolysaccharides. *J. Infect. Dis.* **170**: 135–9.

Nishioka, H., I. Baesso, G. Semenzato *et al.* 2003. The neutrophil-activating protein of *Helicobacter pylori* (HP-NAP) activates the MAPK pathway in human neutrophils. *Eur. J. Immunol.* **33**: 840–9.

Norgaard, A., L. P. Andersen, and H. Nielsen. 1995. Neutrophil degranulation by *Helicobacter pylori* proteins. *Gut* **36**: 354–7.

O'Toole, P. W., L. Janzon, P. Doig *et al.* 1995. The putative neuraminyllactose-binding hemagglutinin HpaA of *Helicobacter pylori* CCUG 17874 is a lipoprotein. *J. Bacteriol.* **177**: 6049–57.

Oddo, M., T. Renno, A. Attinger *et al.* 1998. Fas ligand-induced apoptosis of infected human macrophages reduces the viability of intracellular *Mycobacterium tuberculosis. J. Immunol.* **160**: 5448–54.

Odenbreit, S., B. Wieland, and R. Haas. 1996. Cloning and genetic characterization of *Helicobacter pylori* catalase and construction of a catalase-deficient mutant strain. *J. Bacteriol.* **178**: 6960–7.

Odenbreit, S., J. Puls, B. Sedlmaier *et al.* 2000. Translocation of *Helicobacter pylori* CagA into gastric epithelial cells by type IV secretion. *Science* **287**: 1497–500.

Odenbreit, S., B. Gebert, J. Puls, W. Fischer, and R. Haas. 2001. Interaction of *Helicobacter pylori* with professional phagocytes: role of the cag pathogenicity island and translocation, phosphorylation and processing of CagA. *Cell. Microbiol.* **3**: 21–31.

Olczak, A. A., J. W. Olson, and R. J. Maier. 2002. Oxidative-stress resistance mutants of *Helicobacter pylori. J. Bacteriol.* **184**: 3186–93.

Oshima, H., M. Oshima, K. Inaba, and M. M. Taketo. 2004. Hyperplastic gastric tumors induced by activated macrophages in COX-2/mPGES-1 transgenic mice. *EMBO J.* **23**: 1669–78.

Papini, E., B. Satin, C. Bucci *et al.* 1997. The small GTP binding protein rab7 is essential for cellular vacuolation induced by *Helicobacter pylori* cytotoxin. *EMBO J.* **16**: 15–24.

Paziak-Domanska, B., M. Chmiela, A. Jarosinska, and W. Rudnicka. 2000. Potential role of CagA in the inhibition of T cell reactivity in *Helicobacter pylori* infections. *Cell. Immunol.* **202**: 136–9.

Pece, S., D. Fumarola, G. Giuliani, E. Jirillo, and A. P. Moran. 1995. Activity in the *Limulus* amebocyte lysate assay and induction of tumor necrosis factor-alpha by diverse *Helicobacter pylori* lipopolysaccharide preparations. *J. Endotoxin Res.* **2**: 455–62.

Peiser, L., and S. Gordon. 2001. The function of scavenger receptors expressed by macrophages and their role in the regulation of inflammation. *Microbes Infect.* **3**: 149–59.

Peiser, L., S. Mukhopadhyay, and S. Gordon. 2002. Scavenger receptors in innate immunity. *Curr. Opin. Immunol.* **14**: 123–8.

Peppoloni, S., S. Mancianti, G. Volpini *et al.* 2002. Antibody-dependent macrophage-mediated activity against *Helicobacter pylori* in the absence of complement. *Eur. J. Immunol.* **32**: 2721–5.

Phadnis, S. H., M. H. Parlow, M. Levy *et al.* 1996. Surface localization of *Helicobacter pylori* urease and a heat shock protein homolog requires bacterial autolysis. *Infect. Immun.* **64**: 905–12.

Pignatelli, B., B. Bancel, J. Esteve *et al.* 1998. Inducible nitric oxide synthase, anti-oxidant enzymes and *Helicobacter pylori* infection in gastritis and gastric precancerous lesions in humans. *Eur. J. Cancer Prev.* **7**: 439–47.

Pignatelli, B., B. Bancel, M. Plummer *et al.* 2001. *Helicobacter pylori* eradication attenuates oxidative stress in human gastric mucosa. *Am. J. Gastroenterol.* **96**: 1758–66.

Pomorski, T., T. F. Meyer, and M. Naumann. 2001. *Helicobacter pylori*-induced prostaglandin E-2 synthesis involves activation of cytosolic phospholipase A(2) in epithelial cells. *J. Biol. Chem.* **276**: 804–10.

Prakobphol, A., F. Xu, V. M. Hoang *et al.* 2000. Salivary agglutinin, which binds *Streptococcus mutans* and *Helicobacter pylori*, is the lung scavenger receptor cysteine-rich protein gp-340. *J. Biol. Chem.* **275**: 39860–6.

Pruul, H., P. C. Lee, C. S. Goodwin, and P. J. McDonald. 1987. Interaction of *Campylobacter pyloridis* with human immune defence mechanisms. *J. Med. Microbiol.* **23**: 233–8.

Putsep, K., C. I. Branden, H. G. Boman, and S. Normark. 1999. Antibacterial peptide from *H-pylori. Nature* **398**: 671–2.

Raeder, E. M. B., P. J. Mansfield, V. Hinkovska-Galcheva, J. A. Shayman, and L. A. Boxer. 1999. Syk activation initiates downstream signaling events during human polymorphonuclear leukocyte phagocytosis. *J. Immunol.* **163**: 6785–93.

Ramarao, N., and T. F. Meyer. 2001. *Helicobacter pylori* resists phagocytosis by macrophages: Quantitative assessment by confocal microscopy and fluorescence-activated cell sorting. *Infect. Immun.* **69**: 2604–11.

Ramarao, N., S. D. Gray-Owen, S. Backert, and T. F. Meyer. 2000a. *Helicobacter pylori* inhibits phagocytosis by professional phagocytes involving type IV secretion components. *Mol. Microbiol.* **37**: 1389–04.

Ramarao, N., S. D. Gray-Owen, and T. F. Meyer. 2000b. *Helicobacter pylori* induces but survives the extracellular release of oxygen radicals from professional phagocytes using its catalase activity. *Mol. Microbiol.* **38**: 103–13.

Rautelin, H., B. Blomberg, H. Fredlund, G. Jarnerot, and D. Danielsson. 1993. Incidence of *Helicobacter pylori* strains activating neutrophils in patients with peptic ulcer disease. *Gut* **34**: 599–603.

Rautelin, H., G. Blomberg, G. Jarnerot, and D. Danielsson. 1994a. Nonopsonic activation of neutrophils and cytotoxin production by *Helicobacter pylori*: ulcerogenic markers. *Scand. J. Gastroenterol.* **29**: 128–32.

Rautelin, H., C. H. von Bonsdorff, B. Blomberg, and D. Danielsson. 1994b. Ultrastructural study of two patterns in the interaction of *Helicobacter pylori* with neutrophils. *J. Clin. Pathol.* **47**: 667–9.

Rautemaa, R., H. Rautelin, P. Puolakkainen *et al.* 2001. Survival of *Helicobacter pylori* from complement lysis by binding of GPI-anchored protectin (CD59). *Gastroenterology* **120**: 470–9.

Ringner, M., K. H. Valkonen, and T. Wadstrom. 1994. Binding of vitronectin and plasminogen to *Helicobacter pylori. FEMS Immunol. Med. Microbiol.* **9**: 29–34.

Rittig, M. G., B. Shaw, D. P. Letley *et al.* 2003. *Helicobacter pylori*-induced homotypic phagosome fusion in human monocytes is independent of the bacterial vacA and cag status. *Cell. Microbiol.* **5**: 887–99.

Roche, N., J. Angstrom, M. Hurtig *et al.* 2004. *Helicobacter pylori* and complex gangliosides. *Infect. Immun.* **72**: 1519–29.

Rokita, E., A. Makristathis, E. Presterl, M. L. Rotter, and A. M. Hirschl. 1998. *Helicobacter pylori* urease significantly reduces opsonization by human complement. *J. Infect. Dis.* **178**: 1521–5.

Sabarth, N., S. Lamer, U. Zimny-Arndt *et al.* 2002. Identification of surface proteins of *Helicobacter pylori* by selective biotinylation, affinity purification, and two-dimensional gel electrophoresis. *J. Biol. Chem.* **277**: 27896–902.

Sakai, T., H. Fukui, F. Franceschi *et al.* 2003. Cyclooxygenase expression during *Helicobacter pylori* infection in mongolian gerbils. *Dig. Dis. Sci.* **48**: 2139–46.

Sasaki, M., T. Joh, T. Tada *et al.* 1998. Altered expression of membrane inhibitors of complement in human gastric epithelium during *Helicobacter*-associated gastritis. *Histopathology* **33**: 554–60.

Satin, B., N. Norais, J. L. Telford *et al.* 1997. Effect of *Helicobacter pylori* vacuolating cytotoxin on maturation and extracellular release of procathepsin D and on epidermal growth factor degradation. *J. Biol. Chem.* **272**: 25022–8.

Satin, B., G. Del Giudice, V. Della Bianca *et al.* 2000. The neutrophil activating protein (HP-NAP) of *Helicobacter pylori* is a protective antigen and a major virulence factor. *J. Exp. Med.* **191**: 1467–76.

Saunders, B. M., S. L. Fernando, R. Sluyter, W. J. Britton, and J. S. Wiley. 2003. A loss-of-function polymorphism in the human P2X(7) receptor abolishes ATP-mediated killing of mycobacteria. *J. Immunol.* **171**: 5442–6.

Schmauber, B., M. Andrulis, S. Endrich *et al.* 2004. Expression and subcellular distribution of toll-like receptors TLR4, TLR5 and TLR9 on the gastric epithelium in *Helicobacter pylori* infection. *Clin. Exp. Immunol.* **136**: 521–6.

Schraw, W., M. S. McClain, and T. L. Cover. 1999. Kinetics and mechanisms of extracellular protein release by *Helicobacter pylori. Infect. Immun.* **67**: 5247–52.

Segal, E. D., J. Cha, J. Lo, S. Falkow, and L. S. Tompkins. 1999. Altered states: Involvement of phosphorylated CagA in the induction of host cellular growth changes by *Helicobacter pylori. Proc. Natl. Acad. Sci. USA* **96**: 14559–64.

Selbach, M., S. Moese, R. Hurwitz *et al.* 2003. The *Helicobacter pylori* CagA protein induces cortactin dephosphorylation and actin rearrangement by c-Src inactivation. *EMBO J.* **22**: 515–28.

Seto, K., Y. Hayashikuwabara, T. Yoneta, H. Suda, and H. Tamaki. 1998. Vacuolation induced by cytotoxin from *Helicobacter pylori* is mediated by the egf receptor in HelA cells. *FEBS Lett.* **431**: 347–50.

Seyler, R. W., J. W. Olson, and R. J. Maier. 2001. Superoxide dismutase-deficient mutants of *Helicobacter pylori* are hypersensitive to oxidative stress and defective in host colonization. *Infect. Immun.* **69**: 4034–40.

Shimoyama, T., S. Fukuda, Q. Liu *et al.* 2003. *Helicobacter pylori* water soluble surface proteins prime human neutrophils for enhanced production of reactive oxygen species and stimulate chemokine production. *J. Clin. Pathol.* **56**: 348–51.

Spiegelhalder, C., B. Gerstenecker, A. Kersten, E. Schlitz, and M. Kist. 1993. Purification of *Helicobacter pylori* superoxide dismutase and cloning and sequencing of the gene. *Infect. Immun.* **61**: 5315–25.

St-Denis, A., V. Cauras, F. Gervais, and A. Descoteaux. 1999. Role of protein kinase C-α in the control of infection by intracellular pathogens in macrophages. *J. Immunol.* **163**: 5505–11.

Standaert, M. L., L. Galloway, P. Karnam *et al.* 1997. Protein kinase C-zeta is a downstream effector of phosphatidylinositol 3-kinase during insulin stimulation in rat adipocytes. *J. Biol. Chem.* **272**: 30075–82.

Stein, M., R. Rappuoli, and A. Covacci. 2000. Tyrosine phosphorylation of the *Helicobacter pylori* CagA antigen after cag-driven host cell translocation. *Proc. Natl. Acad. Sci. USA* **97**: 1263–8.

Suda, Y., T. Ogawa, W. Kashihara *et al.* 1997. Chemical structure of lipid a from *Helicobacter pylori* strain 206–1 lipopolysaccharide. *J. Biochem.* **121**: 1129–33.

Sundrud, M. S., V. J. Torres, D. Unutmaz, and T. L. Cover. 2004. Inhibition of primary human T cell proliferation by *Helicobacter pylori* vacuolating toxin (VacA) is independent of VacA effects on IL-2 secretion. *Proc. Natl. Acad. Sci. USA* **101**: 7727–32.

Suzuki, H., M. Mori, K. Seto *et al.* 1999. Polaprezinc, a gastroprotective agent: attenuation of monochloramine-evoked gastric DNA fragmentation. *J. Gastroenterol.* **34**: 43–6.

Suzuki, M., S. Miura, M. Mori *et al.* 1994. Rebamipide, a novel antiulcer agent, attenuates *Helicobacter pylori* induced gastric mucosal cell injury associated with neutrophil derived oxidants. *Gut* **35**: 1375–8.

Suzuki, T., K. Ina, T. Nishiwaki *et al.* 2004. Differential roles of interleukin-1 beta and interleukin-8 in neutrophil transendothelial migration in patients with *Helicobacter pylori* infection. *Scand. J. Gastroenterol.* **39**: 313–21.

Taki, C., S. Kitajima, K. Sueyoshi *et al.* 2002. MUC1 mucin expression in follicular dendritic cells and lymphoepithelial lesions of gastric mucosa-associated lymphoid tissue lymphoma. *Pathol. Internat.* **52**: 691–701.

Tatsuguchi, A., C. Sakamoto, K. Wada *et al.* 2000. Localisation of cyclooxygenase 1 and cyclooxygenase 2 in *Helicobacter pylori* related gastritis and gastric ulcer tissues in humans. *Gut* **46**: 782–9.

Teneberg, S., H. Miller-Podraza, H. C. Lampert *et al.* 1997. Carbohydrate binding specificity of the neutrophil-activating protein of *Helicobacter pylori*. *J. Biol. Chem.* **272**: 19067–71.

Teneberg, S., M. Jurstrand, K. A. Karlsson, and D. Danielsson. 2000. Inhibition of nonopsonic *Helicobacter pylori*-induced activation of human neutrophils by sialylated oligosaccharides. *Glycobiology* **10**: 1171–81.

Thorensen, A.-C. E., A. Hamlet, J. Celik *et al.* 2000. Differences in surface-exposed antigen expression between *Helicobacter pylori* strains isolated from duodenal ulcer patients and from asymptomatic subjects. *J. Clin. Microbiol.* **38**: 3436–41.

Tonello, F., W. G. Dundon, B. Satin *et al.* 1999. The *Helicobacter pylori* neutrophil-activating protein is an iron-binding protein with dodecameric structure. *Mol. Microbiol.* **34**: 238–46.

Trottein, F., S. Nutten, J. P. Papin *et al.* 1997. Role of adhesion molecules of the selectin-carbohydrate families in antibody-dependent cell-mediated cytoxicity to schistosome targets. *J. Immunol.* **159**: 804–11.

Trust, T. J., P. Doig, L. Emody *et al.* 1991. High-affinity binding of the basement membrane proteins collagen type IV and laminin to the gastric pathogen *Helicobacter pylori. Infect. Immun.* **59**: 4398–404.

Underhill, D. M., and A. Ozinsky. 2002a. Toll-like receptors: key mediators of microbe detection. *Curr. Opin. Immunol.* **14**: 103–10.

Underhill, D. M., and A. Ozinsky. 2002b. Phagocytosis of microbes: complexity in action. *A. Rev. Immunol.* **20**: 825–52.

Valkonen, K. H., T. Wadstrom, and A. P. Moran. 1994. Interaction of lipopolysaccharides of *Helicobacter pylori* with the basement membrane protein laminin. *Infect. Immun.* **62**: 3640–8.

Valkonen, K. H., T. Wadstrom, and A. P. Moran. 1997. Identification of the N-acetylneuraminyllactose-specific laminin-binding protein of *Helicobacter pylori. Infect. Immun.* 65: 916–23.

van Kooyk, Y., and T. B. H. Geijtenbeek. 2003. DC-sign: Escape mechanism for pathogens. *Nat. Rev. Immunol.* **3**: 697–709.

Vinogradov, E., M. B. Perry, and J. W. Conlan. 2002. Structural analysis of *Francisella tularensis* lipopolysaccharide. *Eur. J. Biochem.* **269**: 6112–18.

Wang, J., E. G. Brooks, K. B. Bamford *et al.* 2001. Negative selection of T cells by *Helicobacter pylori* as a model for bacterial strain selection by immune evasion. *J. Immunol.* **167**: 926–34.

Ways, D. K., P. P. Cook, C. Webster, and P. J. Parker. 1992. Effect of phorbol esters on protein kinase C-zeta. *J. Biol. Chem.* **267**: 4799–805.

Wen, S., C. P. Felley, H. Bouzourene *et al.* 2004. Inflammatory gene profiles in gastric mucosa during *Helicobacter pylori* infection in humans. *J. Immunol.* **172**: 2592–606.

Williams, M. A., and J. S. Solomkin. 1999. Integrin-mediated signaling in human neutrophil functioning. *J. Leukoc. Biol.* **65**: 725–36.

Wilson, K. T., K. S. Ramanujam, H. L. T. Mobley *et al.* 1996. *Helicobacter pylori* stimulates inducible nitric oxide synthase expression and activity in a murine macrophage cell line. *Gastroenterology* **111**: 1524–33.

Wirth, H. P., M. Q. Yang, M. Karita, and M. J. Blaser. 1996. Expression of the human cell surface glycoconjugates lewis x and lewis y by *Helicobacter pylori* isolates is related to caga status. *Infect. Immun.* **64**: 4598–605.

Wu, K. C., L. M. Jackson, A. M. Galvin *et al.* 1999. Phenotypic and functional characterization of the myofibroblasts, macrophages, and lymphocytes migrating out of the human gastric lamina propria following loss of epithelial cells. *Gut* **44**: 323–30.

Xiang, Z., S. Censini, P. F. Bayeli *et al.* 1995. Analysis of expression of CagA and VacA virulence factors in 43 strains of *Helicobacter pylori* reveals that clinical isolates can be divided into two major types and that CagA is not necessary for expression of the vacuolating cytotoxin. *Infect. Immun.* **63**: 94–8.

Yahiro, K., T. Niidome, M. Kimura *et al.* 1999. Activation of *Helicobacter pylori* VacA toxin by alkaline or acid conditions increases its binding to a 250-kDa receptor protein-tyrosine phosphatase beta. *J. Biol. Chem.* **274**: 36693–9.

Yahiro, K., A. Wada, M. Nakayama *et al.* 2003. Protein-tyrosine phosphatase alpha, RPTP alpha, is a *Helicobacter pylori* VacA receptor. *J. Biol. Chem.* **278**: 19183–9.

Zevering, Y., L. Jacob, and T. F. Meyer. 1999. Naturally acquired human immune responses against *Helicobacter pylori* and implications for vaccine development. *Gut* **45**: 465–74.

Zheng, P. Y., and N. L. Jones. 2003. *Helicobacter pylori* strains expressing the vacuolating cytotoxin interrupt phagosome maturation in macrophages by recruiting and retaining TACO (coronin 1) protein. *Cell. Microbiol.* **5**: 25–40.

Zu, Y., N. D. Cassai, and G. S. Sidhu. 2000. Light microscopic and ultrastructural evidence of *in vivo* phagocytosis of *Helicobacter pylori* by neutrophils. *Ultrastruct. Pathol.* **24**: 319–23.

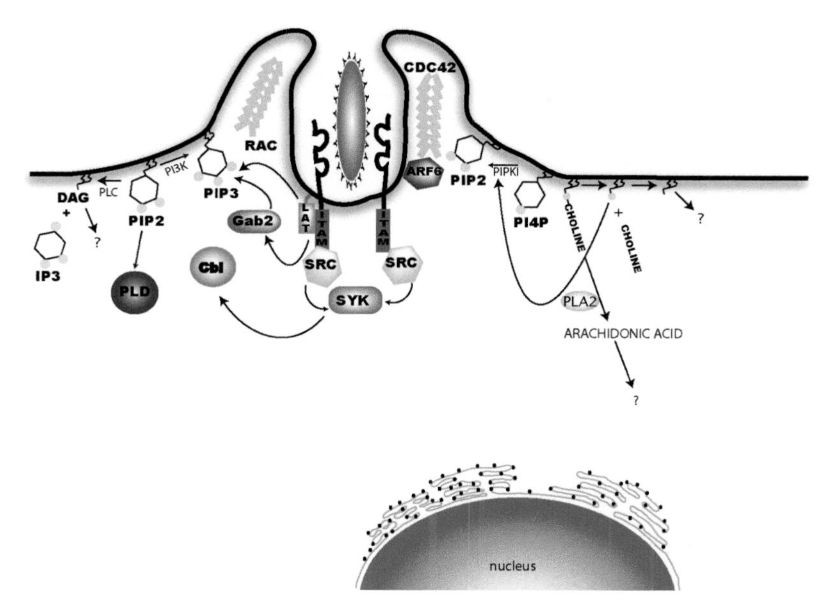

Figure 3.2 Fc receptor-mediated signaling. IgG stimulation results in Fc receptor clustering followed by phosphorylation of the ITAM motif by Src kinases. Phosphorylated tyrosines serve as docking sites for Syk kinase, initiating a cascade of responses involving adaptor proteins, serine/threonine kinases, phosphoinositides, and proteins that modify these lipids. These molecules generate synergistic signals culminating in pseudopodial extension and particle internalization. See text for details.

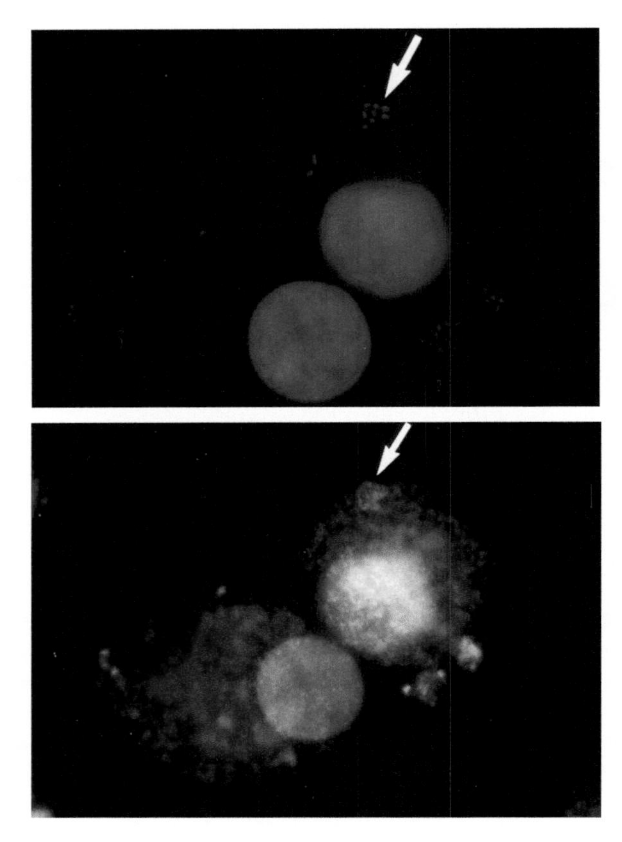

Figure 5.7 Immunofluorescence of human alveolar macrophages 80 min after uptake of opsonized *Streptococcus pneumoniae*. External bacteria are labeled with FITC (green); internalized bacteria are observed as blue (DAPI-stained) elements within the cell. Some internalized bacteria (marked with an arrow in both micrographs) are co-localizing with a marker (rhodamine-red) labeling up the lysosomal membrane protein LAMP-1, which is enriched in late endosomal–lysosomal compartments.

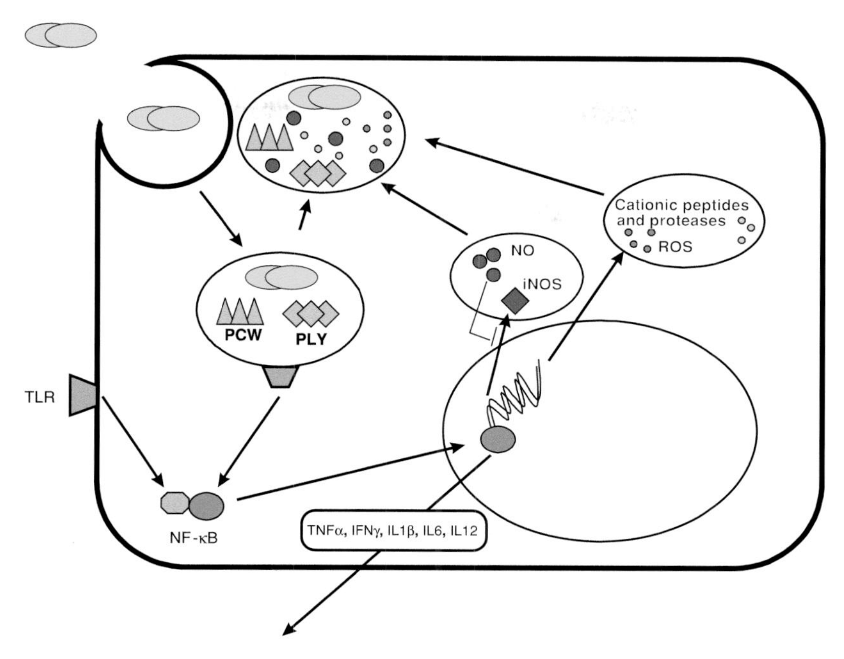

Figure 5.8 Microbicidal mechanisms: antimicrobial defenses of phagocytes against pneumococci. Internalized pneumococci enter a phagosome. Cell-surface signals mediated by Toll-like receptors (TLR) result in cytokine signaling with distinct patterns for macrophages (as shown) and PMN. Microbicidal molecules include those located in lysosomes such as reactive oxygen species (ROS), proteases and cationic peptides (many of which are distinct for PMN), and those located in other organelles such as inducible nitric oxide synthase (iNOS), derived from nitric oxide (NO), which are relatively more important in macrophages. All of these organelles fuse with the phagolysosome to mediate bacterial killing and release of factors such as pneumococcal cell wall (PCW) and pneumolysin (PLY), which of themselves may contribute to further production of antimicrobial host defense molecules.

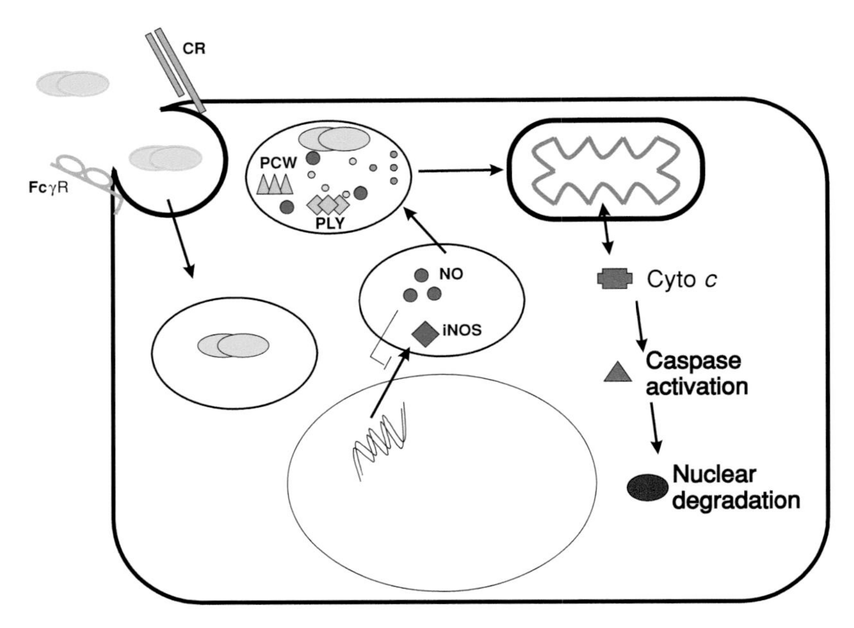

Figure 5.9 Induction of apoptosis in macrophages by pneumococci. Pneumococci opsonized by complement and immunoglobulin are internalized via complement receptors (CR) and Fc gamma receptors (FcγR) and enter a phagosome. The phagosome fuses with lysosomes and potentially with other organelles containing nitric oxide generated from inducible nitric oxide synthase (iNOS). Killing of bacteria in a phagolysosome results in release of pneumococcal cell wall (PCW) and pneumolysin (PLY), which contribute to mitochondrial membrane permeabilization and release of cytochrome *c* (Cyto *c*). Caspase activation results in nuclear degradation and other morphologic features of apoptosis.

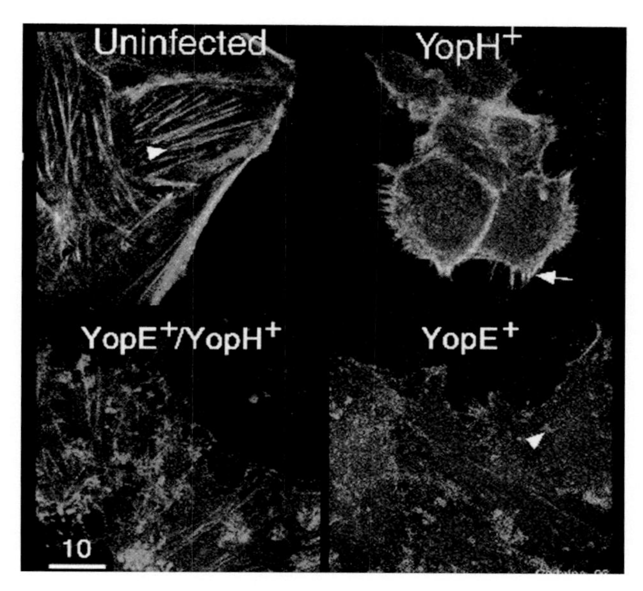

Figure 6.4 YopH and YopE have distinct effects on host cell cytoskeleton. Confocal images of HeLa cells, not infected, or infected with a *Yersinia* multiple *Yop* mutant strain expressing either YopH or YopE, or infected with *Yersinia* wild type (wt). YopE causes fragmentation of the F-actin cytoskeleton, whereas YopH affects the integrity of focal adhesions and associated stress fibers. The combined effect is seen with the *Yersinia* wt strain. (Note that this strain translocates a smaller amount of the effectors compared with the multiple mutant strain.) Cellular F-actin was visualized by staining with fluorescein-conjugated phalloidin (green); vinculin-containing focal adhesions were visualized by indirect immunofluorescence (red). The yellow color represents co-localization of microfilaments and vinculin. Vinculin-containing focal adhesions (arrowheads) and vinculin-containing retraction fibers (arrows) are shown. All sections were scanned under identical conditions and show the basolateral side of the cells. Scale bar: 10 μm.

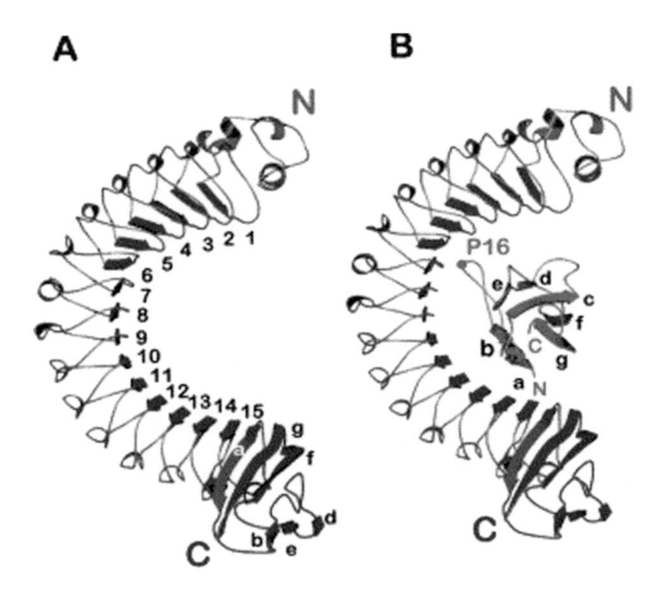

Figure 7.2 Structure of InlA. (A) Uncomplexed InlA′: cap domain, pink; LRR-domain, violet; Ig-like inter-repeat domain, blue. Strands of the LRR are numbered; strands of the Ig-like domain are indicated by letters. (B) The complex InlA′/hEC1 viewed as in (A). hEC1 is rendered in green; strands indicated by letters. (From Schubert, W. D. *et al.* 2002. *Cell* **111**: 825–36.)

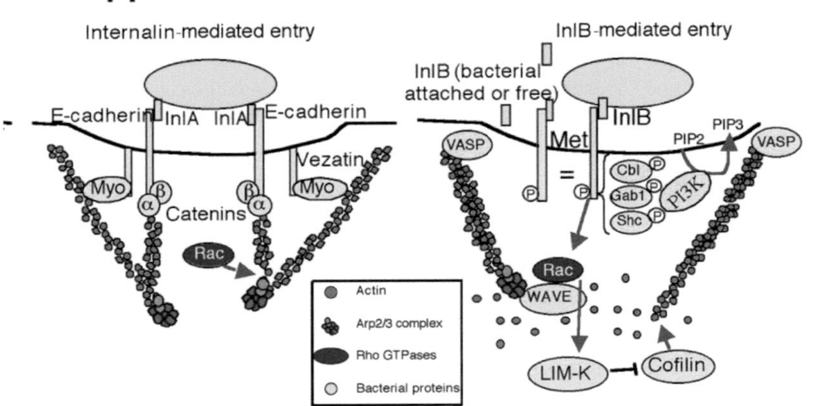

Figure 7.3 Zipper mechanisms of phagocytosis induced by InlA (internalin) and InlB.

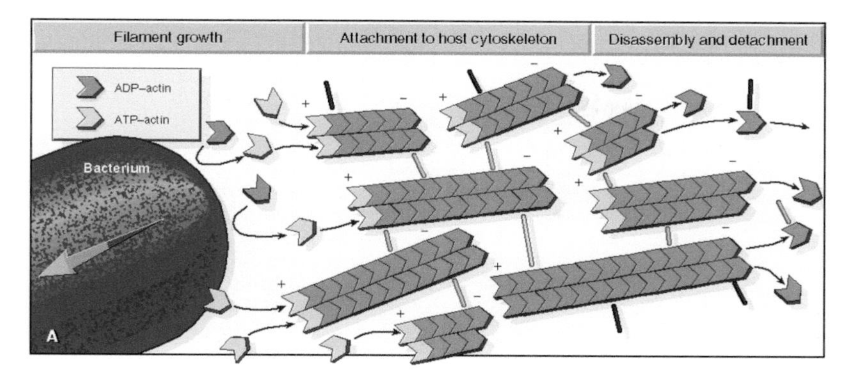

Figure 7.6 Schematic image of *Listeria*-induced actin rocket tail assembly. New ATP-actin monomers are added in the polymerization zone, located at the back surface of the motile bacteria. As the monomers are incorporated onto the barbed or fast-growing ends of actin filaments, ATP is hydrolyzed to ADP. The actin filaments in the tail steadily depolymerize over time and the regions of the tail farthest from the bacteria disappear as monomers steadily dissociate from the filaments as ADP-actiin. Actin filaments in the tails are linked to the host cell actin cytoskeleton by actin filament bundling-proteins such as alpha-actinin.

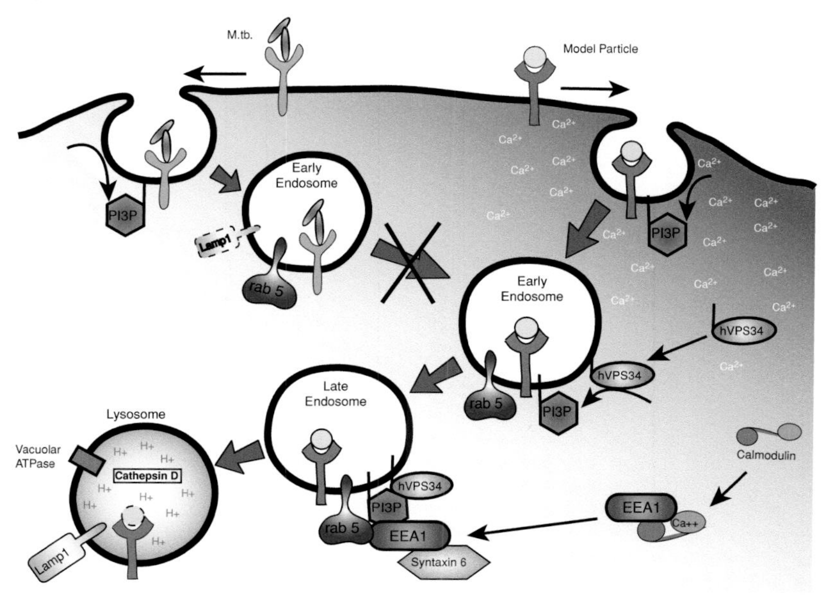

Figure 8.1 Maturation of phagosomes containing live *Mycobacterium tuberculosis* (M. tb.) compared with that of phagosomes containing a model particle. Phagosomes containing live M.tb. retain markers of early endosomes such as Rab5 and transferrin receptor, whereas phagosomes containing model particles acquire hVPS34, EEA1, and syntaxin 6, which promote phagosome acquisition of the vacuolar ATPase, LAMP-1, and mature cathepsin D. An M.tb.-associated defect in calcium signaling may be the most proximal determinant of the fate of distinct phagosomes. See the text for fetails.

CHAPTER 5

Phagocytosis of *Streptococcus pneumoniae*

Dominic L. Jack, David H. Dockrell, Stephen B. Gordon,
and Robert C. Read

INTRODUCTION

Streptococcus pneumoniae is a major cause of morbidity and mortality worldwide, and commonly colonizes the upper respiratory tract. In some colonized individuals the organism translocates to other tissues and causes life-threatening diseases including pneumonia, bacteremia, and meningitis. Rates of disease are especially high in the very young and old and in patients with predisposing conditions including HIV infection, cardiopulmonary or co-morbidities, renal diseases including nephrotic syndrome, and sickle cell disease and other causes of hyposplenia.

Approximately 5 million children under the age of 5 years die with a respiratory tract infection every year; the major causative pathogen in these cases is *S. pneumoniae* (Williams *et al.* 2002). Pneumococcal infection caused approximately 45,000 deaths in adults in the United States in 1998, about two years prior to the introduction of the 7-valent pneumococcal conjugate vaccine for infants (Robinson *et al.* 2001). *Streptococcus pneumoniae* also causes less serious but extremely common diseases such as otitis media, sinusitis and exacerbations of chronic obstructive pulmonary disease. It has been estimated that there are 7 million cases of otitis media in the United States every year (Stool & Field 1989).

OVERVIEW OF PATHOGENESIS

Nasopharyngeal carriage

The first step in the pathogenesis of pneumococcal disease is nasopharyngeal colonization, with individuals often carrying more than one serotype

Phagocytosis and Bacterial Pathogenicity, ed. J. D. Ernst and O. Stendahl. Published by Cambridge University Press.

at a time. Asymptomatic nasopharyngeal carriage is established when surface components of the pneumococcus bind to nasal epithelium (Tuomanen & Masure 1997; Weiser *et al.*, 1996). Pneumococci of a given genotype exist in opaque and transparent phenotypes: the transparent phase exhibits less polysaccharide capsule and more prominent surface proteins than the capsule-rich opaque phase and as such has a greater capacity to colonize host epithelium. Transparent pneumococci have relatively scant capsule, however, and are hence ill-equipped to invade other host sites better defended by phagocytes (Kim & Weiser 1998). The nasopharyngeal mucosal surface is devoid of macrophages and colonized by pneumococci in the transparent phase. Nasopharyngeal colonization typically ends after several months and this is accompanied by the detection of capsule-specific mucosal and circulating antibody (Smith *et al.* 1993; Musher *et al.* 1997). The emergence of anticapsular antibody after colonization suggests that antigen-presenting cells, probably in association with local lymphoid tissue, have been exposed to antigen. Adults experience less nasopharyngeal colonisation than children, and are colonized by pneumococci with a smaller range of capsular types, probably owing to the development of persistent capsule-specific mucosal antibody (Lloyd-Evans *et al.* 1996; Simell *et al.* 2002). Most episodes of disease develop during the first week of carriage prior to the generation of a humoral antibody response (Musher *et al.* 1990). In normal individuals, without concurrent viral infections and with intact mucociliary clearance, the incidence of invasive disease is extremely low, and pneumococci reside in a stable relationship with the respiratory tract mucosa of the nasopharynx. Traffic into sinuses, the middle ear, and the lower respiratory tract is prevented by the normal activity of mucociliary clearance.

From carriage to disease

Disease of the lower respiratory tract follows micro-aspiration of organisms from the nasopharynx into the gas-exchanging areas of the lungs (Kiluchi *et al.*, 1994). Following aspiration into the lower airways, pneumococci bind to type II alveolar epithelial cells via receptors including PAF receptors (Cundell *et al.*, 1995), and are then internalized to the basal surface of the cell and extruded (Ring *et al.*, 1998). In the lung parenchyma there are large numbers of interstitial macrophages, dendritic cells, and pulmonary intravascular macrophages (Lohmann-Matthes *et al.* 1994). These cells modulate local pulmonary inflammation and are responsible for antigen presentation in the regional lymph nodes (Kunkel & Butcher 2003). Alveolar macrophages are the dominant phagocyte of the non-inflamed alveolar space; they patrol the respiratory epithelial surface and constitute 90% of the cell population obtained by bronchoalveolar lavage, distributed at a density of approximately one per alveolus and are critical for resistance to pneumonia in animal models. Depletion of alveolar macrophages with administration of liposomal dichloromethylene-biphosphonate induces exquisite susceptibility to pneumococcal infection (Knapp *et al.* 2003). Failure of macrophages to control pneumococci within this compartment and subsequent binding of pneumococcal cell walls to epithelia and endothelia results in the release of TNF, IL1, and chemokines, separation of endothelial cells, and flooding of alveoli by a serous exudate (Tuomanen *et al.* 1995). The activated endothelium expresses integrins, tissue factor, and platelet-activating factor, which begin the massive recruitment of polymorphonuclear leukocytes (PMN) to the consolidating lung. Activation of the alternative complement pathway by bacterial components further amplifies the recruitment of leukocytes. In successful control of pneumonia, there is phagocytic destruction of pneumococci by neutrophils; dying bacteria release cell walls, pneumolysin, and other components, which induce further inflammation and tissue damage. Finally, repair begins with the onset of programmed cell death (apoptosis) of neutrophils. Pneumococci and pneumococcal fragments that enter the bloodstream are rapidly cleared in experimental models by the action of macrophages of the reticuloendothelial system, particularly in the liver and spleen (Hosea *et al.* 1980). The spleen has a further important role in generating capsule-specific antibody responses by the presentation of antigen to marginal-zone B cells in the presence of follicular dendritic cells.

VIRULENCE DETERMINANTS OF *STREPTOCOCCUS PNEUMONIAE*

Streptococcus pneumoniae is a human pathogen with no other known hosts or reservoirs. Selective pressure for virulence determinants arises from the requirement for stable nasopharyngeal colonization; disease is probably an accidental consequence of pneumococcal colonization. The major virulence factors (see Figure 5.1) of the pneumococcus include the polysaccharide capsule, structural components of the cell wall including peptidoglycan and lipoteichoic acid, the thiol-activated toxin pneumolysin, surface proteins (which may be attached to the surface via an interaction with choline [choline-binding proteins] or via a Gram-positive attachment motif, LPXTG), pneumococcal enzymes including autolysin, neuraminidases, hyaluronidase and superoxide dismutase, and the 2-component signal transduction systems. The availability of the genome sequence has accelerated the discovery of additional virulence mechanisms. The genomes of most pneumococcal serotypes contain approximately 2,200 genes, including at least 70 that are

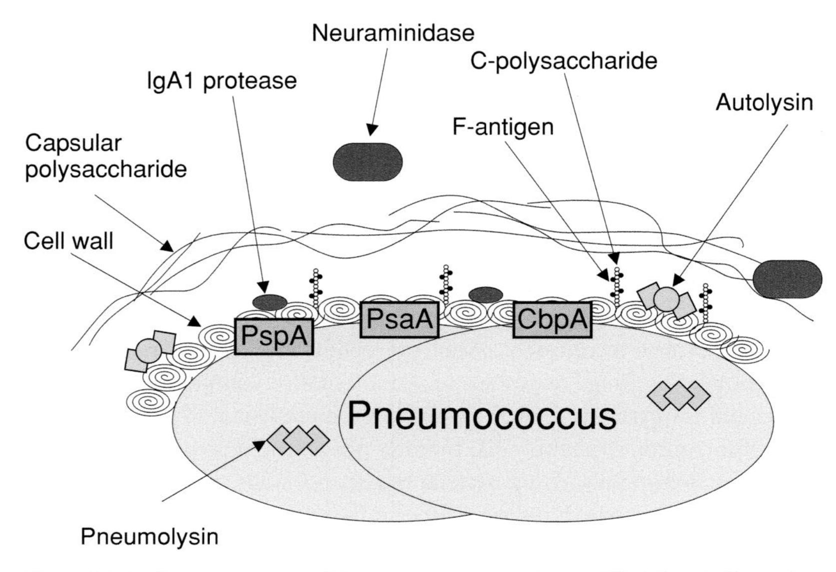

Figure 5.1 Surface components of *Streptococcus pneumoniae* (modified from Briles *et al.* 1998).

predicted to encode surface proteins, in addition to genes encoding many other proteins of hitherto unknown function (Tettelin *et al.* 2001).

Polysaccharide capsule

The polysaccharide capsule is the best known virulence determinant of the pneumococcus. There are 90 different serotypes with structural differences of the repeating units of oligosaccharides that alter chemical structure to an extent that influences disease epidemiology. Some serotypes commonly cause disease in children (types 6A, 14, 19F, 23F) whereas others rarely cause disease in children and affect predominantly adults (e.g. types 3 and 8) (Scott *et al.* 1996). The capsular gene clusters (*cap/cps*) of most disease-related pneumococcal types have been sequenced and are located between *dexB* and *aliA*. The genes within the capsular cluster encode various proteins involved in the synthesis of lipid-linked saccharides on the intracellular face of the cell membrane and the export of these to the surface and subsequent polymerization. There is great variation between serotypes in terms of the number of genes within the capsular gene cluster: for example, the capsule cluster of serotype 3 contains 3 genes whereas that of serotype 23F contains 18. Except in the cases of serotypes 3 and 37, the first four open reading frames of the clusters are well conserved among serotypes, although only the first gene of the cluster (*cap/cpsA*) is more than 90% identical in all the gene clusters sequenced. Otherwise all other genes have considerable heterogeneity and can be considered serotype-specific (Lopez & Garcia 2004). Although the capsule undergoes phase variation (Weiser *et al.* 2001), the molecular basis of this is yet to be determined.

Cell wall

Peptidoglycan, lipoteichoic acid, and teichoic acid are the main constituents of the cell wall, which is rich in polysaccharides and choline. The cell wall's integrity is maintained by numerous cross-linking peptides. Enzymatic degradation of the cell wall (for example, following exposure to β-lactams) releases components that are highly proinflammatory, particularly lipoteichoic acid (Tuomanen *et al.* 1995); exposure of *S. pneumoniae* to penicillin has been shown to enhance Toll-like receptor 2 activation in human cells (Moore *et al.* 2003).

Pneumolysin

Pneumolysin is a 53 kDa cytoplasmic protein produced by all clinical isolates of the pathogen. It is released on autolysis. The virulence properties of pneumolysin are therefore directly dependent on the action of autolysin (Paton *et al.* 1993). Pneumolysin has multiple biological effects that influence pathogenicity, including lytic activity (which is due to the binding of pneumolysin to cell-membrane cholesterol and subsequent pore formation) and activation of complement. Purified pneumolysin is capable of inducing the salient features of pneumonia in animal models. Its biologic activity includes cytotoxicity to epithelial cells and disruption of mucociliary clearance, inhibition of phagocyte and immune cell function, inhibition of neutrophil chemotaxis and respiratory burst, and suppression of proinflammatory cytokine and chemokine release (Mitchell 2000; Jedrzejas 2001).

Surface proteins

Surface proteins that bind to choline

Pneumococcal surface protein A (PspA)

Pneumococcal surface protein A is present on the surface of all pneumococci and is a lactoferrin-binding protein (Hammerschmidt *et al.* 1999). It also inhibits complement activation by *S. pneumoniae* (Tu *et al.* 1999). Choline binding protein A (CbpA, otherwise known as PspC) promotes adherence to a range of cells including pneumocytes and endothelial cells and nasal epithelial cells by mediating binding to glycoconjugates containing lacto-*N*-neotetraose and sialic acids on the surface of target cells (Jedrzejas 2001).

Surface proteins that bind to LPXTG

Hyaluronidase

The enzyme hyaluronidase degrades hyaluronic acid, which is a component of connective tissue. Hyaluronan is a target for the receptor molecule CD44, which is present on many different cell types including macrophages, neutrophils, T cells, B cells, and various epithelial cells (Haynes *et al.* 1989).

Neuraminidase

Neuraminidase is present on all strains of pneumococci and cleaves terminal sialic acid from cell surface glycans such as mucin, glycolipids and glycoproteins. This property probably facilitates successful colonization of the nasopharynx (Jedrzejas 2001).

Pneumococcal surface adhesin A (PsaA)

Pneumococcal surface adhesin A (PsaA) probably mediates adherence of pneumococci to type II pneumocytes and is critical for virulence of the organism. However, the major function of the protein is transport of manganese and other metals into the cytoplasm of pneumococci (Dintilhac *et al.* 1997).

IgA protease

Pneumococci possess two specific defences against IgA. CbpA is able to bind to the secretory component of sIgA, which presumably antagonizes IgA binding via its antigen-specific binding sites (Hammerschmidt *et al.* 1997). This property of CbpA is independent of its complement regulatory role (Dave *et al.* 2004). Pneumococci also posses a surface-bound IgA1 protease, which apart from cleaving IgA may increase adherence of IgA1-ligated organisms (Weiser *et al.* 2003). The second subclass of IgA, IgA2, is resistant to the IgA1 protease.

DEVELOPMENT OF IMMUNITY FOLLOWING INFECTION WITH *STREPTOCOCCUS PNEUMONIAE*

Antibody to pneumococcal components can be detected in the serum of individuals following both colonization and disease. Development of antibody against the capsular polysaccharide will only provide protection against an isolate of homologous serotype. About two-thirds of people who become colonized by a pneumococcal strain develop serotype-specific antibody (Musher *et al.* 1993). However, during experimental colonization of humans, there is also development of antibody against proteins, including both PspA and CbpA (McCool *et al.* 2002, 2003). Studies of children naturally colonized with pneumococci, or subject to otitis media, have shown that even very young children can produce antibodies to the pneumococcal proteins PspA, PsaA, and pneumolysin.

Regarding protective immunity, there is good evidence that antibody against PspA can protect humans against pneumococcal colonization. Anticapsular polysaccharide antibodies are less protective against experimental pneumococcal colonization (McCool *et al.* 2002), but studies with conjugate polysaccharide vaccines in the field have shown that anticapsular antibodies reduce nasopharyngeal colonization by pneumococci of serotypes included in the vaccine (Veenhoven *et al.* 2003).

HOST DEFENSE AGAINST *STREPTOCOCCUS PNEUMONIAE*

The humoral and cellular arms of the immune system are both employed in protection against the pneumococcus with the involvement of adaptive and innate mechanisms. The primary immune defense against the pneumococcus is phagocytosis, which is enhanced by the deposition of opsonins derived from soluble host components. In the naïve host, opsonization and phagocytosis are effected through innate mechanisms but, with age, the importance of adaptive immunity increases.

In serum, the primary component of innate defense is the complement system activated by natural IgM antibody, which is then augmented by the adaptive response of specific IgG. At the respiratory mucosa, locally produced complement with a number of other components, which are currently poorly characterized, make up innate defense. The primary adaptive defense of the mucosa is secretory IgA (sIgA).

INNATE HUMORAL IMMUNITY

The complement system

The complement system is one of the most essential components of defense against pneumococci (Hosea *et al.* 1980). The complement system is a complex collection of over 35 proteins, which make up one of the main effector functions of both the innate and the adaptive immune system (Walport

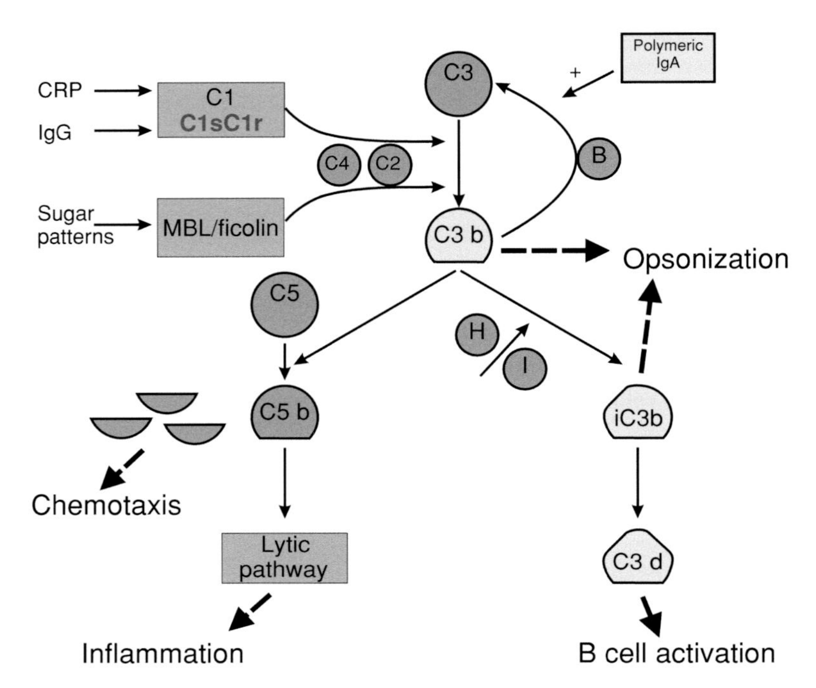

Figure 5.2 The activation of complement. Only the parts of the pathway most relevant to the pneumococcal disease are shown here. The classical pathway is activated by antibody or CRP bound to the cell wall. MBL binds to the pneumococcus, but the ligand or ligands are not known. Ficolin may be able to bind to lipoteichoic acid in the cell wall. The alternative pathway is activated by hydrolyzed C3 bound to bacterial components. All three ways generate a C3 convertase to cleave C3. Further complement reaction can generate the more potent opsonin iC3b and the B-cell ligand C3d, or promote inflammation through the generation of C5a fragments and the lytic pathway. Opsonic components are shown in blue (See Plate 00).

2001a,b) (Figure 5.2). The complement system has a number of different functions, which differ in importance for different organisms. In the pneumococcus, the most important process is the cleavage of the major serum protein, C3, to produce two fragments, C3b and a smaller C3a fragment.

C3

C3 is produced mainly in the liver, but is synthesized by a number of other cell types, including immune cells such as monocytes/macrophages and neutrophils. It is present in serum at a concentration of 1–1.5 mg ml^{-1} with concentrations approximately doubling in an acute-phase response (Botto 2000). The elevation of concentrations is a less impressive fold-increase than other proteins such as C-reactive protein (CRP), but the overall effect on the immune system is probably far more profound.

C3 is cleaved by a C3 convertase derived from one of the three activation pathways: the classical pathway, the lectin pathway, or the alternative pathway, which will be described later. Cleavage of C3 exposes a thioester bond, which can form either an amino or an ester bond with nearby surfaces within a minimum life of about 60 μs (Sim *et al.* 1981). The result is C3b covalently attached to a potential pathogen surface with the C3a fragment released into the fluid phase as a chemotactic anaphylatoxin. On pneumococci, C3b is found attached to the capsular polysaccharide and teichoic acid in the cell wall (Hostetter 1986, 1999).

C3b is an opsonin interacting with complement receptor 1 (CR1, CD35) on phagocytes. However, C3b is susceptible to further proteolytic cleavage to iC3b and then C3d via two host-expressed soluble proteins, Factor I and its co-factor, Factor H. The protein iC3b is also an opsonin, which is detected by complement receptor 3 (CR3, CD11b/CD18) found on neutrophils and macrophages. In contrast, C3d is not opsoninic but is the ligand for complement receptor 2 (CR2) on B cells, and enhances B cell antibody production and maturation (Dempsey *et al.* 1996).

C5 and the lytic pathway

The activity of C3 convertases results in the generation of a number of C3b molecules, which can bind to the activating surface, but the C3b can also be incorporated into the C3 convertase that created it. This creates an enzyme with a different specificity for the protein C5. Cleavage of C5 generates C5b, which is bound to the organism surface, and C5a, which is released into the fluid phase.

C5b can interact with further complement components to generate the membrane attack complex, which is not able to lyse or kill pneumococci but could influence the inflammatory response of host cells, in particular vascular endothelial cells (Ward 1996). The inability of complement to kill pneumococci directly, in contrast to Gram-negative organisms such as *Neisseria meningitidis*, explains why complement-mediated opsonization is so critical in host defense against pneumococci. C5a is a potent anaphylatoxin, which in combination with C3a recruits immune cells to the sites of local inflammation or infection (Walport 2001b).

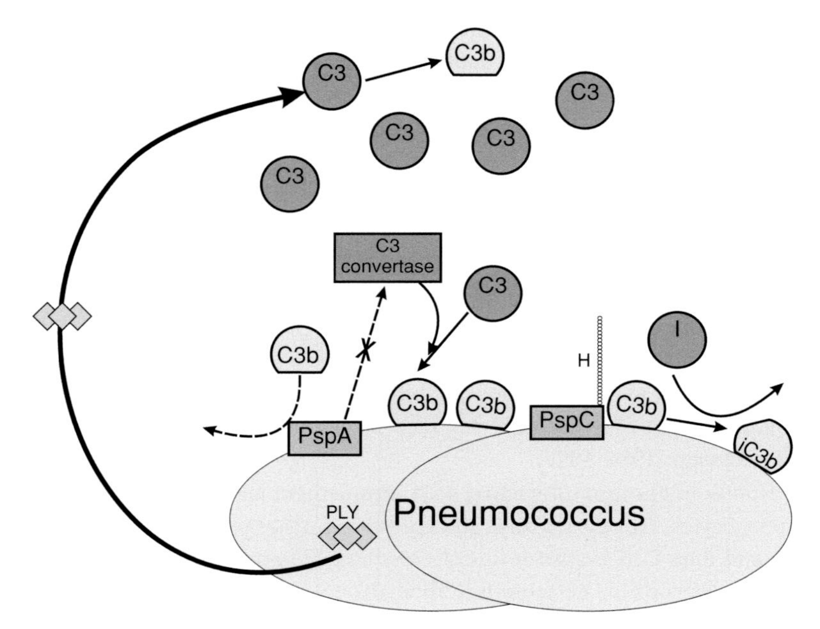

Figure 5.3 Regulation of complement by *Streptococcus pneumoniae*. Pneumococci posses a number of specific elements to modify complement activation. Pneumolysin (PLY) activates complement at a distance from the organism to prevent deposition on the bacterial surface. PspA either prevents the deposition of C3b or interferes with the C3 convertase enzymes that produce it. PspC (or the alternative gene product, Hic) binds factor H from human serum to promote the cleavage of C3b to iC3b, which is then susceptible to further proteolytic cleavage to C3d. The possible mechanisms are shown by broken lines. Promotion of C3b decay reduces alternative pathway activation. Not shown here is the fact that capsule can mask opsonic C3 fragments from detection by phagocytes.

Pneumococci and C3

Pneumococcal Cbpa (Dave *et al.* 2001), and a closely related protein, Hic (Jarva *et al.* 2002), bind Factor H and thereby enhance Factor I cleavage of C3b (Figure 5.3). Therefore the pneumococcus can manipulate innate immunity by modifying the rate of decay of C3b to iC3b to C3d. Hostetter (1986) had initially proposed that C3 on the surface of serotype 3 and 4 organisms, which are highly resistant to phagocytosis but very immunogenic, was mostly in the form of C3d. Serotypes 6A and 14, which are poorly immunogenic but easily phagocytosed and most commonly found in childhood otitis media, bacteremia, or meningitis, mostly bind iC3b. Capsule reduces binding of Factor H to Hic but the effect of capsule type has not been described (Jarva *et al.* 2002). Capsule type does not seem to influence the binding of Factor

H to CbpA (Dave *et al.* 2001). The location of bound opsonic C3 is important in determining the accessibility of these proteins to their receptors on phagocytic cells (Brown *et al.* 1983b). For an encapsulated bacterium like the pneumococcus, C3 bound to cell wall components is less accessible to phagocytic receptors than C3 bound to the capsular polysaccharide. The location of C3 deposition on the bacterial surface is largely determined by the pathway of complement activation.

Activators of complement

C3 convertases are generated by the three different pathways of complement activation: the classical pathway, the lectin pathway, and the alternative pathway. The classical pathway involves the activation of the C1 complex, which comprises the recognition element C1q and the serine proteases C1r and C1s. Activation occurs by the binding of C1q to either the paired Fc regions of two or more IgG molecules or the grouped Fc regions of pentameric IgM. C1q can also bind to C-reactive protein (CRP), which in the presence of calcium binds to phosphocholine, a substantial component of the cell-wall C-polysaccharide of *S. pneumoniae* (Szalai 2002). On binding of the C1 complex, the serine proteases attached to C1q activate and C1s cleaves C4 and then C2 to produce the C3 convertase, C4b2a (Gaboriaud *et al.* 2004).

The lectin pathway is activated by two different complexes: mannose-binding lectin (MBL) complexed with MBL-asssociated serine proteases (MASPs), and L-ficolin complexed with MASPs. MBL recognizes particular sugar patterns and some bacterial proteins; in a manner analogous to that of C1, the MASPs become activated and cleave C4 and C2 to produce a C3 convertase (Jack *et al.* 2001). Pneumococci have been shown to bind MBL directly (van Emmerik *et al.* 1994), although the levels appear to be very low compared with those seen in other bacteria (Neth *et al.* 2000). L-Ficolin is less well characterized, but is able to bind to lipoteichoic acids to activate the lectin complement pathway (Holmskov *et al.* 2003). A possible role for L-ficolin in pneumococcal infection has not yet been explored.

The alternative pathway is not usually thought of as directly activated. C3 is thought to be hydrolyzed normally in serum at a low rate. The form of C3b produced, C3b•H_2O, can bind to any surfaces close to the site of activation, including host cells but also potential pathogens. Host cells have membrane-bound mechanisms to deal with the consequences of such binding, but on pathogens that lack such mechanisms C3b can associate with Factor B to produce another C3 convertase, C3bBb. This mechanism apparently works with quite low efficiency; probably a more effective source of C3b for alternative pathway activation comes from C3b produced by the classical or lectin pathways.

Human subjects deficient in components of the classical and lectin pathways (C4, C2, and C3) and the classical pathway specifically (C1q) are more susceptible to pyogenic infections, including those caused by pneumococci (Walport 2001a), showing that complement activation leading to opsonization by C3 is critical in defense against the pneumococcus. C3- or C4-deficient mice are more susceptible to experimental pneumococcal infection (Brown *et al.* 2002). However, the relative importance of the activation pathways is the subject of some debate. Data from knockout mice would suggest that the classical pathway activation by naturally occurring IgM is probably the most important activation pathway (Brown *et al.* 2002), although considerable augmentation in the amount of C3 deposited occurs via the alternative pathway (Xu *et al.* 2001). Naturally occurring IgM is mainly directed against phosphocholine, which is present in the pneumococcal cell wall (Briles *et al.* 1981). Other models suggest that human CRP can protect mice from lethal pneumococcal injection via complement activation (Mold *et al.* 2002). In contrast, the lectin pathway appears to be relatively unimportant (Brown *et al.* 2002), which is surprising since humans deficient in MBL appear to be more susceptible to pneumococcal infection (Kronborg & Garred 2002; Roy *et al.* 2002), although the relative risk factor of around 2 suggests that the susceptibility is not strong. Interspecies differences in the relative efficiencies of the classical and lectin pathway (Zhang *et al.* 1999) and the presence of two forms of MBL in mice compared with a single form in humans (Shi *et al.* 2004) may explain the apparent discrepancy. It is interesting to note that MBL deficiency occurs at such a high frequency in most human populations (Turner & Hamvas 2000) that it may be a far more substantial cause of pneumococcal disease than much rarer genetic deficiencies of the classical pathway components.

Direct activation of the alternative pathway by capsule-specific polymeric IgA has been described (Janoff *et al.* 1999). The mechanism and importance of this is not clear. Bound IgA may provide a protected site on the organism surface that is not susceptible to complement regulation, or IgA may provide a binding site for MBL that can bind polymeric IgA (Roos *et al.* 2001). Results such as these, combined with the interdependence of the complement pathways in normal hosts, suggest that there may be a number of different approaches to complement-mediated killing of pneumococci.

Complement regulation by pneumococci

Pneumococci actively bind Factor H via CbpA and Hic to enhance the cleavage of C3b to iC3b via Factor I (Figure 5.3). This will have the consequence of altering the phagocytosis of the organism. However, it will also have the function of modifying the activation of the alternative pathway by converting C3b, which would otherwise combine with Factor B into iC3b, the "i" denoting inactive for further participation in the complement pathway. PspA can inhibit the deposition of C3b and/or inhibits the formation of the C3 convertase produced by the alternative pathway, although the mechanism is not known (Tu *et al.* 1999). The site of C3b deposition also appears to be important. Anticapsular antibody deposits C3b on the capsule and is opsonic. In contrast, anti-cell-wall antibodies do not successfully bind C3b to the capsule and are not opsonic (Brown *et al.* 1983a).

Pneumolysin is one of the major toxins of pneumococcus, released on autolysis of bacterial cells. Pneumolysin is able to directly activate complement in the fluid phase away from the surface of intact bacteria (Paton *et al.* 1984). Pneumococcal pneumolysin mutants that lack complement-activating function are less virulent than wild-type organisms and appear to invoke reduced inflammatory responses, likely to be mediated through reduced C3a and C5a production (Rubins & Janoff 1998). Changes to complement activation may have complex effects in the timing of bacteremia in animal models and may affect the recruitment of cells such as T cells to the lung (Jounblat *et al.* 2003).

INNATE PHAGOCYTIC MEDIATORS

SP-A/SP-D

The lung surfactant proteins (SP)-A and -D are synthesized by alveolar type II and Clara cells in the respiratory tract and lung (Crouch & Wright 2001; Madsen *et al.* 2000). These proteins are related to MBL of the lectin pathway and are members of the collectin (collagenous lectin) family. Unlike MBL they do not activate complement, but like MBL they can modify phagocyte function. SP-A enhances uptake of *S. pneumoniae* by murine alveolar macrophages (Kuronuma *et al.* 2004; Tino & Wright 1996). Although SP-A binds to the organism, SP-A does not have to be bound to the organism to enhance phagocytosis: "priming" of macrophages with the protein seems to be enough (Kuronuma *et al.* 2004). This occurs via the upregulation of scavenger receptor-A on the macrophage surface in response to SP-A. SP-A also enhances uptake by neutrophils, but in this experiment SP-A was bound to organisms before exposure to the cells (Hartshorn *et al.* 1998).

SP-D also binds to and agglutinates pneumococci (Hartshorn *et al.* 1998; Jounblat *et al.* 2004). The composition of the capsule influences binding, but the relationship between capsule composition and binding is not yet apparent (Jounblat *et al.* 2004). Although SP-D enhanced uptake by human

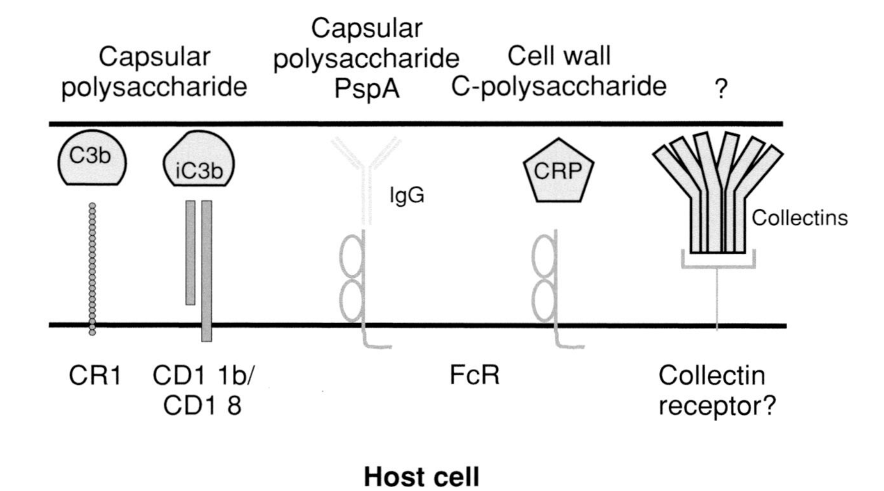

Figure 5.4 Opsonins of *Streptococcus pneumoniae*. The major serum and mucosal fluid opsonins for the organism are shown with their receptors on host cells and the major known ligands on the organism.

macrophages (Hartshorn *et al.* 1998), different experiments found no increase in killing of organisms to which the collectin bound (Jounblat *et al.* 2004). The major opsonins for *S. pneumoniae*, together with their microbial and host receptors, are briefly summarized in Figure 5.4.

MACROPHAGE PHAGOCYTOSIS OF PNEUMOCOCCI

Alveolar macrophages

Early experiments with animal models showed that alveolar macrophages have an important role in ingesting inhaled bacteria and that this defense can be overwhelmed by an excess of bacteria in the form of droplet inocula (Stillman & Branch 1924). The sterility of the distal respiratory tract in humans was therefore initially attributed to the action of alveolar macrophages, which are mature, mobile, long-lived tissue macrophages capable of repeated episodes of phagocytosis. Subsequent work by Coonrod and others demonstrated, however, that there are important interspecies differences in the degree of extracellular killing of pneumococci in the lung (Coonrod 1986; Coonrod *et al.*

1990) and the role of alveolar macrophages in defense against pneumococci (Coonrod 1989; Nguyen *et al.* 1982). In addition, it is now known that several innate factors – soluble antibacterial compounds such as α- and β-defensins (Bals *et al.* 1998), lysozyme, and lactoferrin (Jeffery 1987) – as well as complement and pulmonary immunoglobulin play a large role in creating this sterile environment and that the role of the alveolar macrophage in lung defense against pneumococci is more complex and subtle than that of an avid but simple phagocyte (Boyton & Openshaw 2002).

Binding and phagocytosis of non-opsonized pneumococci

Pneumococcal interactions with phagocytes are dominated by the antiphagocytic properties of the polysaccharide capsule (Alonso De Velasco *et al.* 1995). Non-opsonized bacteria can interact with both the mannose receptor and the scavenger receptor of alveolar macrophages. Polysaccharide is recognized by the mannose receptor and bound in a calcium-dependent manner (Zamze *et al.* 2002). Lipoteichoic acid (LTA) interacts with macrophage scavenger receptors in a manner critically dependent on negative charge on the LTA structure, but *S. pneumoniae* LTA is a zwitterion and as such very poorly binding (Greenberg *et al.* 1996). Despite these mechanisms, assays of phagocytosis using fully capsulate pneumococci have shown that non-opsonized bacteria resist phagocytosis almost completely (Hof *et al.* 1980; Gardner *et al.* 1982). Macrophages are activated in acute infection, and capable of increased phagocytosis after direct activation or incubation with lung fluid from infected lungs (Johnson *et al.* 1975). Phagocytosis of pneumococci in animal models has been shown to be increased by activation with GM-CSF and IL-1 (Hebert *et al.* 1994; Hebert & O'Reilly 1996). It has been suggested that activation and release of critical cytokines is also stimulated by pneumococcal polysaccharide (Simpson *et al.* 1994), and by pneumolysin (Cockeran *et al.* 2002b).

We studied the interaction of human alveolar macrophages with type 1 *S. pneumoniae in vitro* (Gordon *et al.* 2000). The number of cell-associated bacteria after 60 min binding was greater for opsonized bacteria than for non-opsonized, but the binding of non-opsonized bacteria could be increased by increased multiplicity of infection (Figure 5.5). Following binding of non-opsonized pneumococci to the alveolar macrophages, less than 5% of bound bacteria were seen to be internalized over a 3 h warm incubation with no detectable bacterial killing (Figure 5.6). The very small numbers of pneumococci that were internalized were shown to traffic to compartments containing LAMP-1 (lysosome-associated membrane protein) (Figure 5.7). We

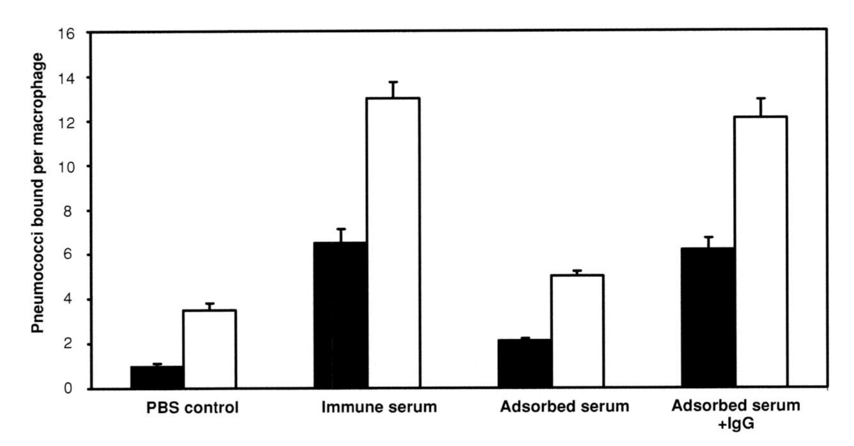

Figure 5.5 Influence of opsonization on binding of *Streptococcus pneumoniae* serotype 1 to human monocyte-derived macrophages (MDMs) (Gordon *et al.* 2000). Bacteria were opsonized and inoculated onto MDMs at two inocula: – 1×10^7 CFU (multiplicity of infection (MOI) = 10) (closed columns) and 2×10^7 CFU (MOI = 20) (open columns). Bacteria were opsonized with immune serum pooled from immune individuals, or the same serum after adsorption with Protein A (to remove immunoglobulin), or from adsorbed serum to which eluted immunoglobulin had been restored. Relative bacterial binding to MDMs was measured by immunofluorescence.

therefore concluded that opsonization is critical for optimal phagocytosis of pneumococci by human alveolar macrophages.

Binding and phagocytosis of opsonized pneumococci

Alveolar lining fluid has long been known to be effective at optimizing phagocytosis by alveolar macrophages, probably by a combined contribution of immunoglobulin (Hof *et al.* 1981), complement (Coonrod & Yoneda 1981) and surfactant proteins (McNeely & Coonrod 1993) in the fluid. Studies of alveolar lung fluid are complicated, however, by the dilution that results from the extraction of these fluids – typically 60-fold in human bronchoalveolar lavage studies – and by the lower concentrations of opsonins found in bronchoalveolar lavage compared with serum, even after correction for dilution. Studies of alveolar macrophages have therefore often used diluted serum as a source of opsonins and have demonstrated that different pathogens have different opsonic requirements, but fully capsulate pneumococci are consistently more resistant to phagocytosis than are other bacteria (Hof *et al.* 1980; Jonsson *et al.* 1985).

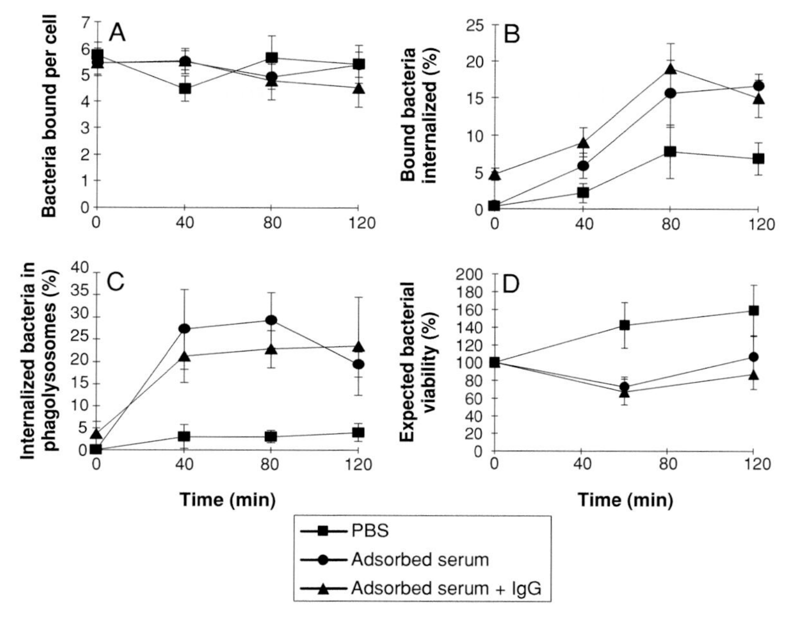

Figure 5.6 Binding (A), internalization (B), percentage of internalized bacteria within terminal phagolysosomes (C), and loss of viability of (D) *Streptococcus pneumoniae* under various conditions of opsonization and inoculation of one million human alveolar macrophages with a MOI of 10. (A) Association of bacteria with human alveolar macrophages. (B) Percentage of associated bacteria that were internalized (Kruskal–Wallis analysis for area under the curve (AUC), $p = 0.02$): adsorbed serum versus no opsonization, $p = 0.05$; IgG-replete adsorbed serum versus no opsonization, $p = 0.03$ (Mann–Whitney U test). (C) Percentage of internalized bacteria that were seen within terminal phagolysosomes (Kruskal–Wallis analysis for AUC, $p = 0.01$): IgG-depleted serum versus no opsonization, $p = 0.03$; IgG-replete adsorbed serum versus no opsonization, $p = 0.04$ (Mann–Whitney U test). (D) Loss of viability of *S. pneumoniae*: non-opsonized versus adsorbed serum, $z = -1.95$, $p = 0.05$; non-opsonized versus replete serum, $z = 2.17$, $p = 0.03$; adsorbed versus IgG-replete adsorbed serum, $z = -1.12$, $p = 0.26$ (Mann–Whitney U test, $n = 6$ different human donors). (Data are means ± standard errors of the means for six different human donors.)

We found that alveolar macrophages are dependent on FcγR or CR-dependent binding for optimal ingestion of pneumococci and processing to terminal phagolysosomes (Figure 5.6). Binding to FcγRIIa has been shown by others to result in Syk-kinase-dependent, proinflammatory cytokine-inducing, pseudopod-dependent phagocytosis, whereas CR binding results in a non-inflammatory “sinking” into the macrophage membrane

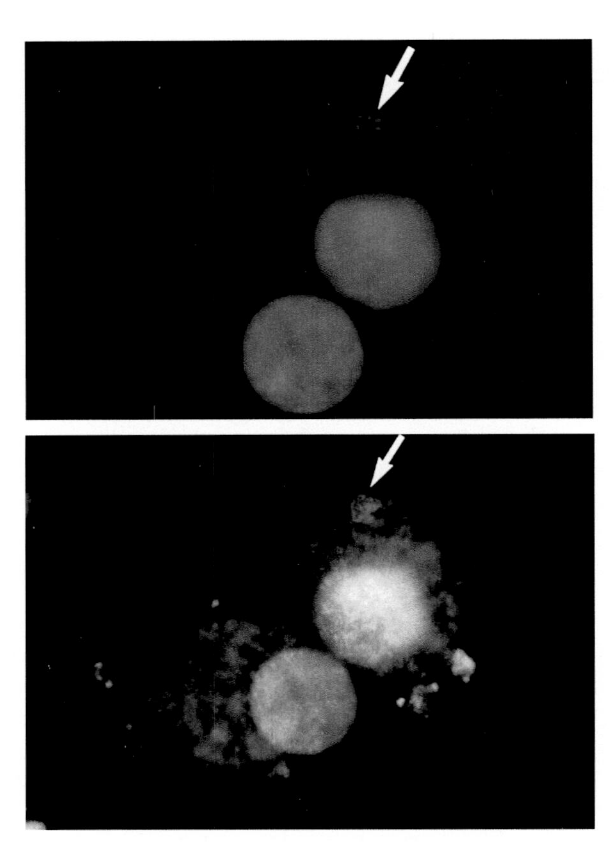

Figure 5.7 Immunofluorescence of human alveolar macrophages 80 min after uptake of opsonized *Streptococcus pneumoniae*. External bacteria are labeled with FITC (green) (see Plate 2); internalized bacteria are observed as blue (DAPI-stained) elements within the cell. Some internalized bacteria (marked with an arrow in both micrographs) are co-localizing with a marker (rhodamine-red) labeling up the lysosomal membrane protein LAMP-1, which is enriched in late endosomal–lysosomal compartments.

(Underhill & Ozinsky 2002). In our studies, bacterial killing was correlated with the processing observed at microscopy but Fcγ R-dependent processing was faster and more efficient. These observations are consistent with the clinical correlation between Fcγ IIa (IgG receptor) polymorphisms and recurrent respiratory tract infections (Sanders *et al.* 1994). We also observed that internalization reached a maximum level of only 20% of bound bacteria at 2 h and then declined, despite the presence of persistent viable bacteria on the macrophage surface. This is consistent with the observation of Shaw and Griffin, who reported that maintained macrophage phagocytosis required repeated new stimuli (Shaw & Griffin 1981). Our observations suggested that alveolar macrophages ingest and kill a limited number of pneumococci and that further bacterial killing must be dependent on other mechanisms.

Neutrophil recruitment by macrophage chemokine production

Alveolar macrophages in the process of being overwhelmed by pneumococci must recruit assistance from neutrophils to clear the infection (Hunninghake *et al.* 1978). Neutrophils are highly effective phagocytes of opsonized bacteria by Fcγ R-dependent phagocytosis (Hoidal *et al.* 1981). Further, neutrophils are concentrated in the pulmonary vasculature at three times the concentration found in circulating blood, owing to the relative size and lack of deformability of the neutrophil and the diameter of the pulmonary microvessels (Doerschuk 1981). Large numbers of neutrophils are therefore poised just outside the alveolar space and can be recruited rapidly by CD18-dependent or CD18-independent signals, including the beta-chemokine IL-8 (Mizgerd *et al.* 1997). Alveolar macrophages (Nielsen *et al.* 1994) and respiratory epithelium (Murdoch *et al.* 2002) are both responsible for generating the proinflammatory signal necessary; blocking of this signal results in impaired neutrophil recruitment and poor survival in animal models (Esposito 1984, 1985).

In mice, neutrophils infiltrate into the lung between 4 and 24 h after pneumococcal instillation (Bergeron *et al.* 1998). This is associated with increasing concentrations of TNF, IL-6, IL-1 and leukotriene B_4 in the lung fluid and a transient rise in serum IL-1 concentrations. Chemokines that recruit leukocytes can be produced by lung epithelium in response to the pneumococcal surface protein CbpA (Murdoch *et al.* 2002). The influx of neutrophils is influenced by pneumolysin (Kadioglu *et al.* 2000); the presence of both the complement-activating and the cytolytic activity of the protein is necessary for the complete effect (Jounblat *et al.* 2003). The release of C3a and C5a during complement activation is probably responsible for cellular infiltration of the lungs (Guo & Ward 2002), but the specific effect of pneumolysin appears to be more potent in recruiting in T cells than in recruiting neutrophils (Jounblat *et al.* 2003). Loss of the cytotoxic portion of pneumolysin reduces neutrophil influx into the lungs in mice (Jounblat *et al.* 2003). Pneumolysin is also responsible for the release of IL-8 by neutrophils, which could contribute to ongoing inflammation through recruitment of more neutrophils and autocrine activation of the cells (Cockeran *et al.* 2002a).

We recently measured the chemotactic cytokine profile of human alveolar macrophages stimulated *in vitro* with opsonized type 1 pneumococci. Increases in IL-1 and IL-8 were seen over the first 16 h with a subsequent decline in production (Gordon *et al.* 2005). In a mouse model, the cytokine kinetics of an initially proinflammatory and later anti-inflammatory response to pulmonary pneumococcal infection have been described (Bergeron *et al.* 1998; Knapp *et al.* 2003). In models of sepsis, an aberrant anti-inflammatory response can be beneficially corrected by exogenous TNF (Simpson *et al.* 1991; Chen *et al.* 2000). Alveolar macrophage phagocytosis of pneumococci is therefore combined with an important immunoregulatory role (Gordon & Read 2002).

Alveolar macrophage modulation of inflammation

Alveolar macrophages differ substantially from other macrophage populations. They are less effective at presenting antigen than are dendritic cells (Nicod *et al.* 2000; Chelen *et al.* 1995; Ettensohn & Roberts 1983), produce downregulation of CD4 T-cells (Blumenthal *et al.* 2001), and produce a cytokine repertoire in response to stimuli that is different from that produced by monocyte-derived macrophages (Toossi *et al.* 1996; Steffen *et al.* 1993; Wewers *et al.* 1984). In addition, alveolar macrophages are much less avid phagocytes than either polymorphonuclear cells or monocyte-derived macrophages (Hoidal *et al.* 1981). The obvious advantage to the host of an anti-inflammatory milieu in the alveolus is that inhaled particles will not result in chronic pulmonary inflammation and hence impaired gas exchange as seen in sarcoidosis (Muller *et al.* 1992; Venet *et al.* 1985) or asthma (Gant *et al.* 1992). In order to maintain a sterile lung, however, alternative non-inflammatory mechanisms for infection control must be developed (Rubins 2003), including opsonin-dependent mechanisms using the inhibitory FcγIIb receptor (Underhill & Ozinsky 2002; Clatworthy & Smith 2004). Alveolar macrophages are capable of both ingestion of pneumococci followed by apoptosis (Dockrell *et al.* 2003) and recruitment of neutrophils with subsequent induction of anti-inflammatory mechanisms (Knapp *et al.* 2003).

Lung dendritic cells

Acquired immune responses, including production of pulmonary immunoglobulin, require antigen presentation at the regional lymph node and in the spleen. Alveolar macrophages are known to be poor antigen-presenting cells (Nicod *et al.* 2000; Chelen *et al.* 1995) so, although some alveolar macrophages may migrate to the regional lymph node, antigen transfer from alveolar macrophages to dendritic cells or direct sampling of antigen by dendritic cells is necessary (Ramirez & Sigal 2002; Vermaelen *et al.*

2001). Dendritic cells in the lung occupy a subepithelial location with end-processes able to sample alveolar contents. Dendritic-cell antigen challenge by direct phagocytosis or antigen transfer is followed by chemokine-gradient-dependent migration of the dendritic cells to regional lymph nodes (Lambrecht *et al.* 2001). During this process, activation of dendritic cells occurs to facilitate optimal antigen presentation and co-receptor expression (Soilleux *et al.* 2002).

At the local lymph node, T-lymphocyte-dependent responses to protein and peptide antigens are made, and effector lymphocytes (T cells and plasma cells) traffic back to the pulmonary mucosa (Kunkel & Butcher 2003; Out *et al.* 2002). This process of mucosal immune response stimulation can be manipulated to achieve pulmonary responses by oral (Zuercher *et al.* 2002) or nasal (Arulanandam *et al.* 2001) presentation of antigen. Immune responses to T-independent antigens such as pure pneumococcal capsular polysaccharide have been shown to require both follicular dendritic cells and marginal zone B cells in the specialized environment of the spleen (Garg *et al.* 1994; Peset Llopis *et al.* 1996). Pulsed dendritic cells participated in both protein and polysaccharide responses in a murine model (Colino *et al.* 2002).

Splenic macrophages and follicular dendritic cells

The spleen is rich in macrophages and lymphocytes and has two major functions in the defence against pneumococcal disease (Zandvoort & Timens 2002). First, immunoglobulin- and complement-mediated phagocytosis of bacteria in the blood is carried out largely by splenic macrophages and dendritic cells (Brown *et al.* 1981, 1983b). Second, the close juxtaposition of follicular dendritic cells (Kang *et al.* 2004) and macrophage-regulated (Karlsson *et al.* 2003) marginal-zone B lymphocytes is necessary for the development of acquired polysaccharide immunoglobulin responses (Harms *et al.* 1996; Kruetzmann *et al.* 2003). The susceptibility of post-splenectomy patients to invasive pneumococcal disease is largely explained by the failure of phagocytosis, but lack of immunoglobulin response to antigen is a problem too (Hosea *et al.* 1981a,b).

Alterations of macrophage function in populations at risk from pneumococcal disease

Cigarette smoking

Tobacco smoke has been reported to alter the bactericidal function of alveolar macrophages (Huber *et al.* 1997) by several mechanisms (Green, 1985), but none in pneumococcal models. The excess invasive pneumococcal disease observed in otherwise immunocompetent tobacco smokers (Nuorti *et al.* 2000) is likely to be multi-factorial and include effects on both ciliary function and bronchial mucus secretion.

HIV infection

HIV is known to infect macrophages in lung and spleen and to affect macrophage functions including phagocytosis (Ieong *et al.* 2000), oxidative burst (Koziel *et al.* 2000), and cytokine production (Agostini *et al.* 1995; Twigg *et al.* 1992). We compared binding, internalization and killing of opsonized type 1 pneumococci by alveolar macrophages from HIV-uninfected and HIV-infected Malawian volunteers and found no difference in these parameters at any stage of HIV infection (Gordon *et al.* 2001). It is likely that the susceptibility of HIV-infected adults to pneumococcal infections and pneumonia relates more to alterations of pulmonary and systemic antipneumococcal immunoglobulin function (Gordon *et al.* 2003) and to lack of T-lymphocyte support (Kadioglu *et al.* 2004) than to altered macrophage function.

Alcohol abuse and liver cirrhosis

Alcohol abuse and alcoholic liver cirrhosis are known to affect macrophage function (Wallaert *et al.* 1991). This has been demonstrated in non-pneumococcal challenges in animal models to be due to altered phagocytosis, cytokine release, superoxide release of TNF, and expression of TNF receptors in alveolar macrophages (Greenberg *et al.* 1999; Omidvari *et al.* 1998; Antony *et al.* 1993). In a pneumococcal model, however, the major defect was in lung neutrophil phagocytosis, with concomitant increases in alveolar macrophage function (Gentry *et al.* 1995).

NEUTROPHIL PHAGOCYTOSIS OF PNEUMOCOCCI

Neutrophils are often portrayed as the villains of pneumococcal disease but it is useful to remember that patients with neutropenia are more susceptible to pneumococcal infection, suggesting that these cells are required for full immunity (Alonso de Velasco *et al.* 1995). However, excessive stimulation of neutrophils with release of their toxic contents can contribute substantially to host damage in disease. Neutrophils are involved in many of the classic hallmarks of pneumococcal disease: pneumonia, pneumococcal sepsis, and meningitis. In murine models of pneumococcal disease, neutrophils infiltrate into the lung between 4 and 24 h after pneumococcal instillation (Bergeron *et al.* 1998). This is associated with increasing concentrations of TNF, IL-6, IL-1 and leukotriene B_4 in the lung fluid and a transient rise in serum IL-1 concentrations.

The primary function of neutrophils is to kill opsonized bacteria. The pneumococcal capsule is probably effective in reducing the effectiveness of antibodies to subcapsular antibodies (Brown *et al.* 1983a). Unopsonized organisms are poorly phagocytosed by neutrophils (Mold *et al.* 1982). Opsonization with CRP can improve phagocytosis irrespective of capsule (Mold *et al.* 1982). Despite this, opsonization via CRP and FcR appears to be redundant for defense in mouse models (Mold *et al.* 2002).

Pneumococci are able to activate neutrophils quite specifically. Organisms alone can upregulate superoxide production by neutrophils (Kragsbjerg & Fredlund 2001). The expression of CD18, one component of complement receptor 3 (CD11b/CD18), is increased by neutrophil exposure to pneumococci, but the effect on CD11b is not significant (Kragsbjerg & Fredlund 2001). This is in contrast with other bacteria, which can substantially influence the expression of CD11b (Klein *et al.* 1996). CD18 combines with other members of the CD11 family that have roles in neutrophil adhesion to endothelial cells (Hogg 1992). Anti-CD18 administered to mice in experimental pneumococcal meningitis ameliorates the inflammation (Tuomanen *et al.* 1989). This result suggests that CD18 could be important in localizing neutrophils to sites of focal inflammation in disease. The production of C5a during complement activation might be expected to have a more potent effect in stimulating the expression of both adhesion molecules and complement receptors (Guo & Ward 2002).

The deployment of killing mechanisms such as superoxide is essential to control proliferating pneumococci. However, in sensitive tissues such as the lung or meninges, excessive activation of killing mechanisms with the release of neutrophil components can have catastrophic effects on host function (Lee & Downey 2001). This is thought to occur through the release of reactive oxygen species used by the cell to kill infectious agents (Shasby *et al.* 1982). Proteolytic enzymes such as neutrophil elastase and metalloproteinases are also likely to be important in the recruitment of neutrophils into the lung, because these are involved in other models of the pathogenesis of lung injury (Tkalcevic *et al.* 2000). Other infectious models suggest that neutrophil influx alone is not responsible for injury to the lung (Delclaux *et al.* 1997) and so it may be that the combination of neutrophil influx and the presence of pneumococcus is relatively unique in the damage observed.

MICROBICIDAL MOLECULES

Phagocytes utilize a number of molecules, including reactive oxygen species (ROS) and nitric oxide (NO), to facilitate microbial killing (Figure 5.8). Microbicidal molecules also modulate the inflammatory response and

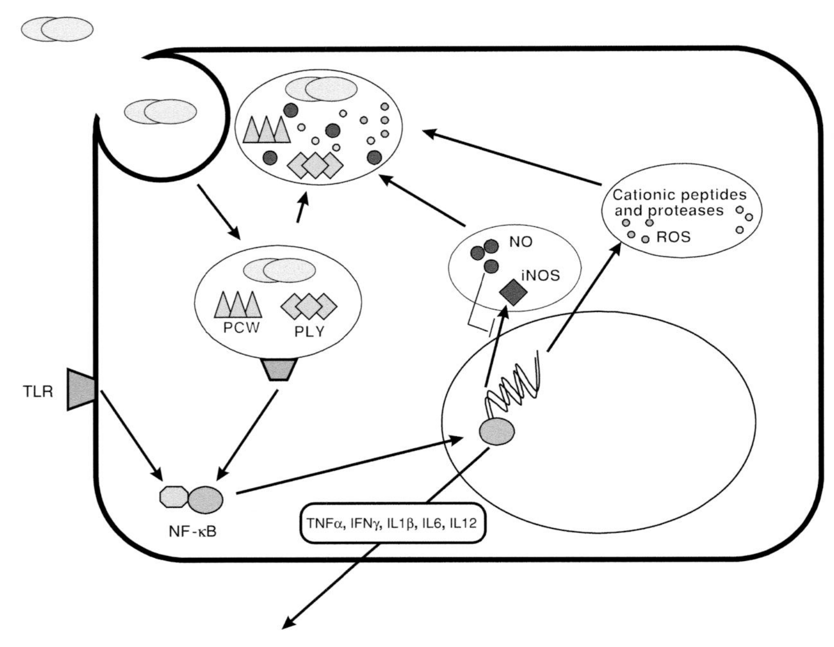

Figure 5.8 Microbicidal mechanisms: antimicrobial defenses of phagocytes against pneumococci. Internalized pneumococci enter a phagosome. Cell-surface signals mediated by Toll-like receptors (TLR) result in cytokine signaling with distinct patterns for macrophages (as shown) and PMN. Microbicidal molecules include those located in lysosomes such as reactive oxygen species (ROS), proteases and cationic peptides (many of which are distinct for PMN), and those located in other organelles such as inducible nitric oxide synthase (iNOS), derived from nitric oxide (NO), which are relatively more important in macrophages. All of these organelles fuse with the phagolysosome to mediate bacterial killing and release of factors such as pneumococcal cell wall (PCW) and pneumolysin (PLY), which of themselves may contribute to further production of antimicrobial host defense molecules.

tissue injury. ROS are among the key microbicidal molecules used by neutrophils to kill bacteria, although recent elegant work has suggested that ROS play an indirect role in facilitating activation of cationic granule proteases such as cathepsin G and elastase, which are ultimately responsible for the antimicrobial action (Reeves *et al.* 2002). Phagocyte nicotinamide adenine dinucleotide phosphate (NADPH) oxidase contains membrane components ($gp91^{phox}$ and $p22^{phox}$), cytosolic components ($p40^{phox}$, $p47^{phox}$ and $p67^{phox}$), Rac, and Rap1A and 2 (Leusen *et al.* 1996). ROS generated by NADPH oxidase are combined with halides by myeloperoxidase (MPO) to produce more potent antimicrobicidals. MPO activity is most pronounced in neutrophils and is relatively lacking in macrophages. ROS such as superoxide and hydrogen pyroxide can also combine with NO or its derivatives to generate reactive nitrogen species (RNS). These include peroxynitrite ($ONOO^-$) formed by the interaction of NO and superoxide, nitrous acid (HNO_2) and hydrogen peroxide, or nitroxyl (NO^-) and oxygen. Other important RNS include *S*-nitrosothiols, nitrogen dioxide, dinitrogen trioxide, dinitrogen tetroxide and dinitrosyl–iron complexes (Fang 2004).

Pneumococci and ROS

Pneumococci are killed by ROS (Pruul *et al.* 1988). Neutrophils and mononuclear phagocytes produce ROS in experimental models of pneumococcal meningitis (Bottcher *et al.* 2000). Some studies have suggested that pneumococci are less effective stimulators of neutrophil ROS than chemical stimuli such as phorbol myristate acetate (PMA) and that they inhibit spontaneous ROS production by PMN (Perry *et al.* 1993). Pneumococci express manganese superoxide dismutase, which limits ROS-mediated killing (Yesilkaya *et al.* 2000). However, pneumococci are catalase-negative and generate H_2O_2. In the absence of catalase H_2O_2 is not degraded by bacteria and substitutes for host-mediated H_2O_2. Individuals with chronic granulomatous disease (CGD) have defective H_2O_2 production, owing to defects in NADPH oxidase, but have normal MPO activity. In neutrophils of CGD patients bacterial H_2O_2 is used as a substrate by MPO and allows for successful generation of peroxidized halides (Allen *et al.* 1981). In keeping with these observations, mice that lack the $gp91^{phox}$ component of NADPH oxidase show no defect in bacterial clearance in a meningitis model compared with wild-type mice, although this observation may also reflect a relative lack of importance of ROS in the control of bacterial replication in cerebrospinal fluid (Schaper *et al.* 2003). Generation of ROS by phagocytes in response to group B streptococci has been found to be dependent on the TLR adaptor protein myeloid differentiation factor 88 (MyD88); similar mechanisms may be operative for ROS induction in response to pneumococci, although this awaits clarification (Henneke *et al.* 2002). In murine models of meningitis ROS production also contributes to inflammation, in part owing to activation of matrix metalloproteinase (MMP)-9 (Meli *et al.* 2003). However, the effects of ROS on modulating inflammation may be complex: pneumolysin, which is responsible for some of the inflammatory effects of the pneumococcus, is oxidized by the combination of hydrogen peroxide and halide ions, for example hypochlorite, which results in the loss of biological activity (Clark 1986). This suggests that ROS may also detoxify proinflammatory microbial products and limit microbe-induced inflammation but at the expense of inducing host-mediated tissue dysfunction.

Pneumococci and nitric oxide

The role of NO in host defence against bacteria is best established for intracellular bacteria (Vazquez-Torres *et al.* 2000). NO both synergizes with ROS to induce bactericidal killing and also contributes to a later ROS-independent bacterostatic effect. NO is also stimulated by macrophages exposed to pneumococci (Orman *et al.* 1998) both pneumococcal cell wall and pneumolysin stimulate NO production (Orman *et al.* 1998; Braun *et al.* 1999). Pneumococcal-cell-wall-stimulated NO production results from p38 kinase-stimulated inducible nitric oxide synthase (iNOS) (Monier *et al.* 2002). Although NO production by macrophages is more prominent in rodent macrophages, human macrophages also produce NO in response to pneumococci; the NO contributes to both bacterial killing and induction of apoptosis in macrophages (Marriott *et al.* 2004). In murine models NO aids bacterial clearance from the lung but may actually worsen the outcome in models of pneumococcal bacteremia (Kerr *et al.* 2004). As with ROS, however, NO may contribute to tissue injury; in murine models, iNOS inhibition has been associated with greater inflammatory injury and death (Bergeron *et al.* 1999). There is no evidence that pneumococci can detoxify or metabolize NO.

Role of other microbicidal molecules

Much less is known of the specific role of proteases such as cathepsins or of cationic peptides such as the defensins or cathelicidins against pneumococci. Cathepsin E gene expression is upregulated in monocytes exposed to pneumolysin (Rogers *et al.* 2003), cathepsin B is upregulated in a rat model of pneumococcal sepsis (Ruff & Secrist 1984), and cathepsin B gene expression is upregulated in a rat model of pneumococcal otitis media, but the microbiologic consequences require delineation (Li-Korotky *et al.* 2004). β-Defensins have antimicrobial activity against pneumococci *in vitro*. The effect is more marked for β-defensin-2 than β-defensin-1, and lysozyme synergizes with β-defensins to optimize antimicrobial activity (Lee *et al.* 2004).

APOPTOSIS OF PHAGOCYTES

Programmed cell death or apoptosis is essential for tissue homeostasis, but altered susceptibility to apoptosis is a feature of many infectious diseases

(Hetts 1998). Pathogen-induced apoptosis may remove phagocytes required for host defense functions, or alternatively pathogen-induced inhibition of apoptosis can result in intracellular persistence of pathogens and prolonged proinflammatory effects (Dockrell 2001). Apoptosis may in certain circumstances contribute to microbial clearance (Lammas *et al.* 1997), modulate the inflammatory response (Lucas *et al.* 2003), or facilitate development of adaptive immunity when apoptotic bodies are used as a source of antigens for presentation by dendritic cells (Yrlid & Wick 2000). Ultimately, regulation of apoptosis is a critical determinant of phagocyte host defense function. Tissue macrophages are long-lived cells, surviving in the human lung for approximately 90 days (Sibille & Reynolds 1990). This phenotype is associated with a relative resistance to apoptosis (Liu *et al.* 2001). In contrast, neutrophils are short-lived cells in tissues (Sibille & Reynolds 1990). Apoptosis and clearance of apoptotic bodies is essential to limit tissue injury during inflammation in organs such as the lung (Haslett 1999).

Phagocytosis of pneumococci results in macrophage apoptosis (Figure 5.9) (Dockrell *et al.* 2001). Internalization of bacteria is required for maximal apoptosis induction; opsonization of bacteria with type-specific immunoglobulin and complement enhances the degree of apotosis. Apoptosis induction is delayed and, in contrast to that associated with enteric Gram-negative bacteria (Zychlinsky & Sansonetti 1997), is not associated with immune evasion (Dockrell *et al.* 2001). It is the intracellular bacterial burden, rather than the pattern of complement or Fc gamma receptors engaged, that determines the level of apoptosis (Ali *et al.* 2003). Polysaccharide capsule impedes both internalization and apoptosis. In contrast, exogenous pneumolysin and cell-wall extract induce apoptosis but the levels observed are low in comparison with that induced by live bacteria. In the absence of pneumolysin, internalized bacteria are much less potent inducers of apoptosis. This observation suggests that pneumolysin may make its contribution to apoptosis induction only in the context of release from internalized pneumococci (Marriott *et al.* 2004). Heat-inactivated pneumococci, or those exposed to protein-synthesis inhibitors, also induce significantly lower levels of apoptosis than do live bacteria. The induction of apoptosis involves a mitochondrially mediated pathway of apoptosis induction with release of cytochrome *c* and downstream activation of caspases (Dockrell *et al.* 2001; Marriott *et al.* 2004). Nitrosative stress contributes to apoptosis induction by inducing downregulation of the macrophage pro-survival factor and anti-apoptotic Bcl-2 family member myeloid cell leukemia sequence 1 (Mcl-1) (Marriott *et al.* 2004). Inhibition of bacterial killing results in reduced apoptosis and development of a necrotic form of cell death. In addition to apoptosis in directly infected

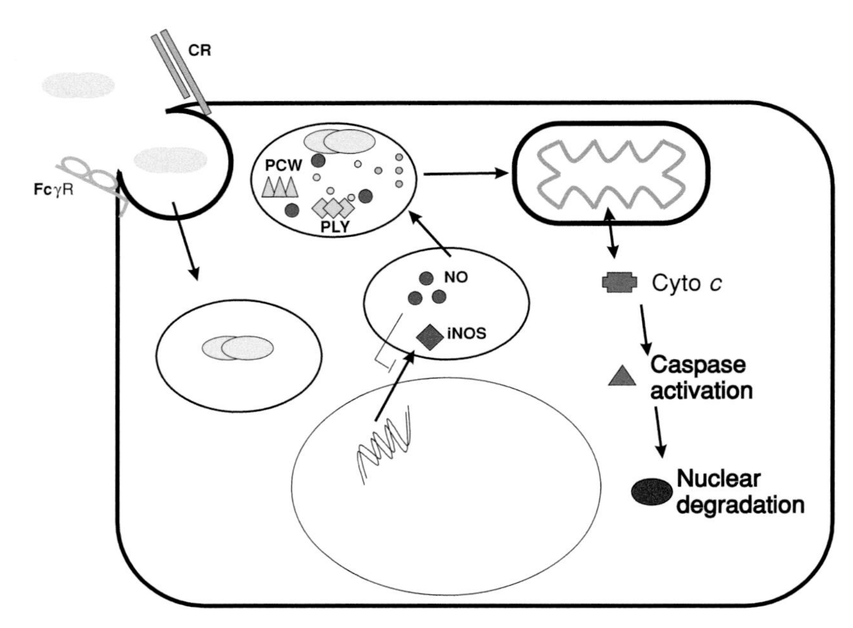

Figure 5.9 Induction of apoptosis in macrophages by pneumococci. Pneumococci opsonized by complement and immunoglobulin are internalized via complement receptors (CR) and Fc gamma receptors (FcγR) and enter a phagosome. The phagosome fuses with lysosomes and potentially with other organelles containing nitric oxide generated from inducible nitric oxide synthase (iNOS). Killing of bacteria in a phagolysosome results in release of pneumococcal cell wall (PCW) and pneumolysin (PLY), which contribute to mitochondrial membrane permeabilization and release of cytochrome *c* (Cyto *c*). Caspase activation results in nuclear degradation and other morphologic features of apoptosis.

macrophages, uninfected recruited inflammatory macrophages are susceptible to Fas-mediated apoptosis, which is induced by infected differentiated macrophages after pneumococcal infection (Dockrell *et al.* 2001).

Importance of phagocyte apoptosis to host defense against pneumococci

Phagocytosis of pneumococci by neutrophils can induce either apoptosis or necrosis (Zysk *et al.* 2000). Apoptosis of neutrophils is believed to play an important role in ensuring the clearance of the inflammatory exudate that is so characteristic of pneumococcal pneumonia; it is the successful clearance of these apoptotic cells that ensures resolution without consequent tissue injury (Haslett 1999). Heat-killed pneumococci enhance neutrophil apoptosis. However, viable pneumococci induce necrosis. Pneumococcal production of hydrogen peroxide, pneumococcal cell wall, or pneumolysin each contribute to the loss of cell viability observed (Zysk *et al.* 2000). Furthermore, induction of neutrophil apoptosis by the macrolide antimicrobial azithromycin is reversed by co-incubation in the presence of pneumococci, suggesting that pneumococci can inhibit some neutrophil apoptotic pathways (Koch *et al.* 2000). Co-incubation of neutrophils with pneumococci and influenza A virus, however, results in enhanced apoptosis, which is mediated by caspase 3 activation and associated with enhanced production of reactive oxygen species (ROS). Hence, just as in macrophages, neutrophil apoptosis may be host-mediated but pneumococci may cause a more inflammatory necrotic cell death. The predominant form of neutrophil death may be influenced by environmental conditions, including extracellular pH, as has been shown for phagocytosis by neutrophils of *Escherichia coli* (Coakley *et al.* 2002).

Alveolar macrophage apoptosis is a feature of both subclinical murine pneumococcal infection models and established pneumococcal pneumonia (Dockrell *et al.* 2003). Inhibition of macrophage apoptosis *in vivo* by caspase inhibition is associated with decreased bacterial clearance and enhanced rates of invasive pneumococcal disease. These experiments suggest that the induction of macrophage apoptosis may contribute to bacterial killing, although the mechanisms underlying this association remain to be defined. Neutrophil apoptosis is also observed *in vivo* following pneumococcal infection. In a murine model, levels of neutrophil apoptosis 6 h after challenge with *Streptococcus pneumoniae* exceeded those observed following infection with *E. coli* (Wang *et al.* 2002). Alveolar macrophages are required to phagocytose and clear apoptotic neutrophils *in vivo*; alveolar macrophage depletion in a murine pneumococcal pneumonia model resulted in enhanced levels of inflammation in association with a failure of clearance of apoptotic neutrophils (Knapp *et al.* 2003).

PNEUMOCOCCAL VACCINATION

Current vaccination protocols use capsular polysaccharides to generate specific antibody to pneumococcus, which is predominantly of the IgG class and is protective through the activation of complement and deposition of C3 on capsular polysaccharide (Goldblatt 2000). Serum concentrations of specific Ig have been used as correlates of protection in the past, but there is accumulating evidence that affinity maturation of immunoglobulin is also important. Vaccination can also stimulate the production of IgA, with up to a third of capsule-specific immunoglobulin being of this class (Johnson *et al.* 1996). In contrast to the generally high concentrations of monomeric IgA in serum, the majority of capsule-specific IgA is polymeric. The polymeric form may play a role in systemic clearance through alternative pathway activation (Janoff *et al.* 1999), but its major role is probably at the mucosal surface in blocking adherence or inflammation, although the mechanism is unclear.

Until recently, the most widely used polysaccharide vaccine was the 23-valent preparation. Such vaccines are not immunogenic in infants. Overall, data suggest that the simple polysaccharide vaccines fail to prevent pneumococcal pneumonia in adults, although they do reduce pneumococcal bacteremia and invasive disease by 55%–60% overall (Hirschmann & Lipsky 1994). The development of the 7-valent protein–polysaccharide pneumococcal conjugate vaccine in early 2000 reduced the burden of invasive disease in young children: in 2001 the rate of invasive disease among children under 2 years of age was 69% lower than in 1998 and 1999. There was also a concomitant decline in disease rate among adults (a decline of 32% in those aged 20–39 years, with lower reductions in individuals older than this) (Whitney *et al.* 2003), likely due to a reduction in nasopharyngeal carriage among children.

SUMMARY

Studies of phagocytosis of *Streptococcus pneumoniae* have identified the critical role of the pneumococcal capsule in resistance to innate and acquired humoral host defense. The pneumococcus is resistant to complement-mediated killing and so opsonin-mediated phagocytosis and killing by microbicidal molecules of neutrophils and macrophages is of prime importance. The pneumococcus has a number of virulence determinants other than capsule that thwart opsonization by complement, collectins and immunoglobulins and intracellular killing. Following successful phagocytosis and killing of pneumococci, macrophages undergo host-initiated apoptosis, to control infection and regulate inflammation at the site of infection. Future work will be needed to identify further mechanisms whereby the pneumococcus is able to resist successful control by macrophages within the alveolus, an environment replete with complement and collectins in the non-immune host.

REFERENCES

Agostini, C., R. Sancetta, A. Cerutti, and G. Semenzato. 1995. Alveolar macrophages as a cell source of cytokine hyperproduction in HIV-related interstitial lung disease. *J Leukoc Biol* **58**(5): 495–500.

Ali, F., M. E. Lee, F. Iannelli *et al.* 2003. *Streptococcus pneumoniae*-associated human macrophage apoptosis after bacterial internalization via complement and Fcgamma receptors correlates with intracellular bacterial load. *J Infect Dis* **188**: 1119–31.

Allen, R. C., E. L. Mills, T. R. McNitt, and P. G. Quie. 1981. Role of myeloperoxidase and bacterial metabolism in chemiluminescence of granulocytes from patients with chronic granulomatous disease. *J Infect Dis* **144**: 344–8.

Alonso de Velasco, E., A. F. Verheul, J. Verhoef, and H Snippe. 1995. *Streptococcus pneumoniae*: virulence factors, pathogenesis, and vaccines. *Microbiol Rev* **59**: 591–603.

Antony, V. B., S. W. Godbey, J. W. Hott, and S. F. Queener. 1993. Alcohol-induced inhibition of alveolar macrophage oxidant release *in vivo* and *in vitro*. *Alcohol Clin Exp Res* **17**(2): 389–93.

Arulanandam, B. P., J. M. Lynch, D. E. Briles, S. Hollingshead, and D. W. Metzger. 2001. Intranasal vaccination with pneumococcal surface protein A and interleukin-12 augments antibody-mediated opsonization and protective immunity against *Streptococcus pneumoniae* infection. *Infect Immun* **69**(11): 6718–24.

Bals, R., X. Wang, Z. Wu *et al.* 1998. Human beta-defensin 2 is a salt-sensitive peptide antibiotic expressed in human lung. *J Clin Invest* **102**(5): 874–80.

Bergeron, Y., N. Ouellet, A. M. Deslauriers *et al.* 1998. Cytokine kinetics and other host factors in response to pneumococcal pulmonary infection in mice. *Infect Immun* **66**(3): 912–22.

Bergeron, Y., N. Ouellet, M. Simard, M. Olivier, and M. G. Bergeron. 1999. Immunomodulation of pneumococcal pulmonary infection with N(G)-monomethyl-L-arginine. *Antimicrob Agents Chemother* **43**: 2283–90.

Blumenthal R. L., D. E. Campbell, P. Hwang *et al.* 2001. Human alveolar macrophages induce functional inactivation in antigen-specific CD4 T cells. *J Allergy Clin Immunol* **107**(2): 258–64.

Bottcher, T., J. Gerber, A. Wellmer *et al.* 2000. Rifampin reduces production of reactive oxygen species of cerebrospinal fluid phagocytes and hippocampal neuronal apoptosis in experimental *Streptococcus pneumoniae* meningitis. *J Infect Dis* **181**: 2095–8.

Botto, M. 2000. C3. In *The Complement Factsbook*, 1st edn, B. J. Morley & M. J. Walport, eds, pp. 88–94. London: Academic Press.

Boyton, R. J., and P. J. Openshaw. 2002. Pulmonary defences to acute respiratory infection. *Br Med Bull* **61**: 1–12.

Braun, J. S., R. Novak, G. Gao, P. J. Murray, and J. L. Shenep. 1999. Pneumolysin, a protein toxin of *Streptococcus pneumoniae*, induces nitric oxide production from macrophages. *Infect Immun* **67**: 3750–6.

Briles, D. E., M. Nahm, K. Schroer *et al.* 1981. Antiphosphocholine antibodies found in normal mouse serum are protective against intravenous infection with type 3 *Streptococcus pneumoniae*. *J Exp Med* **153**: 694–705.

Briles, D. E., R. C. Tart, E. Swiatlo *et al.* 1998. Pneumococcal diversity: considerations for new vaccine strategies with emphasis on pneumococcal surface protein A (PspA). *Clin Microbiol Rev* **11**: 645–57.

Brown, E. J., S. W. Hosea, and M. M. Frank. 1981. The role of the spleen in experimental pneumococcal bacteremia. *J Clin Invest* **67**(4): 975–82.

Brown, E. J., S. W. Hosea, and M. M. Frank. 1983a. The role of antibody and complement in the reticuloendothelial clearance of pneumococci from the bloodstream. *Rev Infect Dis* **5** (Suppl 4): S797–S805.

Brown, E. J., K. A. Joiner, R. M. Cole, and Berger, M. 1983b. Localisation of complement component 3 on *Streptococcus pneumoniae*; anti-capsular antibody causes complement deposition on the pneumococcal capsule. *Infect Immun* **39**(1): 403–9.

Brown, J. S., T. Hussell, S. M. Gilliand *et al.* 2002. The classical pathway is the dominant complement pathway required for innate immunity to *Streptococcus pneumoniae* infection in mice. *Proc Natl Acad Sci USA* **99**: 16969–74.

Chelen, C. J., Y. Fang, G. J. Freeman *et al.* 1995. Human alveolar macrophages present antigen ineffectively due to defective expression of B7 costimulatory cell surface molecules. *J Clin Invest* **95**(3): 1415–21.

Chen, G. H., R. C. Reddy, M. W. Newstead *et al.* 2000. Intrapulmonary TNF gene therapy reverses sepsis-induced suppression of lung antibacterial host defense. *J Immunol* **165**(11): 6496–503.

Clark, R. A. 1986. Oxidative inactivation of pneumolysin by the myeloperoxidase system and stimulated human neutrophils. *J Immunol* **136**: 4617–22.

Clatworthy, M. R., and K. G. Smith. 2004. Fc{gamma}RIIb balances efficient pathogen clearance and the cytokine-mediated consequences of sepsis. *J Exp Med* **199**(5): 717–23.

Coakley, R. J., C. Taggart, N. G. McElvaney, and S. J. O'Neill. 2002. Cytosolic pH and the inflammatory microenvironment modulate cell death in human neutrophils after phagocytosis. *Blood* **100**: 3383–91.

Cockeran, R., R. Anderson, and C. Feldman. 2002a. The role of pneumolysin in the pathogenesis of *Streptococcus pneumoniae* infection. *Curr Opin Infect Dis* **15**(3): 235–9.

Cockeran, R., C. Durandt, C. Feldman, T. J. Mitchell, and R. Anderson. 2002b. Pneumolysin activates the synthesis and release of interleukin-8 by human neutrophils *in vitro*. *J Infect Dis* **186**: 562–565.

Colino, J., Y. Shen, and C. M. Snapper. 2002. Dendritic cells pulsed with intact *Streptococcus pneumoniae* elicit both protein- and polysaccharide-specific immunoglobulin isotype responses *in vivo* through distinct mechanisms. *J Exp Med* **195**(1): 1–13.

Coonrod, J. D. 1986. The role of extracellular bactericidal factors in pulmonary host defense. *Semin Respir Infect* **1**: 118–29.

Coonrod, J. D. 1989. Role of leukocytes in lung defenses. *Respiration* **55** (Suppl 1): 9–13.

Coonrod, J. D., and K. Yoneda. 1981. Complement and opsonins in alveolar secretions and serum of rats with pneumonia due to *Streptococcus pneumoniae*. *Rev Infect Dis* **3**(2): 310–22.

Coonrod, J. D., R. Varble, and M. C. Jarrells. 1990. Species variation in the mechanism of killing of inhaled pneumococci. *J Lab Clin Med* **116**(3): 354–62.

Crouch, E., and J. R. Wright. 2001. Surfactant proteins A and D and pulmonary host defense. *Rev Physiol* **63**: 521–54.

Cundell, D. R., N. P. Gerard, C. Gerard, H. I. Idanpaan, and E. I. Tuomanen. 1995. *Streptococcus pneumoniae* anchor to activated human cells by the receptor for platelet-activating factor. *Nature* **377**(6548): 435–8.

Dave, S., A. Brooks-Walter, M. K. Pangburn, and L. S. McDaniel. 2001. PspC, a pneumococcal surface protein, binds human factor H. *Infect Immun* **69**: 3435–7.

Dave, S., S. Carmicle, S. Hammerschmidt, M. K. Pangburn, and L. S. McDaniel. 2004. Dual roles of PspC, a surface protein of *Streptococcus pneumoniae*, in binding human secretory IgA and factor H. *J Immunol* **173**: 471–4.

Delclaux, C., S. Rezaiguia-Delclaux, C. Delacourt *et al.* 1997. Alveolar neutrophils in endotoxin-induced and bacteria-induced acute lung injury in rats. *Am J Physiol* **273**: L104–2.

Dempsey, P. W., M. E. Allison, S. Akkaraju, C. C. Goodnow, and D. T. Fearon. 1996. C3d of complement as a molecular adjuvant: bridging innate and acquired immunity. *Science* **271**: 348–50.

Dintilhac, A., G. Alloing, C. Granadel, and J. P. Claverys. 1997. Competence and virulence of *Streptococcus pneumoniae*: Adc and PsaA mutants exhibit a requirement for zinc and manganese resulting from an activation of putative ABC metal permeases. *Molec Microbiol* **25**: 727–39.

Dockrell, D. H. 2001. Apoptotic cell death in the pathogenesis of infectious diseases. *J Infect* **42**: 227–34.

Dockrell, D. H., M. Lee, D. H. Lynch, and R. C. Read. 2001. Immune-mediated phagocytosis and killing of *Streptococcus pneumoniae* are associated with direct and bystander macrophage apoptosis. *J Infect Dis* **184**: 713–22.

Dockrell, D. H., H. M. Marriott, L. R. Prince *et al.* 2003. Alveolar macrophage apoptosis contributes to pneumococcal clearance in a resolving model of pulmonary infection. *J Immunol* **171**(10): 5380–8.

Doerschuk, C. M. 2001. Mechanisms of leukocyte sequestration in inflamed lungs. *Microcirculation* **8**: 71–8.

Emmerik, L. C. v., E. J. Kuijper, C. A. Fijen, J. Dankert, and S. Thiel. 1994. Binding of mannan-binding protein to various bacterial pathogens of meningitis. *Clin Exp Immunol* **97**: 411–6.

Esposito, A. L. 1984. Aspirin impairs antibacterial mechanisms in experimental pneumococcal pneumonia. *Am Rev Respir Dis* **130**(5): 857–62.

Esposito, A. L. 1985. Digoxin disrupts the inflammatory response in experimental pneumococcal pneumonia. *J Infect Dis* **152**(1): 14–23.

Ettensohn, D. B., and N. J. Roberts, Jr. 1983. Human alveolar macrophage support of lymphocyte responses to mitogens and antigens. Analysis and comparison with autologous peripheral-blood-derived monocytes and macrophages. *Am Rev Respir Dis* **128**(3): 516–22.

Fang, F. C. 2004. Antimicrobial reactive oxygen and nitrogen species: concepts and controversies. *Nat Rev Microbiol* **2**: 820–32.

Gaboriaud, C., N. M. Thielens, L. A. Gregory *et al.* 2004. Structure and activation of the C1 complex of complement: unraveling the puzzle. *Trends Immunol* **25**: 368–73.

Gant, V., M. Cluzel, Z. Shakoor *et al.* 1992. Alveolar macrophage accessory cell function in bronchial asthma. *Am Rev Respir Dis* **146**(4): 900–4.

Gardner, S. E., D. C. Anderson, B. J. Webb *et al.* 1982. Evaluation of *Streptococcus pneumoniae* type XIV opsonins by phagocytosis-associated chemiluminescence and a bactericidal assay. *Infect Immun* **35**(3): 800–8.

Garg, M., A. M. Kaplan, and S. Bondada. 1994. Cellular basis of differential responsiveness of lymph nodes and spleen to 23-valent Pnu-Imune vaccine. *J Immunol* **152**(4): 1589–96.

Gentry, M. J., M. U. Snitily, and L. C. Preheim. 1995. Phagocytosis of *Streptococcus pneumoniae* measured *in vitro* and *in vivo* in a rat model of carbon tetrachloride-induced liver cirrhosis. *J Infect Dis* **171**(2): 350–5.

Goldblatt, D. 2000. Conjugate vaccines. *Clin Exp Immunol* **119**: 1–3.

Gordon, S. B., and R. C. Read. 2002. Macrophage defences against respiratory tract infections. *Br Med Bull* **61**: 45–61.

Gordon, S. B., G. R. Irving, R. A. Lawson, M. E. Lee, and R. C. Read. 2000. Intracellular trafficking and killing of *Streptococcus pneumoniae* by human alveolar macrophages are influenced by opsonins. *Infect Immun* **68**(4): 2286–93.

Gordon, S. B., M. E. Molyneux, M. J. Boeree *et al.* 2001. Opsonic phagocytosis of *Streptococcus pneumoniae* by alveolar macrophages is not impaired in human immunodeficiency virus-infected Malawian adults. *J Infect Dis* **184**(10): 1345–9.

Gordon, S. B., D. E. Miller, R. B. Day *et al.* 2003. Pulmonary immunoglobulin responses to *Streptococcus pneumoniae* are altered but not reduced in Human Immunodeficiency Virus-infected Malawian adults. *J Infect Dis* **188**(5): 666–70.

Gordon, S. B., E. R. Jarman, S. Kanyanda *et al.* 2005. Reduced interleukin-8 response to *Streptococcus pneumoniae* by alveolar macrophages from adults with HIV/AIDS. *AIDS* **19**(11): 1197–200.

Green, G. M. 1985. Mechanisms of tobacco smoke toxicity on pulmonary macrophage cells. *Eur J Respir Dis* (Suppl) **139**: 82–5.

Greenberg, J. W., G. W. Fischer, and K. A. Joiner. 1996. Influence of lipoteichoic acid structure on recognition by the macrophage scavenger receptor. *Infect Immun* **64**(8): 3318–25.

Greenberg, S. S., X. Zhao, L. Hua *et al.* 1999. Ethanol inhibits lung clearance of *Pseudomonas aeruginosa* by a neutrophil and nitric oxide-dependent mechanism, *in vivo*. *Alcohol Clin Exp Res* **23**(4): 735–44.

Guo, R. F., and P. A. Ward. 2002, Mediators and regulation of neutrophil accumulation in inflammatory responses in lung: insights from the IgG immune complex model. *Free Radical Biol Med* **33**: 303–10.

Hammerschmidt, S., S. R. Talay, P. Brandtzaeg, and G. S. Chatwall. 1997. SpsA, a novel pneumococcal surface protein with specific binding to secretory immunoglobulin A and secretory component. *Molec Microbiol* **35**: 1113–24.

Hammerschmidt, S., G. Bethe, P. H. Remane, and G. S. Chatwall. 1999. Identification of pneumococcal surface protein A as a lactoferrin-binding protein of *Streptococcus pneumoniae*. *Infect Immun* **67**: 1683–7.

Harms, G., M. J. Hardonk, and W. Timens. 1996. *In vitro* complement-dependent binding and *in vivo* kinetics of pneumococcal polysaccharide TI-2 antigens in the rat spleen marginal zone and follicle. *Infect Immun* **64**(10): 4220–5.

Hartshorn, K. L., E. Crouch, M. R. White *et al.* 1998. Pulmonary surfactant proteins A and D enhance neutrophil uptake of bacteria. *Am J Physiol* **274**: L958–69.

Haslett, C. 1999. Granulocyte apoptosis and its role in the resolution and control of lung inflammation. *Am J Respir Crit Care Med* **160**: S5–11.

Haynes, B. F., M. J. Telen, L. P. Hale, and S. N. Denning. 1989. CD44 – a molecule involved in leucocyte adherence and T cell activation. *Immunol Today* **10**: 423–8.

Hebert, J. C., and M. O'Reilly. 1996. Granulocyte-macrophage colony-stimulating factor (GM-CSF) enhances pulmonary defenses against pneumococcal infections after splenectomy. *J Trauma* **41**(4): 663–6.

Hebert, J. C., M. O'Reilly, K. Yuenger *et al.* 1994. Augmentation of alveolar macrophage phagocytic activity by granulocyte colony stimulating factor and interleukin-1: influence of splenectomy. *J Trauma* **37**: 909–12.

Henneke, P., O. Takeuchi, R. Malley *et al.* 2002. Cellular activation, phagocytosis, and bactericidal activity against group B streptococcus involve parallel myeloid differentiation factor 88-dependent and independent signaling pathways. *J Immunol* **169**: 3970–7.

Hetts, S. W. 1998. To die or not to die: an overview of apoptosis and its role in disease. *JAMA* **279**: 300–7.

Hirschmann, J. V., and B. A. Lipsky. 1994. The pneumococcal vaccine after 15 years of use. *Arch Intern Med* **154**: 373–7.

Hof, D. G., J. E. Repine, P. K. Peterson, and J. R. Hoidal. 1980. Phagocytosis by human alveolar macrophages and neutrophils: qualitative differences in the opsonic requirements for uptake of *Staphylococcus aureus* and *Streptococcus pneumoniae in vitro. Am Rev Respir Dis* **121**: 65–71.

Hof, D. G., J. E. Repine, G. S. Giebink, and J. R. Hoidal. 1981. Production of opsonins that facilitate phagocytosis of *Streptococcus pneumoniae* by human alveolar macrophages or neutrophils after vaccination with pneumococcal polysaccharide. *Am Rev Respir Dis* **124**: 193–5.

Hogg, N. 1992. Roll, roll, roll your leukocyte gently down the vein. *Immunol Today* **13**: 113–5.

Hoidal, J. R., D. Schmeling, and P. K. Peterson. 1981. Phagocytosis, bacterial killing, and metabolism by purified human lung phagocytes. *J Infect Dis* **144**(1): 61–71.

Holmskov, U., S. Thiel, and J. C. Jensenius. 2003. Collectins and ficolins: Humoral lectins of the innate immune defense. *A Rev Immunol* **21**: 547–78.

Hosea, S. W., E. J. Brown, and M. M. Frank. 1980. The critical role of complement in experimental pneumococcal sepsis. *J Infect Dis* **142**: 903–9.

Hosea, S. W., E. J. Brown, M. I. Hamburger, and M. M. Frank. 1981a. Opsonic requirements for intravascular clearance after splenectomy. *New Engl J Med* **304**: 245–50.

Hosea, S. W, C. G. Burch, E. J. Brow, R. A. Berg, and M. M. Frank. 1981b. Impaired immune response of splenectomised patients to polyvalent pneumococcal vaccine. *Lancet* 1(8224): 804–7.

Hostetter, M. K. 1986. Serotypic variations among virulent pneumococci in deposition and degradation of covalently bound C3b: implications for phagocytosis and antibody production. *J Infect Dis* **153**: 682–93.

Hostetter, M. K. 1999. Opsonic and nonopsonic interactions of C3 with *Streptococcus pneumoniae. Microb Drug Resist Mech Epidem Dis* **5**: 85–9.

Huber, G. L., G. C. Sornberger, V. Mahajan, M. E. Cutting, and C. R. McCarthy. 1997. Impairment of alveolar macrophage bactericidal function by cigar smoke. *Bull Eur Physiopathol Respir* **13**(4): 513–21.

Hunninghake, G. W., J. I. Gallin, and A. S. Fauci. 1978. Immunologic reactivity of the lung: the *in vivo* and *in vitro* generation of a neutrophil chemotactic factor by alveolar macrophages. *Am Rev Respir Dis* **117**(1): 15–23.

Ieong, M. H., C. C. Reardon, S. M. Levitz, and H. Kornfeld. 2000. Human immunodeficiency virus type 1 infection of alveolar macrophages impairs their innate fungicidal activity. *Am J Resp Crit Care Med* **162**(3 Pt 1): 966–70.

Jack, D. L., N. J. Klein, and M. W. Turner. 2001. Mannose-binding lectin: targetting the microbial world for complement attack and opsonophagocytosis. *Immunol Rev* **180**: 86–99.

Janoff, E. N., C. Fasching, J. M. Orenstein *et al.* 1999. Killing of *Streptococcus pneumoniae* by capsular polysaccharide-specific polymeric IgA, complement, and phagocytes. *J Clin Invest* **104**: 1139–47.

Jarva, H., R. Janulczyk, J. Hellwage *et al.* 2002. *Streptococcus pneumoniae* evades complement attack and opsonophagocytosis by expressing the pspC locus-encoded Hic protein that binds to short consensus repeats 8–11 of factor H. *J Immunol* **168**: 1886–94.

Jedrzejas, M. J. 2001. Pneumococcal virulence factors: structure and function. *Microbiol Molec Biol Rev* **65**: 187–207.

Jeffery, P. K. 1987. The origins of secretions in the lower respiratory tract. *Eur J Respir Dis* (Suppl) **153**: 34–42.

Johnson, J. D., W. L. Hand, N. L. King, and C. G. Hughes. 1975. Activation of alveolar macrophages after lower respiratory tract infection. *J Immunol* **115**: 80–4.

Johnson, S., N. L. Opstad, J. M. Jr. Douglas, and E. N. Janoff. 1996. Prolonged and preferential production of polymeric immunoglobulin A in response to *Streptococcus pneumoniae* capsular polysaccharides. *Infect Immun* **64**: 4339–44.

Jonsson, S., D. M. Musher, A. Chapman, A. Goree, and E. C. Lawrence. 1985. Phagocytosis and killing of common bacterial pathogens of the lung by human alveolar macrophages. *J Infect Dis* **152**(1): 4–13.

Jounblat, R., A. Kadioglu, T. J. Mitchell, and P. W. Andrew. 2003. Pneumococcal behavior and host responses during bronchopneumonia are affected differently by the cytolytic and complement-activating activities of pneumolysin (vol **71**: pg 1813, 2003). *Infect Immun* **71**: 7239.

Jounblat, R., A. Kadioglu, F. Iannelli *et al.* 2004. Binding and agglutination of *Streptococcus pneumoniae* by human surfactant protein D (SP-D) vary between strains but, SP-D fails to enhance killing by neutrophils. *Infect Immun* **72**: 709–16.

Kadioglu, A., N. A. Gingles, K. Grattan *et al.* 2000. Host cellular immune response to pneumococcal lung infection in mice. *Infect Immun* **68**: 492–501.

Kadioglu, A., W. Coward, M. J. Colston, C. R. A. Hewitt, and P. W. Andrew. 2004. CD4-T-lymphocyte interactions with pneumolysin and pneumococci suggest a crucial protective role in the host response to pneumococcal infection. *Infect Immun* **72**(5): 2689–97.

Kang, Y.-S., J. Y. Kim, S. A. Bruening *et al.* 2004. The C-type lectin SIGN-RI mediates uptake of the capsular polysaccharide of *Streptococcus pneumoniae* in the marginal zone of mouse spleen. *Proc Natl Acad Sci USA* **101**(1): 215–20.

Karlsson, M. C. I., R. Guinamard, S. Bolland *et al.* 2003. Macrophages control the retention and trafficking of B lymphocytes in the splenic marginal zone. *J Exp Med* **198**(2): 333–40.

Kerr, A. R., X. Q. Wei, P. W. Andrew, and T. J. Mitchell. 2004. Nitric oxide exerts distinct effects in local and systemic infections with *Streptococcus pneumoniae*. *Microb Pathog* 36:303–310.

Kiluchi, R., N. Watabe, T. Konno, N. Mishina, K. Sekizawa, and H. Sasaki. 1994. High incidence of silent aspiration in elderly individuals with community acquired pneumonia. *Am J Resp Crit Care Med* **150**: 251–3.

Kim, J. O., and J. N. Weiser. 1998. Association of intrastrain phase variation in quantity of capsular polysaccharide and teichoic acid with the virulence of *Streptococcus pneumoniae*. *J Infect Dis* **177**(2): 368–77.

Klein, N. J., C. A. Ison, M. Peakman *et al.* 1996. The influence of capsulation and lipooligosaccharide structure on neutrophil adhesion molecule expression and endothelial injury by *Neisseria meningitidis*. *J Infect Dis* **173**: 172–9.

Knapp, S., J. C. Leemans, S. Florquin *et al.* 2003. Alveolar macrophages have a protective antiinflammatory role during murine pneumococcal pneumonia. *Am J Respir Crit Care Med* **167**(2): 171–9.

Koch, C. C., D. J. Esteban, A. C. Chin *et al.* 2000. Apoptosis, oxidative metabolism and interleukin-8 production in human neutrophils exposed to azithromycin: effects of *Streptococcus pneumoniae*. *J Antimicrob Chemother* **46**: 19–26.

Koziel, H., X. Li, M. Y. Armstrong, F. F. Richards, and R. M. Rose. 2000. Alveolar macrophages from human immunodeficiency virus-infected persons demonstrate impaired oxidative burst response to *Pneumocystis carinii in vitro*. *Am J Respir Cell Mol Biol* **23**(4): 452–9.

Kragsbjerg, P., and H. Fredlund. 2001. The effects of live *Streptococcus pneumoniae* and tumor necrosis factor-α on neutrophil oxidative burst and β2-integrin expression. *Clin Microbiol Infect* **7**: 125–9.

Kronborg, G., and P. Garred. 2002. Mannose-binding lectin genotype as a risk factor for invasive pneumococcal infection. *Lancet* **360**: 1176.

Kruetzmann, S., M. M. Rosado, H. Weber *et al.* 2003. Human immunoglobulin M memory B cells controlling *Streptococcus pneumoniae* infections are generated in the spleen. *J Exp Med* **197**(7): 939–45.

Kunkel, E. J., and E. C. Butcher. 2003. Plasma-cell homing. *Nat Rev Immunol* **3**(10): 822–829.

Kuronuma, K., H. Sano, K. Kato *et al.* 2004. Pulmonary surfactant protein A augments the phagocytosis of *Streptococcus pneumoniae* by alveolar macrophages through a casein kinase 2-dependent increase of cell surface localization of scavenger receptor A. *J Biol Chem* **279**: 21421–30.

Lambrecht, B. N., J. B. Prins, and H. C. Hoogsteden. 2001. Lung dendritic cells and host immunity to infection. *Eur Respir J* **18**(4): 692–704.

Lammas, D. A., C. Stober, C. J. Harvey *et al.* 1997. ATP-induced killing of mycobacteria by human macrophages is mediated by purinergic P2Z(P2X7) receptors. *Immunity* **7**: 433–44.

Lee, W. L., and G. P. Downey. 2001. Neutrophil activation and acute lung injury. *Curr Opin Crit Care* **7**: 1–7.

Lee, H. Y., A. Andalibi, P. Webster *et al.* 2004. Antimicrobial activity of innate immune molecules against *Streptococcus pneumoniae, Moraxella catarrhalis* and nontypeable *Haemophilus influenzae. BMC Infect Dis* **4**: 12.

Leusen, J. H. W., A. J. Verhoeven, and D. Roos. 1996. Interactions between the components of the human NADPH oxidase: A review about the intrigues in the phox family. *Front Biosci* **1**: d72–90.

Li-Korotky, H. S., J. D. Swarts, P. A. Hebda, and W. J. Doyle. 2004. Cathepsin gene expression profile in rat acute pneumococcal otitis media. *Laryngoscope* **114**: 1032–6.

Liu, H., H. Perlman, L. J. Pagliari, and R. M. Pope. 2001. Constitutively activated Akt-1 is vital for the survival of human monocyte-differentiated macrophages. Role of Mcl-1, independent of nuclear factor (NF)-kappaB, Bad, or caspase activation. *J Exp Med* **194**: 113–26.

Lloyd-Evans, N., T. J. O'Dempsey, I. Baldeh *et al.* 1996. Nasopharyngeal carriage of pneumococci in Gambian children and in their families. *Pediatr Infect Dis J* **15**(10): 866–71.

Lohmann-Matthes, M. L., C. Steinmuller, and G. Franke-Ullmann. 1994. Pulmonary macrophages. *Eur Respir J* **7**(9): 1678–89.

Lopez, R., and E. Garcia. 2004. Recent trends in the molecular microbiology of pneumococcal capsules, lytic enzymes and bacteriophage. *FEMS Microbiol Rev* **28**: 553–80.

Lucas, M., L. M. Stuart, J. Savill, and A. Lacy-Hulbert. 2003. Apoptotic cells and innate immune stimuli combine to regulate macrophage cytokine secretion. *J Immunol* **171**: 2610–15.

Madsen, J., A. Kliem, I. Tornøe *et al.* 2000. Localization of lung surfactant protein D on mucosal surfaces in human tissues. *J Immunol* **164**: 5866–70.

Marriott, H. M., F. Ali, R. C. Read *et al.* 2004. Nitric oxide levels regulate macrophage commitment to apoptosis or necrosis during pneumococcal infection. *FASEB J* **18**: 1126–8.

McCool, T. L., T. R. Cate, G. Moy, and J. N. Weiser. 2002. The immune response to pneumococcal proteins during experimental human carriage. *J Exp Med* **195**: 359–65.

McCool, T. L., T. R. Cate, E. I. Tuomanen *et al.* 2003. Serum immunoglobulin G response to candidate vaccine antigens during experimental pneumococcal colonisation. *Infection and Immunity* **71**: 5724–32.

McNeely, T. B., and J. D. Coonrod. 1993. Comparison of the opsonic activity of human surfactant protein A for *Staphylococcus aureus* and *Streptococcus pneumoniae* with rabbit and human macrophages. *J Infect Dis* **167**(1): 91–7.

Meli, D. N., S. Christen, and S. L. Leib. 2003. Matrix metalloproteinase-9 in pneumococcal meningitis: activation via an oxidative pathway. *J Infect Dis* **187**: 1411–5.

Mitchell, T. J. 2000. Virulence factors in the pathogenesis of disease caused by *Streptococcus pneumoniae*. *Research Microbiol* **151**: 413–19.

Mizgerd, J. P, H. Kubo, G. J. Kutkosk *et al.* 1997. Neutrophil emigration in the skin, lungs, and peritoneum: different requirements for CD11/CD18 revealed by CD18-deficient mice. *J Exp Med* **186**(8): 1357–64.

Mold, C., K. M. Edwards, and H. Gewurz. 1982. Effect of C-reactive protein on the complement-mediated stimulated of human neutrophils by *Streptococcus pneumoniae* serotypes 3 and 6. *Infect Immun* **37**: 987–92.

Mold, C., B. Rodic-Polic, B., and T. W. Du Clos. 2002. Protection from *Streptococcus pneumoniae* Infection by C-reactive protein and natural antibody requires complement but not Fc{gamma} receptors. *J Immunol* **168**: 6375–81.

Monier, R. M., K. L. Orman, E. A. Meals, and B. K. English. 2002. Differential effects of p38- and extracellular signal-regulated kinase mitogen-activated protein kinase inhibitors on inducible nitric oxide synthase and tumor necrosis factor production in murine macrophages stimulated with *Streptococcus pneumoniae*. *J Infect Dis* **185**: 921–6.

Moore, L. J., A. C. Pridmore, S. K. Dower, and R. C. Read. 2003. Penicillin enhances the Toll-like Receptor 2-mediated proinflammatory activity of *Streptococcus pneumoniae*. *J Infect Dis* **188**: 1040–8.

Muller, Q. J., S. Pfeifer, D. Mannel, J. Strausz, and R. Ferlinz. 1992. Lung-restricted activation of the alveolar macrophage/monocyte system in pulmonary sarcoidosis. *Am Rev Respir Dis* **145**(1): 187–92.

Murdoch, C., R. C. Read, Q. Zhang, and A. Finn. 2002. Choline-binding protein A of *Streptococcus pneumoniae* elicits chemokine production and expression of intercellular adhesion molecule 1 (CD54) by human alveolar epithelial cells. *J Infect Dis* **186**(9): 1253–60.

Musher, D. M., D. A. Watson, and R. E. Baughn. 1990. Does naturally-acquired IgG antibody to cell wall polysaccharide protect human subjects against pneumococcal infection. *J Infect Dis* **161**: 736–40.

Musher, D. M., J. E. Groover, J. M. Roland *et al.* 1993. Antibody to capsular polysaccharides of *Streptococcus pneumoniae* in adults: prevalence, persistence, relation to carriage and resistance to infection. *Clin Infect Dis* **17**: 66–73.

Musher, D. M., J. E. Groover, M. R. Reichler *et al.* 1997. Emergence of antibody to capsular polysaccharides of *Streptococcus pneumoniae* during outbreaks of pneumonia: association with nasopharyngeal colonization. *Clin Infect Dis* **24**(3): 441–6.

Neth, O., D. L. Jack, A. W. Dodds *et al.* 2000. Mannose-binding lectin binds to a range of clinically relevant microorganisms and promotes complement deposition. *Infect Immun* **68**: 688–93.

Nguyen, B. Y., P. K. Peterson, H. A. Verbrugh, P. G. Quie, and J. R. Hoidal. 1982. Differences in phagocytosis and killing by alveolar macrophages from humans, rabbits, rats, and hamsters. *Infect Immun* **36**(2): 504–9.

Nicod, L. P., L. Cochand, and D. Dreher. 2000. Antigen presentation in the lung: dendritic cells and macrophages. *Sarcoidosis Vasc Diffuse Lung Dis* **17**(3): 246–55.

Nielsen, B. W., N. Mukaida, and K. Matsushima. 1994. Macrophages as producers of chemotactic proinflammatory cytokines. In *Macrophage-Pathogen Interactions*, 1st edn. B. S. Zwilling & T. K. Eisenstein, eds, pp. 131–42. New York: Marcel Dekker.

Nuorti, J. P., J. C. Butler, M. M. Farley *et al.* 2000. Cigarette smoking and invasive pneumococcal disease. Active Bacterial Core Surveillance Team. *N Engl J Med* **342**(10): 681–9.

Omidvari, K., R. Casey, S. Nelson, R. Olariu, and J. E. Shellito. 1998. Alveolar macrophage release of tumor necrosis factor-alpha in chronic alcoholics without liver disease. *Alcohol Clin Exp Res* **22**(3): 567–72.

Orman, K. L., J. L. Shenep, and B. K. English. 1998. Pneumococci stimulate the production of the inducible nitric oxide synthase and nitric oxide by murine macrophages. *J Infect Dis* **178**: 1649–57.

Out, T. A., S. Z. Wang, K. Rudolph, and D. E. Bice. 2002. Local T-cell activation after segmental allergen challenge in the lungs of allergic dogs. *Immunology* **105**(4): 499–508.

Paton, J. C., B. Rowan-Kelly, and A. Ferrante. 1984. Activation of human complement by the pneumococcal toxin pneumolysin. *Infect Immun* **43**: 1085–7.

Paton, J. C., P. W. Andrew, G. J. Boulnois, and T. J. Medshaw. 1993. Molecular analysis of the pathogenicity of *Streptococcus pneumoniae*: the role of pneumococcal proteins. *A Rev Microbiol* **47**: 89–115.

Perry, F. E., C. J. Elson, L. W. Greenham, and J. R. Catterall. 1993. Interference with the oxidative response of neutrophils by *Streptococcus pneumoniae. Thorax* **48**: 364–9.

Peset Llopis, M. J., G. Harms, M. J. Hardonk, and W. Timens. 1996. Human immune response to pneumococcal polysaccharides: complement-mediated localization preferentially on CD21-positive splenic marginal zone B cells and follicular dendritic cells. *J Allergy Clin Immunol* **97**(4): 1015–24.

Pruul, H., G. Kriek, and P. J. McDonald. 1988. Enoxacin-induced modification of the susceptibility of bacteria to phagocytic killing. *J Antimicrob Chemother* **21**(Suppl B): 19–27.

Ramirez, M. C., and L. J. Sigal. 2002. Macrophages and dendritic cells use the cytosolic pathway to rapidly cross-present antigen from live, vaccinia-infected cells. *J Immunol* **169**(12): 6733–42.

Reeves, E. P., H. Lu, H. L. Jacobs *et al.* 2002. Killing activity of neutrophils is mediated through activation of proteases by K^+ flux. *Nature* **416**: 291–7.

Ring, A., J. N. Weiser, and E. I. Tuomanen. 1998. Pneumococcal trafficking across the blood-brain barrier. *J Clin Invest* **102**(2): 347–60.

Robinson, K. A., W. Baughman, H. Rothrock *et al.* 2001. Epidemiology of invasive *Streptococcus pneumoniae* infections in the United States 1995–1998. Opportunities for prevention in the conjugate vaccine era. *JAMA* **285**: 1729–35.

Rogers, P. D., J. Thornton, K. S. Barker *et al.* 2003. Pneumolysin-dependent and -independent gene expression identified by cDNA microarray analysis of THP-1 human mononuclear cells stimulated by *Streptococcus pneumoniae. Infect Immun* **71**: 2087–94.

Roos, A., L. H. Bouwman, D. J. Gijlswijk-Janssen *et al.* 2001. Human IgA activates the complement system via the mannan-binding lectin pathway. *J Immunol* **167**: 2861–8.

Roy, S., K. Knox, S. Segal *et al.* 2002. MBL genotype and risk of invasive pneumococcal disease: a case-control study. *Lancet* **359**: 1569–73.

Rubins, J. B. 2003. Alveolar macrophages: wielding the double-edged sword of inflammation. *Am J Respir Crit Care Med* **167**(2): 103–4.

Rubins, J. B., and E. N. Janoff. 1998. Pneumolysin: A multifunctional pneumococcal virulence factor. *J Lab Clin Med* **131**: 21–7.

Ruff, R. L., and D. Secrist. 1984. Inhibitors of prostaglandin synthesis or cathepsin B prevent muscle wasting due to sepsis in the rat. *J Clin Invest* **73**: 1483–6.

Sanders, L. A., J. G. van-de-Winkel, G. T. Rijkers *et al.* 1994. Fc gamma receptor IIa (CD32) heterogeneity in patients with recurrent bacterial respiratory tract infections. *J Infect Dis* **170**: 854–61.

Schaper, M., S. L. Leib, D. N. Meli *et al.* 2003. Differential effect of p47phox and gp91phox deficiency on the course of pneumococcal meningitis. *Infect Immun* **71**: 4087–92.

Scott, J. A., A. J. Hall, R. Dagan *et al.* 1996. Serogroup-specific epidemiology of *Streptococcus pneumoniae*: associations with age, sex, geography in 7,000 episodes of invasive disease. *Clin Infect Dis* **22**: 973–81.

Shasby, D. M., K. M. Vanbenthuysen, R. M. Tate *et al.* 1982. Granulocytes mediate acute edematous lung injury in rabbits and in isolated rabbit lungs perfused with phorbol myristate acetate: role of oxygen radicals. *Am Rev Respir Dis* **125**: 443–7.

Shaw, D. R., and F. M. J. Griffin. 1981. Phagocytosis requires repeated triggering of macrophage phagocytic receptors during particle ingestion. *Nature* **289**: 409–11.

Shi, L., K. Takahashi, J. Dundee *et al.* 2004. Mannose-binding lectin-deficient mice are susceptible to infection with *Staphylococcus aureus*. *J Exp Med* **199**: 1379–90.

Sibille, Y., and H. Y. Reynolds. 1990. Macrophages and polymorphonuclear neutrophils in lung defense and injury. *Am Rev Respir Dis* **141**: 471–501.

Sim, R. B., T. M. Twose, D. S. Paterson, and E. Sim. 1981. The covalent binding reaction of complement component C3. *Biochem J* **193**: 115–27.

Simell, B., T. M. Kilpi, and H. Kayhty. 2002. Pneumococcal carriage and otitis media induce salivary antibodies to pneumococcal capsular polysaccharides in children. *J Infect Dis* **186**(8): 1106–14.

Simpson, S. Q., H. N. Modi, R. A. Balk, R. C. Bone, and L. C. Casey. 1991. Reduced alveolar macrophage production of tumor necrosis factor during sepsis in mice and men. *Crit Care Med* **19**(8): 1060–6.

Simpson, S. Q., R. Singh, and D. E. Bice. 1994. Heat-killed pneumococci and pneumococcal capsular polysaccharides stimulate tumor necrosis factor-alpha production by murine macrophages. *Am J Respir Cell Mol Biol* **10**(3): 284–9.

Smith, T., D. Lehmann, J. Montgomery *et al.* 1993. Acquisition and invasiveness of different serotypes of *Streptococcus pneumoniae* in young children. *Epidemiol Infect* **111**(1): 27–39.

Soilleux, E. J., L. S. Morris, G. Leslie *et al.* 2002. Constitutive and induced expression of DC-SIGN on dendritic cell and macrophage subpopulations *in situ* and *in vitro*. *J Leukoc Biol* **71**(3): 445–57.

Steffen, M., H. C. Reinecker, J. Petersen *et al.* 1993. Differences in cytokine secretion by intestinal mononuclear cells, peripheral blood monocytes and alveolar macrophages from HIV-infected patients. *Clin Exp Immunol* **91**(1): 30–6.

Stillman, E. G., and A. Branch. 1924. Experimental production of pneumococcus pneumonia in mice by the inhalation method. *J Exp Med* **40**: 733–42.

Stool, S. E. and M. J. Field. 1989. The impact of otitis media. *Pediatr Infect Dis J* **8**: S14.

Szalai, A. J. 2002. The antimicrobial activity of C-reactive protein. *Microbes Infect* **4**: 201–5.

Tettelin, H., K. E. Nelson, I. T. Paulsen *et al.* 2001. Complete genome sequence of a virulent isolate of *Streptococcus pneumoniae*. *Science* **293**: 490–505.

Tino, M. J., and J. R. Wright. 1996. Surfactant protein A stimulates phagocytosis of specific pulmonary pathogens by alveolar macrophages. *Am J Physiol* **270**: L677–88.

Tkalcevic, J., M. Novelli, M. Phylactides *et al.* 2000. Impaired immunity and enhanced resistance to endotoxin in the absence of neutrophil elastase and cathepsin G. *Immunity* **12**: 201–10.

Toossi, Z., C. S. Hirsch, B. D. Hamilton *et al.* 1996. Decreased production of TGF-beta 1 by human alveolar macrophages compared with blood monocytes. *J Immunol* **156**(9): 3461–8.

Tu, A. H., R. L. Fulgham, M. A. McCrory, D. E. Briles, and A. J. Szalai. 1999. Pneumococcal surface protein A inhibits complement activation by *Streptococcus pneumoniae*. *Infect Immun* **67**: 4720–4.

Tuomanen, E. I., and H. R. Masure. 1997. Molecular and cellular biology of pneumococcal infection. *Microb Drug Resist* **3**(4): 297–308.

Tuomanen, E. I., K. Saukkonen, S. Sande, C. Cioffe, and S. D. Wright. 1989. Reduction of inflammation, tissue damage, and mortality in bacterial meningitis in rabbits treated with monoclonal antibodies against adhesion-promoting receptors of leukocytes. *J Exp Med* **170**: 959–69.

Tuomanen, E. I., R. Austrian, and H. R. Masure. 1995. Pathogenesis of pneumococcal infection. *New Engl J Med* **332**: 1280–4.

Turner, M. W., and R. M. Hamvas. 2000. Mannose-binding lectin: structure, function, genetics and disease associations. *Rev Immunogenet* **2**: 305–22.

Twigg, H. L. III, G. K. Iwamoto, and D. M. Soliman. 1992. Role of cytokines in alveolar macrophage accessory cell function in HIV-infected individuals. *J Immunol* **149**: 1462–9.

Underhill, D. M., and A. Ozinsky. 2002. Phagocytosis of microbes: complexity in action. *A Rev Immunol* **20**: 825–52.

Vazquez-Torres, A., J. Jones-Carson, P. Mastroeni, H. Ischiropoulos, and F. C. Fang. 2000. Antimicrobial actions of the NADPH phagocyte oxidase and inducible nitric oxide synthase in experimental salmonellosis. I. Effects on microbial killing by activated peritoneal macrophages *in vitro*. *J Exp Med* **192**: 227–36.

Veenhoven, R., D. Bogert, C. Uiterwaal *et al.* 2003. Effect of conjugate pneumococcal vaccine followed by polysaccharide pneumococcal vaccine on current acute otitis media: a randomised study. *Lancet* **361**: 2189–95.

Venet, A., A. J. Hance, C. Saltini, B. W. Robinson, and R. G. Crystal 1985. Enhanced alveolar macrophage-mediated antigen-induced T-lymphocyte proliferation in sarcoidosis. *J Clin Invest* **75**(1): 293–301.

Vermaelen, K. Y., I. Carro-Muino, B. N. Lambrecht, and R. A. Pauwels. 2001. Specific migratory dendritic cells rapidly transport antigen from the airways to the thoracic lymph nodes. *J Exp Med* **193**(1): 51–60.

Wallaert, B., C. Aerts, J. F. Colombel, and C. Voisin. 1991. Human alveolar macrophage antibacterial activity in the alcoholic lung. *Am Rev Respir Dis* **144**(2): 278–83.

Walport, M. J. 2001a. Complement. First of two parts. *N Engl J Med* 344:1058–66.

Walport, M. J. 2001b. Complement. Second of two parts. *New Engl J Med* **344**: 1140–4.

Wang, Q., P. Teder, N. P. Judd, P. W. Noble, and C. M. Doerschuk. 2002. CD44 deficiency leads to enhanced neutrophil migration and lung injury in *Escherichia coli* pneumonia in mice. *Am J Pathol* **161**: 2219–28.

Ward, P. A. 1996. Role of complement, chemokines, and regulatory cytokines in acute lung injury. *Ann N Y Acad Sci* **796**: 104–12.

Weiser, J. N., Z. Markiewicz, E. I. Tuomanen, and J. H. Wani. 1996. Relationship between phase variation in colony morphology, intrastrain variation in cell wall physiology, and nasopharyngeal colonization by *Streptococcus pneumoniae*. *Infect Immun* **64**(6): 2240–5.

Weiser, J. N., D. Bae, H. Epino *et al.* 2001. Changes in availability of oxygen accentuate differences in capsular polysaccharide expression by phenotypic variants and clinical isolates of *Streptococcus pneumoniae*. *Infect Immun* **69**(9): 5430–9.

Weiser, J. N., D. Bae, C. Fasching, R. W. Scamurra, A. J. Ratner, and E. N. Janoff. 2003. Antibody-enhanced pneumococcal adherence requires IgA1 protease. *Proc Natl Acad Sci USA* **100**: 4215–20.

Wewers, M. D., S. I. Rennard, A. J. Hance, P. B. Bitterman, and R. G. Crystal. 1984. Normal human alveolar macrophages obtained by bronchoalveolar lavage have a limited capacity to release interleukin-1. *J Clin Invest* **74**(6): 2208–18.

Whitney, C. G., M. M. Farley, J. Hadler *et al.* 2003. Decline in invasive pneumococcal disease after introduction of protein-polysaccharide conjugate vaccine. *New Engl J Med* **348**: 1737–46.

Williams, B. G., E. Gouws, C. Boschi-Pinto, J. Bryce, and C. Dye. 2002. Estimates of worldwide distribution of child deaths from acute respiratory infections. *Lancet Infect Dis* **2**: 25–32.

Xu, Y., M. Ma, G. C. Ippolito *et al.* 2001. Complement activation in factor D-deficient mice. *Proc Natl Acad Sci USA* **98**: 14577–82.

Yesilkaya, H., A. Kadioglu, N. Gingles *et al.* 2000. Role of manganese-containing superoxide dismutase in oxidative stress and virulence of *Streptococcus pneumoniae*. *Infect Immun* **68**: 2819–26.

Yrlid, U., and M. J. Wick. 2000. Salmonella-induced apoptosis of infected macrophages results in presentation of a bacteria-encoded antigen after uptake by bystander dendritic cells. *J Exp Med* **191**: 613–24.

Zamze, S., L. Martinez-Pomares, H. Jones *et al.* 2002. Recognition of bacterial capsular polysaccharides and lipopolysaccharides by the macrophage mannose receptor. *J Biol Chem* **277**(44): 41613–23.

Zandvoort, A., and W. Timens. 2002. The dual function of the splenic marginal zone: essential for initiation of anti-TI-2 responses but also vital in the general first-line defense against blood-borne antigens. *Clin Exp Immunol* **130**(1): 4–11.

Zhang, Y., C. Suankratay, X. H. Zhang *et al.* 1999. Calcium-independent haemolysis via the lectin pathway of complement activation in the guinea-pig and other species. *Immunology* **97**: 686–92.

Zuercher, A. W., H. Q. Jiang, M. C. Thurnheer, C. F. Cuff, and J. J. Cebra. 2002. Distinct mechanisms for cross-protection of the upper versus lower respiratory tract through intestinal priming. *J Immunol* **169**(7): 3920–5.

Zychlinsky, A., and P. Sansonetti. 1997. Perspectives series: host/pathogen interactions. Apoptosis in bacterial pathogenesis. *J Clin Invest* **100**: 493–5.

Zysk, G., L. Bejo, B. K. Schneider-Wald, R. Nau, and H. Heinz. 2000. Induction of necrosis and apoptosis of neutrophil granulocytes by *Streptococcus pneumoniae*. *Clin Exp Immunol* **122**: 61–6.

CHAPTER 6

Yersinia inhibition of phagocytosis

Maria Fällman and Anna Gustavsson

THE *YERSINIA* INFECTION

There are three human pathogenic *Yersinia* species: *Y. pestis*, *Y. enterocolitica*, and *Y. pseudotuberculosis* (Smego *et al.* 1999; Sulakvelidze 2000). *Y. pestis* is the causative agent of bubonic plague and has been responsible for the deaths of millions of people over the years. This pathogen is transmitted to humans by the bite of an infected rodent flea. Once inside, the bacteria initially invade and proliferate in lymphatic tissue. *Y. enterocolitica* and *Y. pseudotuberculosis* cause enteric infections (yersinosis) in humans. These are transmitted to humans by infected beverages and food or by direct contact with infected mammals; pigs are the major reservoir (Bottone 1999; Smego *et al.* 1999). Despite having a different route of infection, the orally transmitted non-plague *Yersinia* species also exhibit tropism for lymphoid tissue. The infection route occurs through the ileal mucosa in the gastrointestinal tract, where they are taken up into the lymphoid follicles through M-cells. These specialized cells cover the lymphoid follicles of Peyer's patches and engulf bacteria in a way that resembles active phagocytosis (Grassl *et al.*, 2003). The bacteria multiply within the Peyer's patches, which are intestinal lymphoid nodules that contain B and T lymphocytes and phagocytes, and then drain to mesenteric lymph nodes. At this location, *Yersinia* encounters cells of the innate immune system, and can exert a block on the customary antimicrobial functions of these cells, including phagocytosis (Hanski *et al.* 1989; Simonet *et al.* 1990).

Phagocytosis and Bacterial Pathogenicity, ed. J. D. Ernst and O. Stendahl. Published by Cambridge University Press.

YERSINIA VIRULENCE EFFECTORS

Proliferation in lymphoid tissues is an essential virulence property assigned to all three pathogenic *Yersinia* species. A common denominator is a virulence plasmid that encodes the key virulence factors (Portnoy *et al.* 1981). These include components of a type III secretion system (TTSS), translocation machinery, and proteins involved in regulating the TTSS-mediated delivery (injection) of the proteins into host cells. The injected virulence factors, the *Yersinia* outer proteins (Yops), enable the bacteria to survive within the host by interfering with different functions of host cells. The TTSS is a secretion system that is found in many bacterial pathogens of animals and plants, including human pathogenic *Yersinia* spp., *Shigella* sp., *Salmonella typhimurium,* enteropathogenic *Escherichia coli* (EPEC), *Pseudomonas aeruginosa, Xanthomonas campestris,* and *Erwinia* spp. (Hueck 1998).

The secretion system comprises approximately 20 proteins, including a cytoplasmic ATPase, several proteins located in the inner and outer bacterial membranes, those that span the periplasmic space in a composition that resembles the flagellar biosynthesis apparatus, and outer membrane proteins that resembles the secretin of type II secretion systems (Hueck 1998). Proteins secreted by the TTSS are not subjected to amino-terminal processing during secretion. In fact, the secretion signal remains unclear, but resides within the first 15–20 amino acids of the secreted proteins or in their 5′-mRNA. The TTSS system in *Yersinia,* which was the first to be identified visually, consists of a basal-body-like structure, reminiscent of the bacterial extracellular flagellum, and is topped by an extracellular needle-like appendage that protrudes outside the bacterium (Cornelis 2002). In close association with the secretion machinery is the translocation machinery, which couples bacterial secretion to the transport of the proteins from the bacterium into the cytoplasm of the host cell. This translocation machinery is build up of at least LcrV, YopB, and YopD, which are believed to form a pore in the host cell through which the Yops that are injected into the host cell can pass (Bröms *et al.* 2003; Holmström *et al.* 2001; Håkansson *et al.* 1996b; Neyt & Cornelis 1999; Pettersson *et al.* 1999; Tardy *et al.* 1999). The delivery of Yop effectors via the TTSS only occurs upon intimate contact with host cells (Pettersson *et al.* 1996).

The different Yop effectors that are secreted and delivered into interacting host cells via this system are YopH, E, T, J, M, and YpkA. These are thought to serve different functions during infection. Since *Yersinia* spp. cause lethal systemic infections in mice it offers an ideal model for studying pathogenicity. These studies have shown that strains mutated in most of these virulence

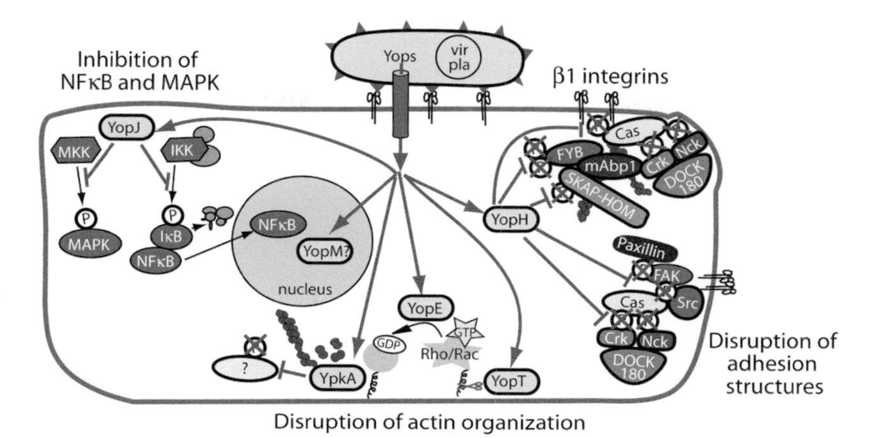

Figure 6.1 Schematic representation of the host cell targets of TTSS delivered *Yersinia* virulence effectors. See text for reviews and references.

effectors are avirulent and cleared by the primary immune defence (Cornelis 2002). **YopH** and **YopE** mediate blocking of phagocytosis by distinct mechanisms (Figure 6.1, and discussed below) (Fällman *et al.* 1995; Rosqvist *et al.* 1988b, 1990). **YopT** and **YpkA** have also been suggested to contribute to this (Figure 6.1, and discussed below), but the effects of these are less prominent (Grosdent *et al.* 2002) and it is possible that the main functions of these effectors are directed towards some other antimicrobial mechanisms not yet identified. **YopJ** has been reported to act as a SUMO-protease, which downregulates host cell stress response signaling, involving suppression of MAP kinases and NF-κB activation (Figure 6.1) (Carter *et al.* 2003; Orth *et al.* 2000; Palmer *et al.* 1998; Ruckdeschel *et al.* 1996; Schesser *et al.* 1998; Zhou *et al.* 2004). **YopM** is an acidic protein with leucine-rich repeats of unknown function (Figure 6.1) (Leung *et al.* 1990). Studies hitherto have revealed that this effector culminates in the host cell nucleus through a vesicle-associated mechanism (Skrzypek *et al.* 1998) and also that it interferes with host cell signaling by forming a complex with the kinases protein kinase C-like 2 (PRK2) and ribosomal S6 protein kinase 1 (RSK1) (McDonald *et al.*, 2003).

ADHESION TO HOST CELLS

The essential, initial stage of many bacterial infections is adherence and penetration of the epithelial barrier of the host ileum. The enteropathogenic *Yersinia* species have three or four proteins with the ability to adhere to host tissues, including *Yersinia* adhesin A (YadA), invasin, and the pH6 Antigen and Attachment invasion locus (Ail) (El Tahir & Skurnik 2001; Isberg & Barnes 2001; Isberg & Leong 1990; Isberg *et al.* 1987; Miller *et al.* 1989). The two main adhesins responsible for adherence to mammalian cells are the outer membrane proteins YadA and invasin (El Tahir & Skurnik, 2001; Isberg & Barnes 2001; Isberg *et al.* 2000). Invasin, which is chromosomally encoded, is more important for initial binding and is the predominant contributor to internalization by M cells although, in the absence of invasin, expression of YadA can mediate cell entry (Clark *et al.* 1998; Eitel & Dersch 2002; Marra & Isberg 1997; Yang & Isberg 1993). Strains that do not express the antiphagocytic factors are therefore capable of entering most types of cell (Bovallius & Nilsson 1975; Cornelis & Wolf-Watz 1997; Grosdent *et al.* 2002; Isberg 1989). However, as mentioned above and further described below, virulent *Yersinia* strains express these factors and cause a block of host cell phagocytosis, resulting in the bacteria remaining extracellular.

Invasin is a chromosomally encoded outer membrane protein expressed by *Y. enterocolitica* and *Y. pseudotuberculosis* but not *Y. pestis* (Rosqvist *et al.* 1988a). It binds to β1 integrins (α3β1, α4β1, α5β1, α6β1, and αVβ1 integrins) on the host cell and triggers uptake of the bacterium; the efficiency of internalization depends on both receptor and ligand density (Isberg *et al.* 1987, 2000; Isberg & Leong 1990; Isberg & Tran Van Nhieu 1994; Plow *et al.* 2000). The binding of invasin to the β1 integrin receptor resembles that of the extracellular matrix protein fibronectin. Both invasin and fibronectin bind to the same or overlapping regions of the β1 integrin, but invasin binds with 100-fold higher affinity than fibronectin (Tran Van Nhieu & Isberg 1991). The integrin-binding domains of fibronectin and invasin show a similarity in the relative position of several residues implicated in integrin interactions (Hamburger *et al.* 1999; Leong *et al.* 1995; Saltman *et al.* 1996; Tran Van Nhieu & Isberg 1991). In addition *Y. pseudotuberculosis*, but not *Y. enterocolitica*, contains a central dimerization region that facilitates multimerization of integrin receptors and contributes to efficient bacterial internalization of the bacteria (Isberg & Barnes 2001).

Invasin is maximally expressed in early stationary phase at 26 °C and at a pH of 8.0, which is a plausible environment for meat or beverages from wherein the bacteria spread. The expression of invasin is also high at 37 °C in pH 5.5, implying that it is expressed in the stomach and intestine of the host (Pepe *et al.* 1994). Strains mutated in the invasin gene exhibit a delayed onset of infection, but the final outcome in the mouse infection model is that invasin mutants are as virulent in mice as are wild-type *Yersinia*. This suggests a role for invasin in the early stages of infection where, possibly together with additional adhesions, it facilitates uptake through intestinal M-cells and subsequent colonization of the Peyer's patches (Isberg *et al.* 2000; Pepe & Miller 1993a,b; Rosqvist *et al.* 1988b). Furthermore, since invasin expression is reduced at 37 °C at a neutral pH, this adhesin may not be needed during later stages of infection (Isberg *et al.* 1988; Pepe *et al.* 1994).

YadA is encoded by the virulence plasmid and it is an essential virulence component of *Y. enterocolitica*. On the other hand, *Y. pseudotuberculosis* mutated in this adhesin is almost as virulent as the wild type (Bölin *et al.* 1982; El Tahir & Skurnik 2001). YadA makes up a fibrillar matrix covering the bacterium, which increases the surface hydrophobicity of the bacterium. It is involved in autoagglutination and contributes to resistance of complement-mediated killing by serum (El Tahir & Skurnik, 2001). This surface protein can bind to mucus and to components of the extracellular matrix, such as collagen, fibronectin, and laminin; it can also promote attachment to cells such as epithelial cells and neutrophils, and mediate invasion (Cornelis 1998; Dersch 2003; El Tahir & Skurnik 2001). Another adhesin is Ail, which, like invasin is encoded on the chromosome (Bliska & Falkow 1992). Although the gene locus is only found in pathogenic strains of *Yersinia*, Ail does not seem to be required for establishing systemic infection in mice (Miller *et al.* 1989; Pierson & Falkow 1993; Wachtel & Miller 1995). Ail mutant strains of *Y. enterocolitica* are more sensitive to killing by serum complement, and are impaired in both adherence and entry into mammalian cells (Bliska & Falkow 1992; Pierson & Falkow 1993). However, Ail of *Y. pseudotuberculosis* does not exhibit adhesive activity, but contributes to serum resistance (Yang *et al.* 1996). There is also a flexible fimbriae structure called the pH6 antigen, which forms on the surface of *Yersinia* species at 37 °C in acidic environments (Lindler & Tall, 1993). Although pH6 antigen mutants of *Y. pestis* exhibit reduced virulence in mice (Lindler *et al.* 1990), the role of this antigen in *Y. pseudotuberculosis* and *Y. enterocolitica* has yet to be shown.

ANTIPHAGOCYTOSIS

The ability to proliferate in the extracellular fluid during infection is an essential property of pathogenic *Yersinia* species. By analogy, most of the *Yersinia* mutants that are attenuated in virulence are cleared in the Peyer's patches, which are a location in which immune cells are abundant (Holmström *et al.* 1995). The immune cells that are primarily considered to constitute the major target cell for the *Yersinia* weaponry are the phagocytes, in which resistance to phagocytosis should be critical. Already in the mid 1950s Burrows and Bacon (1956) had conducted studies demonstrating that virulent strains of *Y. pestis* resisted engulfment by phagocytes. This was later confirmed and extended in studies using *Y. pseudotuberculosis*, where it was shown that the ability to block phagocytosis was linked to expression of the virulence plasmid and that the effectors were YopH and YopE (Bölin & Wolf-Watz 1988; Forsberg & Wolf-Watz 1988; Rosqvist *et al.* 1988a, 1990; Straley & Bowmer 1986). These results emphasize the coupling between virulence and ability to prevent phagocytosis. It has further been shown that *Yersinia* opposes uptake via various phagocytic receptors (Fc- and integrin receptors) by both macrophages and granulocytes (Andersson *et al.* 1999; Fällman *et al.* 1995; Ruckdeschel *et al.* 1996; Visser *et al.* 1995). This indicates that the antiphagocytic effect is general, including uptake via different receptors, and also that it enables the pathogen to overcome the effect of opsonization by both complement and IgG, which are two critical host defense molecules abundant in lymphoid tissue.

It is noteworthy that phagocytosis is a rapid process that is activated instantly as the bacterium interacts with the receptors on the phagocyte. Therefore, the effect exerted by a microbe that can hinder this process has to be extremely rapid and precise. It should be kept in mind that in *Yersinia* antiphagocytosis the effectors, to perform their tasks, have to be translocated from the extracellularly located bacterium into the host cell.

PHAGOCYTOSIS

Phagocytosis is the cellular process of ingesting and degrading large particles (>0.5 μm), including bacteria, senescent cells and cellular debris; in higher eucaryotic organisms, such as mammals, it is associated with host defense by allowing clearance of infectious agents (Chavrier 2001). Cells that have the capacity to phagocytose can be divided into non-professional, paraprofessional, and professional phagocytes, depending on the efficiency with which they phagocytose. The professional phagocytes, which are usually the cells denoted phagocytes, including monocytes/macrophages, neutrophils, and dendritic cells, are equipped with an array of specialized phagocytic receptors, making them very efficient in phagocytosing particles. Non-professional and paraprofessional phagocytes, which can be virtually any kind of cell, also have the capability to ingest particles, albeit less efficiently because they have only a limited number of dedicated receptors (Vieira *et al.* 2002).

Professional phagocytes have so-called opsonic phagocytic receptors, which include the Fc receptors and complement receptors. These receptors bind to and mediate uptake of opsonized (particles opsonins are host-derived proteins that coat the surface of particles and make them more susceptible

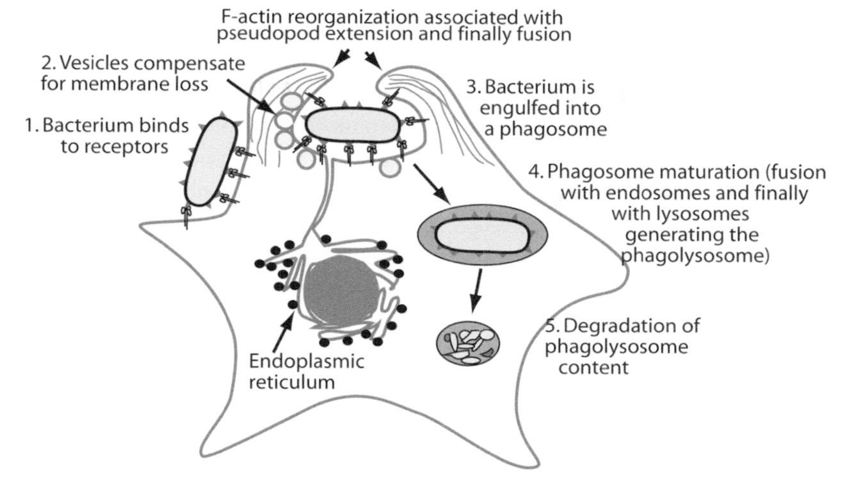

Figure 6.2 Schematic illustration of general steps in phagocytosis. See text for reviews and references.

to phagocytosis). There are also receptors that bind directly to ligands naturally expressed by microorganisms: these include the β1, β3, and β5 integrins, complement receptor 3, Toll-like receptors, mannose receptor, galactose receptor, scavenger receptors, and β-glucan receptor (Tjelle *et al.* 2000). These receptors are found on professional phagocytes, but some, such as integrins, are also found on other cells.

The phagocytic process starts with the interaction of surface receptors on the phagocyte with ligands on the particle (Figure 6.2). This interaction triggers a local reorganization of the submembrane actin cytoskeleton, which provides the force required to internalize the prey. Commonly, cup-like pseudopods are formed around the particle, which subsequently becomes included into the cup, which is sealed by cell-membrane fusion to form a closed vacuole. (Other mechanisms also exist, where cells ingest bacteria without extending pseudopods.) Following internalization, the nascent vacuole, called a phagosome, gradually matures into a phagolysosome through a complex process of fusion and fission. The phagosome initially displays markers characteristic of the plasma membrane, then of early endosomes, followed by late endosomes and finally lysosomes (Desjardins *et al.* 1994). The mature phagolysosome possesses a number of degradative properties including a very low pH, hydrolytic enzymes, defensins, and the ability to generate toxic oxidative compounds, and thus is well equipped to clear bacteria (Hampton *et al.* 1998; Tapper 1996; Tjelle *et al.* 2000). This is a commonly utilized pathway, but the specific fate of the internalized particle very much depends on the ligand–receptor pair. Thus an intracellular pathogen, for example, may follow a different intracellular route (Meresse *et al.* 1999). In comparison with endocytosis, less is known about the specific factors regulating phagosomal maturation, where the many events of fusion and fission to other vesicles gradually modify the composition of both phagosomal content and membrane. A recent proteomic study of phagolysosome-associated proteins revealed that there are more than 500 proteins associated with the phagolysosome. Although the identity of some of these proteins was expected, such as those previously known to participate in microbial killing, also other more unexpected proteins were found: there were raft proteins, ER proteins, actin-associated protein, microtubule proteins, endosome fusion proteins, and different signaling proteins (Brunet *et al.* 2003; Desjardins 2003; Garin *et al.* 2001).

If phagocytes did not replace the plasma membrane that they lose from their surface during pseudopod extension and subsequent "sealing off" of the internalized vesicle, their total surface area would diminish rapidly. However, during phagocytosis the membrane surface area increases (Booth *et al.* 2001); the compensating membrane arises by exocytosis of membranes from internal cellular compartments such as early endosomes and lysosomes (Greenberg & Grinstein 2002). In a recent study, it was demonstrated that the endoplasmic reticulum was a major contributor to phagocytosis, providing new membrane to the cell surface prior to phagocytic cup closure (Gagnon *et al.* 2002). The use of endoplasmic reticulum as a source of membrane appears to be a general mechanism of entry in macrophages, but not neutrophils, and is associated with phagocytosis by FcγRs and complement receptors as well as internalization of the pathogens *Leishmania* and *Salmonella* (Desjardins 2003; Gagnon *et al.* 2002). The neutrophils appear to use another but similar mechanism with local exocytosis of intracellular granules followed by focal pinocytosis near forming phagosomes (Botelho *et al.* 2002).

SIGNAL TRANSDUCTION IN PHAGOCYTOSIS

The major events in phagocytosis, rearrangements of the cytoskeleton and intracellular vesicle fusion and fission, can be regulated by various signaling pathways. The specific signaling mechanism responsible for the initial events of phagocytosis depends on the receptor involved, but common for many receptors is that they trigger kinase activation, alteration in phospholipid metabolism, and activation of GTPases (Chimini & Chavrier 2000; Ernst 2000; Underhill & Ozinsky 2002; Vieira *et al.* 2002). Both tyrosine phosphorylations and activation of Rho-family GTPases are common features for regulation of cytoskeletal activities and to some extent for vesicle trafficking, where the microtubule system, as well as other GTPases such as Arfs and Rabs, plays an important role (Chimini & Chavrier 2000; Ernst 2000).

Tyrosine kinases/phosphatases are important players in many processes involving receptor-mediated changes of the cytoskeleton as well as cell adhesion events. Phosphorylations/dephosphorylations of tyrosine residues in proteins generate or abolish interaction sites for SH2-containing and other phosphotyrosine-binding proteins, or alternatively activate or deactivate proteins involved in cell signaling. Like many other processes involving the cytoskeleton, β1-integrin-mediated, Fc-receptor-mediated and many other types of phagocytosis are blocked by tyrosine kinase inhibitors (Andersson *et al.* 1996; Greenberg *et al.* 1993; Magae *et al.* 1994; Rosenshine *et al.* 1992).

Rho-family GTPases regulate a variety of cellular functions, including actin reorganization (Hall 1998; Hall & Nobes 2000). These proteins cycle between an inactive conformation bound to GDP and an active conformation bound to GTP. This cycling is regulated by guanine nucleotide exchange factors (GEFs) and GTPase-activating proteins (GAPs); the former cause activation through promoting GDP dissociation and GTP binding, whereas the latter counteracts this by stimulating the intrinsic GTPase activity of these proteins. In mammals, the Rho GTPase family consists of 22 members, divided into eight subgroups: Rho (RhoA-C), Rac (Rac1–3, RhoG), Cdc42 (Cdc42, TC10, TCL, Chp. Wrh1), Rnd (Rnd1–3), RhoD (RhoD, Rif), RhoH/TTF, RhoBTB (RhoBTB1–2), and Miro (Miro1–2) (Aspenstrom *et al.* 2004). Most studies so far have focused on three of these, RhoA, Rac1 and Cdc42, whose effects on the actin cytoskeleton were initially characterized in Swiss 3T3 fibroblasts (Kozma *et al.* 1995; Nobes & Hall 1995; Ridley & Hall 1992a,b; Ridley *et al.* 1992). Activation of RhoA induces formation of stress fibers, which are bundles of antiparallel (contractile) actin–myosin filaments, and focal adhesions. The latter are multi-component complexes that act as anchors for the cell and link the extracellular matrix to intracellular cytoskeleton and signaling proteins via integrins (Ridley & Hall 1992b). Rac is required for formation of thin sheet-like processes called lamellipodia, which extend from the cell and contain a dense meshwork of actin filaments, oriented with the growing ends toward the cell front. Cdc42 activation mediates formation of thin protrusions denoted filopodia or microspikes, which contain loose parallel bundles of actin filaments (Kozma *et al.* 1995; Nobes & Hall 1995; Small *et al.* 2002). The individual roles of RhoA, Rac1, and Cdc42 in phagocytosis appear to depend on the receptor involved in triggering the uptake, where these GTPases than operate in different types of phagocytosis. Phagocytosis via complement receptors has been found to require Rho but not Rac and Cdc42; this observation is in agreement with the fact that this phagocytosis occurs without pseudopod extensions (Caron & Hall 1998). Uptake via Fcγ receptors, on the other hand, depends on Rac and Cdc42 but not Rho (Caron & Hall 1998). The β1-integrin-mediated uptake of *Yersinia,* which, like the Fcγ-receptor-mediated uptake involves pseudopod extension, depends on Rac and RhoA, but not Cdc42 (McGee *et al.* 2001; Weidow *et al.* 2000).

YERSINIA EFFECTORS INTERFERING WITH RHO GTPASES

Given that Rho-family GTPases play important roles in regulation of the cytoskeleton, it was not surprising to find that members of this GTPase family or their regulators are commonly targeted by bacterial virulence factors in different ways. Three of the *Yersinia* virulence effectors, YopE, YopT and YpkA, have been suggested to interfere with these regulators.

Upon translocation into host cells, YopE causes fragmentation of F-actin, leading to rounding-up of the cell while still leaving tail-like cytoplasmic membrane remnants that disappear upon prolonged incubation, leading to the detachment of host cells from the substratum (Rosqvist *et al.* 1990, 1991, 1994; Rosqvist & Wolf-Watz 1986). The C-terminal of YopE possesses a GTPase-activating protein (GAP)-like domain, residing in an arginine finger motif that is essential for GAP activity (Von Pawel-Rammingen *et al.* 2000). The YopE-GAP activity has been shown to downregulate Rho, Rac and Cdc42 GTPases *in vitro,* and to be essential for the YopE-mediated cytotoxic effect on HeLa cells, for antiphagocytosis, and for virulence in mice (Aili *et al.* 2002; Black & Bliska 2000; Von Pawel-Rammingen *et al.* 2000). However, even if data from studies *in vitro* indicate that YopE can downregulate Rho, Rac, and Cdc42, the specificity in the situation *in vivo* might be different. A study with *Yersinia enterocolitica* showed that YopE acts as a GAP for Rac, but not for Cdc42 or Rho, in endothelial cells (Andor *et al.* 2001). Moreover, given that the Rho GTPase family consists of many more, but less studied, members, it is likely that some of these can also constitute the critical YopE target. In agreement with this assumption, a recent study showed that YopE point-mutated in the target interaction interface in such a way that it became defective in GAP activity towards Rho, Rac and Cdc42 still elicited cytotoxicity of cells (Aili *et al.* 2003). This result indicates that YopE may well target other members of the Rho family of GTPases, apart from Rho, Rac, or Cdc42.

The *Yersinia* effectors YpkA and YopT also interfere with Rho GTPases, and appear to be involved in preventing phagocytosis of *Y. enterocolitica*

(Grosdent *et al.* 2002). YopT, which is expressed by *Y. enterocolitica* but not all *Y. pseudotuberculosis* strains, confers a cytotoxic effect on cultured cells similar to that induced by YopE. However, unlike YopE, deletion of YopT does not reduce either virulence in mice or colonization of Peyer's patches (Iriarte & Cornelis 1998). Translocated YopT induces cytotoxic effects on the host cell owing to its cysteine protease domain. It localizes to cellular membranes and irreversibly cleaves post-translationally modified RhoA near the C-terminus to release it from the membrane by removing the prenyl group in the CAAX motif (the motif that links RhoA to the membrane), which inactivates RhoA (Aepfelbacher *et al.* 2003; Iriarte & Cornelis 1998; Shao *et al.* 2002). This inactivation results in disruption of actin stress fibres and focal adhesions, causing the rounding up of the affected cell. In addition, YopT traps RhoA in the cytosol by releasing it from GDI (guanine-dissociating factors) (Aepfelbacher *et al.* 2003). Another effector, YpkA, has also been suggested to interfere with Rho GTPases, but its function is still obscure. YpkA is essential for virulence in mice and was identified as an effector that targeted to the inner surface of HeLa cell membranes, causing host cells to round up while still maintaining focal adhesion contacts (Galyov *et al.* 1993, 1994; Håkansson *et al.* 1996a). The N-terminus of YpkA is homologous to eukaryotic Ser/Thr protein kinases (Galyov *et al.* 1993). Although YpkA is inactive in the bacterium, it binds to actin in host cells, and this stimulates the YpkA intrinsic autophosphorylating activity (Galyov *et al.* 1993; Juris *et al.* 2000). Active YpkA potentially phosphorylates proteins that play key roles in the actin cytoskeleton leading to the disruption of F-actin organization and to the rounding up of the cells (Juris *et al.* 2000). YpkA has also been shown to bind to the small GTPases RhoA and Rac-1 independent of its autophosphorylation. However, since this binding is also independent of the activity state of RhoA and Rac-1, the function remains elusive (Barz *et al.* 2000; Dukuzumuremyi *et al.* 2000).

THE *YERSINIA* EFFECTOR YOPH

The virulence effector YopH plays a major role in the mechanism that enables *Yersinia* to resist phagocytosis. This effector is a protein tyrosine phosphatase (PTPase) that is essential for *Yersinia* virulence in mice (Bölin & Wolf-Watz, 1988). Characterizations and structural analyses of YopH have revealed that the C-terminal >200 amino acid PTPase domain shares considerable homology with eucaryotic PTPases (Denu *et al.* 1996; Guan & Dixon 1990; Tonks & Neel 1996). YopH is a highly potent PTPase, exhibiting the highest activity of any PTPase identified to date (Zhang *et al.* 1992). A critical residue for the PTPase activity of YopH is a cysteine, which is situated in the consensus domain found in all PTPases, the P-loop. Mutation of this cysteine to alanine or serine completely disrupts the catalytic activity of YopH (Guan & Dixon 1990); a *Yersinia* strain that carries such a mutation is attenuated in virulence.

Accordingly, since tyrosine kinases/phosphatases were known as important players in receptor-mediated changes of the cytoskeleton, it was generally assumed that *Yersinia*, when blocking its own phagocytosis via YopH, subverted the host cell tyrosine kinase signaling processes that were important for the uptake process. Furthermore, because phagocytosis is initiated almost immediately upon binding of a bacterium to the cell surface, a very rapid effect of YopH was expected. Experiments designed to verify these assumptions indeed showed that YopH interrupted a very early infection-induced phosphotyrosine signal in macrophages, and that certain phosphotyrosine proteins were rapidly dephosphorylated (Andersson *et al.* 1996). It was also found that YopH abrogated a bacterially induced immediate early Ca^{2+}-release in neutrophils (Andersson *et al.* 1999). Importantly, the YopH-mediated blocking of phagocytosis is not restricted to uptake via β1 integrins. In addition, a *Y. pseudotuberculosis* invasin mutant, opsonized with IgG or complement, that thus binds to Fc- or complement-receptors instead of β1 integrins, blocks phagocytosis by macrophages via YopH (Bliska & Black 1995; Fällman *et al.* 1995). Hence, YopH affects a mechanism that is general for different types of receptor-mediated phagocytosis. Moreover, the YopH-mediated resistance of bacterial uptake is not restricted to professional phagocytes. In HeLa cells, for example, invasin-promoted uptake is blocked by YopH, which in this case also acts by interrupting phosphotyrosine signaling induced by bacterial binding to the β1 integrin receptor (Persson *et al.* 1997).

YOPH–SUBSTRATE INTERACTIONS

In many studies of YopH–substrate interactions, a “substrate-trapping” mutant of YopH has been used to find interacting proteins. In this mutant, the cysteine residue of the PTP catalytic signature motif has been replaced with alanine or serine. This mutant retains the ability to bind substrate but is completely devoid of enzyme activity; it thus recognizes and forms stable complexes with its target substrates. These complexes can further be immunoprecipitated with anti-YopH antibodies followed by recognition of associated proteins by Western blot (Fällman *et al.* 2002; Tonks and Neel, 2001). With this method a highly tyrosine phosphorylated form of p130 Crk-associated substrate (p130Cas) was found as a common substrate of YopH in both macrophages and HeLa cells (Black & Bliska 1997; Hamid *et al.*

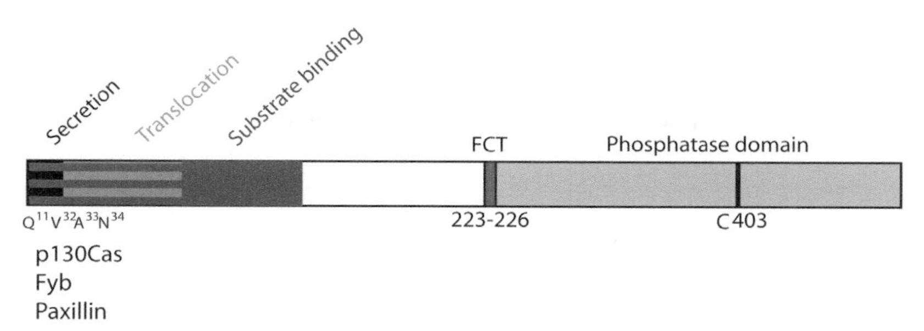

Figure 6.3 Schematic representation of the *Yersinia* tyrosine phosphatase YopH. The N-terminal amino acids 1–130 act as a substrate binding domain. Also indicated are the proteins that have been demonstrated to interact with this region. This part also contains the sequences necessary for secretion and translocation. A central region involved in early focal complex targeting (FCT) includes amino acids 223–226. The conserved cysteine residue located in position 403 is indicated; the mutation of this residue to Ser or Ala abolishes the enzymatic activity of the protein.

1999; Persson *et al.* 1997). p130Cas, however, did not constitute the only or major substrate for YopH in macrophages; instead another protein, namely Fyn-binding protein (Fyb), also called Slp-76 associated protein of 130 kDa (SLAP-130), appeared as the primary substrate in these cells (da Silva *et al.* 1997a,b; Hamid *et al.*, 1999). Both p130Cas and Fyb are known to participate in signal transduction from the β1 integrin receptor to the cytoskeleton; they both become tyrosine phosphorylated upon receptor engagement (Hunter *et al.* 2000; O'Neill *et al.* 2000). Thus, by interfering with these host cell molecules, the signal transduction from the integrin receptor is downregulated, but whether this signaling concerns phagocytic uptake remains to be clarified.

Other host cell proteins such as FAK and paxillin, although in HeLa cells, have also been implicated as YopH substrates (Andersson *et al.* 1996; Black *et al.* 1998; Persson *et al.* 1997). FAK was identified together with p130Cas in the YopH substrate trap assay. However, in contrast to p130Cas and Fyb, FAK does not interact directly with YopH and a direct interaction of paxillin with YopH has only been seen *in vitro* (Black *et al.* 1998; Mogemark, *et al.* 2005). FAK can bind to the SH3 domain of p130Cas via its proline-rich domain (Polte & Hanks 1995); it is this interaction that is behind the detection of FAK in YopH immunoprecipitates (Mogemark *et al.* unpublished).

The first 150 residues of YopH were initially reported to harbor the residues that were critical for recognition of the host cell substrates (Figure 6.3) (Black *et al.* 1998). It was also shown that the interaction with the substrate proteins was phosphotyrosine-dependent, indicating a phosphotyrosine-binding domain within this region. However, when the structure of this region of YopH in *Y. pestis* was solved it showed that the region lacked any similarity to eucaryotic phosphotyrosine-binding domains (Evdokimov *et al.* 2001). Further characterization of this region revealed that binding of YopH to p130Cas, Fyb, and paxillin involved four critical amino acids at the extreme N-terminal end of YopH (Gln^{11}, Val^{32}, $Ala,^{33}$ and Asn^{34} (Deleuil *et al.* 2003; Montagna *et al.* 2001)). The biological significance of this YopH–substrate interaction has recently been confirmed by mouse infection experiments, where a *Y. pseudotuberculosis* strain carrying a YopH variant mutated in Gln^{11} and Asn^{34} was shown to be attenuated in virulence (Deleuil *et al.* 2003).

MOLECULAR MECHANISM OF YOPH

Professional phagocytes are specifically equipped for efficient phagocytosis; therefore bacterial effectors involved in abrogating phagocytosis should exhibit a very prompt and precise mode of action. In support of this, there are several examples of YopH acting immediately upon bacterial binding to host cells. One striking example is that YopH impedes invasin-induced elevations in the intracellular concentration of free Ca^{2+} in human neutrophils (Andersson *et al.* 1999). This Ca^{2+}-signal is β1-integrin-dependent and is triggered at almost the same moment that *Yersinia* binds to receptors on the neutrophil surface. Other examples of early effects on host cells mediated by YopH are that dephosphorylation of phosphotyrosine proteins in cells infected with *Yersinia* can be detected after only 30 s of infection (Andersson *et al.* 1996), and that the association of YopH with substrates is detected within just 2 min of infection (Persson *et al.* 1997).

There are several experiments that indicate that the site of action of YopH is at the integrin adhesion site. An inactive variant of YopH, the substrate trapping mutant, localizes to host cell integrin adhesion structures, i.e. focal adhesions in infected cells (Persson *et al.* 1997, 1999). The YopH target p130Cas and also FAK and paxillin are found at the same subcellular site. These cellular structures are also destroyed by PTPase active YopH upon prolonged infection of cultured cells (Black & Bliska 1997; Persson *et al.* 1997). This is seen as a rounding up and subsequent detachment of infected cells, denoted YopH-mediated cytotoxicity, which is distinct from the well-characterized YopE-mediated cytotoxicity, where F-actin is fragmented (Figure 6.4). Thus, a late effect seen after infecting cells with *Yersinia* expressing YopH is that the substrate–integrin–F-actin connections are destroyed.

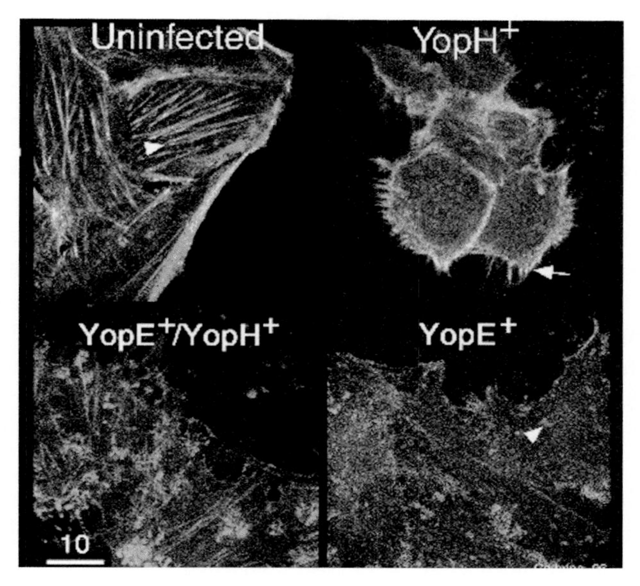

Figure 6.4 YopH and YopE have distinct effects on host cell cytoskeleton. Confocal images of HeLa cells, not infected, or infected with a *Yersinia* multiple *Yop* mutant strain expressing either YopH or YopE, or infected with *Yersinia* wild type (wt). YopE causes fragmentation of the F-actin cytoskeleton, whereas YopH affects the integrity of focal adhesions and associated stress fibers. The combined effect is seen with the *Yersinia* wt strain. (Note that this strain translocates a smaller amount of the effectors compared with the multiple mutant strain.) Cellular F-actin was visualized by staining with fluorescein-conjugated phalloidin (green); vinculin-containing focal adhesions were visualized by indirect immunofluorescence (red). The yellow color represents co-localization of microfilaments and vinculin. Vinculin-containing focal adhesions (arrowheads) and vinculin-containing retraction fibers (arrows) are shown. All sections were scanned under identical conditions and show the basolateral side of the cells. Scale bar: 10 μm.

Given that this bacterium binds to integrin receptors, a similar scenario is expected to take place upon bacterial infection. In the absence of YopH, *Yersinia* interacting with integrins are internalized by the cell via signaling from this receptor to the cytoskeleton, but in the case where YopH is injected this connection is efficiently abrogated.

The biological significance of the binding of YopH to integrin adhesion sites was proven by employing a YopH mutant deficient in this localization but otherwise functional as wild-type YopH (Persson *et al.* 1999). Such a mutant (*YopH*Δ*223–226*), which was affected in early focal complex localization, exhibited reduced capacity to block phagocytosis and, most importantly, was attenuated in virulence (Persson *et al.* 1999). Interestingly, this mutant also failed to block the immediate early Ca^{2+}-signal in neutrophils (Persson *et al.* 1999), suggesting that the Ca^{2+}-signal arises from these signaling complexes. Thus, fully PTPase-active YopH that cannot localize to peripheral focal complex structures fails to block immediate early signaling in the phagocyte and cannot promote phagocytosis and infection. The later-occurring YopH-mediated cytotoxicity was, however, not impaired. This indicated that the targeting of YopH to integrin adhesion complexes was important for blocking of extremely rapidly induced processes induced by the interacting bacteria, and not a prerequisite for later effects. It can therefore be hypothesized that the 223–226 region enables YopH, which enters into the cell via the type III translocation apparatus, immediately to anchor just beneath the bacterium from which it is injected. This then allows YopH to act promptly on the particular integrin-associated signaling complexes that are engaged by the bacterium from the outside and which are involved in mediating the internalization. Hence, YopH can dephosphorylate critical proteins involved in signaling from the integrin complex to the cytoskeleton and only affect those that are important for phagocytosis of the bacterium from where the effector originates, resulting in the bacterium attached to the complex remaining extracellular (Figure 6.5).

THE YOPH TARGETS

As mentioned above, YopH has been reported to target many proteins: p130Cas, FAK, paxillin, Fyb, and SKAP-HOM. Of these, p130Cas, Fyb, and SKAP-HOM are bound by YopH directly; these proteins are also all present in macrophages, the cell type that is generally assumed to be the important target cell for this effector.

p130Cas was identified as a tyrosine phosphorylated protein in v-Src- and v-Crk-transformed cells, and later on shown to be a docking/scaffold protein with several protein–protein interaction sites (Figure 6.6) (Kanner *et al.* 1991; Law *et al.* 1999; Nakamoto *et al.* 1996; Sakai *et al.* 1994). This docking protein is involved in the regulation of cell motility (Panetti 2002), regulation of integrin-mediated cell–matrix adhesions (Honda *et al.* 1998; Nojima *et al.* 1995) and JNK activation (Dolfi *et al.* 1998; Oktay *et al.* 1999). Mouse embryos deficient in p130Cas die 11.5–12.5 dpc showing disorganized myofibrils and Z-discs in the heart and abnormal blood vessels (Honda *et al.* 1998). Fibroblasts isolated from these embryos exhibit changed cellular morphology and changed distribution and organization of the actin cytoskeleton (Honda *et al.* 1998).

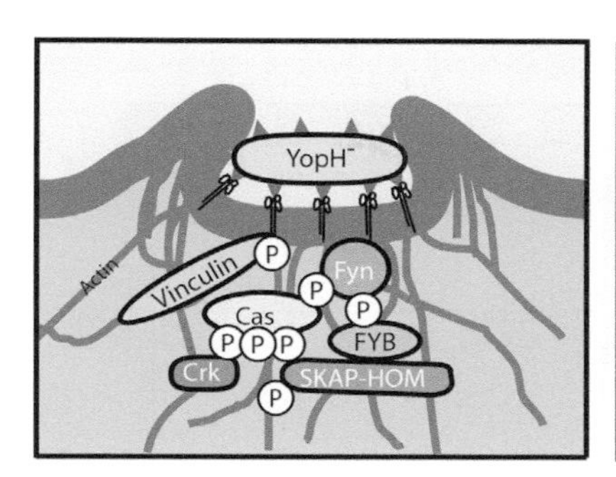

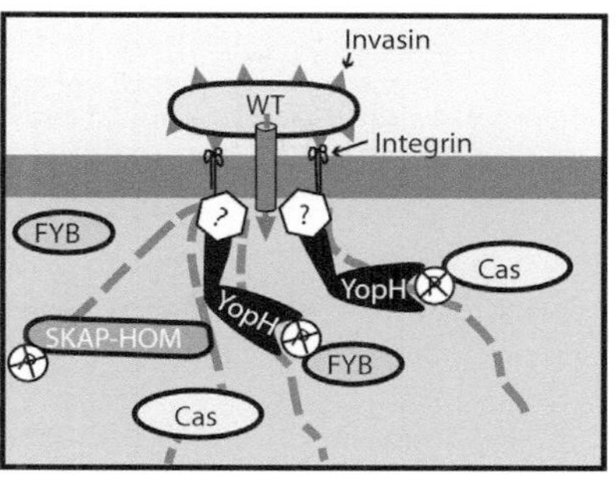

Figure 6.5 Hypothetical model of the molecular mechanism of YopH in antiphagocytosis. In the absence of YopH (left), the *Yersinia* bacterium is internalized as a result of interactions between the bacterial surface protein invasin and β1 integrins on the host cell. These interactions result in clustering of the integrins and subsequent assembly of focal-complex-like structures. The signaling complex transduces signals to the cytoskeleton, resulting in actin reorganization that allows engulfment of the surface-attached bacteria, p130Cas and Fyb are expected to play crucial roles in this event. Upon infection with a YopH-expressing *Yersinia* strain (right), the bacterium binds to the surface of the host cell via invasin–β1 integrin interactions, and then, instead of being internalized, the attached bacterium delivers YopH through the host cell plasma membrane. Upon entering into the host cell, YopH uses its inherent targeting sequence to anchor to the focal-complex-like structures that are engaged in mediating uptake of the bacterium from which the effector originates. At this location YopH then dephosphorylates phosphotyrosine proteins of importance for bacterial engulfment, and thereby abrogates the phagocytic process so the bacterium remains extracellular.

Different types of receptor activation, such as, the engagement of integrins and growth factor receptors, lead to phosphorylation of p130Cas (Casamassima & Rozengurt 1998; Nojima *et al.* 1995; Ojaniemi & Vuori 1997). Phosphorylated p130Cas acts as a docking protein for other signaling molecules, thereby influencing many pathways and hence multiple cellular responses. p130Cas localizes to focal adhesions upon tyrosine phosphorylation; this localization depends on both the SH3 domain and the C-terminus of p130Cas (Harte *et al.* 2000). Integrin-stimulated tyrosine phosphorylation of p130Cas has been suggested to be mediated by activated FAK together with Src family kinases, or as recently shown, by the Bmx/Etk kinase (Abassi *et al.* 2003; Tachibana *et al.* 1997). The phosphorylation status of p130Cas is also controlled through interaction with many tyrosine phosphatases, for example PTP1B, PTP-PEST, and PTP-SH2 (O'Neill *et al.* 2000).

Fyb is a docking/scaffold protein that is specifically expressed in cells of hematopoietic origin, such as T cells, monocytes, platelets, and mast cells, but

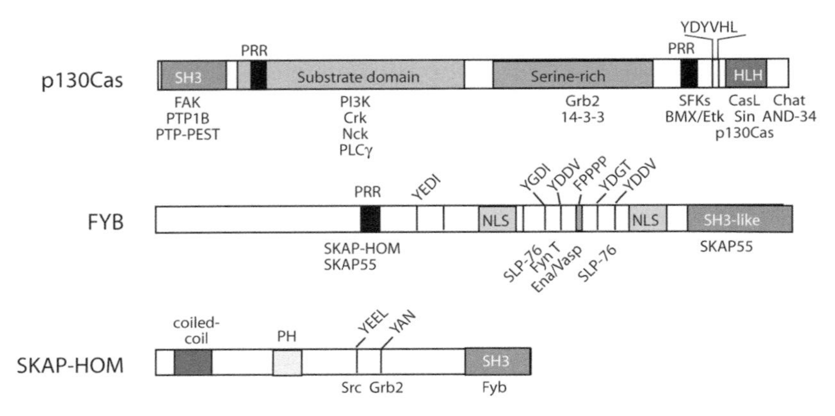

Figure 6.6 Schematic representation of the domain organization of the YopH substrates. Also depicted are proteins with which they have been shown to interact either *in vivo* or *in vitro* (see text for reviews and references). PRR, proline-rich region; FPPPP, VASP/Mena EVH1 binding site; NLS, nuclear localization sequence; PH, pleckstrin homology domain; SH3, Src homology 3 domain; YDYVHL, Src kinase SH2 binding site; YDDV, SLP-76 binding motif; YDGI, Fyn kinase SH2 binding site; YEDI and YEEL, putative binding sites for Src kinase SH2 domain; YAN, putative binding site for Grb2.

not in B cells (da Silva *et al.* 1997a; Fujii *et al.* 2003; Krause *et al.* 2000; Musci *et al.* 1997; Peterson 2003). Fyb exists in two splice variants, 120 and 130 kDa; the larger variant contains a 46 amino acid insert between two tyrosine-based motifs (Veale *et al.* 1999). This docking protein contains several regions with the potential to mediate protein–protein interactions (Figure 6.6): it interacts with the Src family kinase member Fyn, the SH2-domain-containing leukocyte protein of 76 kDa (SLP-76), and via a FPPPP motif to the Ena/VASP homology 1 (EVH1) domain of enabled/vasodilator-stimulated phosphoprotein (Ena/VASP) (da Silva *et al.* 1997a; Krause *et al.* 2000; Musci *et al.* 1997). In T cells, the lymphoid-specific proteins SKAP55 and SKAP-HOM/SKAP55-R also interact with this adaptor (Liu *et al.* 1998).

Most studies on this adaptor protein concern its role in T-cells, the cell type in which Fyb was initially found (da Silva *et al.* 1997a; Musci *et al.* 1997). Fyb becomes tyrosine phosphorylated upon T cell receptor activation as well as through activation of integrins, and this increases its association with the tyrosine kinase Fyn (da Silva *et al.* 1993, 1997a, 1997c; Hunter *et al.* 2000; Musci *et al.* 1997). T-cells deficient in Fyb show impairment in T cell receptor-stimulated adhesion to ligands for β1 and β2 integrins and impaired proliferation (Griffiths *et al.* 2001; Peterson 2003). However, TCR clustering, which is the main T cell signalling pathway in which SLP-76 and gross actin polymerization are involved, is unaffected in the null cells (Griffiths *et al.* 2001; Peterson *et al.* 2001). Based on these studies and on a study showing that Fyb contributes to integrin-mediated adhesion and vasoactive mediator release in mast cells, the protein was redesignated Adhesion- and Degranulation-promoting Adaptor Protein (ADAP) (Geng *et al.* 2001; Griffiths *et al.* 2001; Peterson *et al.* 2001).

SKAP-HOM (Src-kinase-associated protein of 55 kDa homologue/related protein) and SKAP-55 are two structurally related adaptor proteins (Figure 6.6), which can bind Fyb (Geng & Rudd 2002). SKAP-HOM is ubiquitously expressed whereas SKAP-55 is only expressed in T-cells (Geng & Rudd 2002; Timms *et al.* 1999). SKAP-HOM is tyrosine phosphorylated by Src family kinases, and localizes to membrane ruffles in macrophages upon integrin-mediated adhesion, but its function is still unknown (Black *et al.* 2000; Timms *et al.* 1999). SKAP-55 couples to the transmembrane PTPase CD45, which activates Fyn by dephosphorylation and thus leads to T cell receptor activation (Wu *et al.* 2002a). Fyn also phosphorylates SKAP-55, which is suggested to be involved in MAPK activation downstream of the T cell receptor (Wu *et al.* 2002b).

ROLE OF THE YOP SUBSTRATES IN PHAGOCYTOSIS

Common to all identified host cell target proteins of YopH is the fact that they participate in integrin signaling. Both p130Cas and Fyb, which are the ones that will be discussed here, have in addition been implicated to participate in signaling regulating cytoskeletal dynamics.

A role for p130Cas in phagocytosis is highly realistic, given that this protein has been attributed a central role in initial cell protrusive events, where it is suggested to mediate local activation of Rac1 and probably also other signaling pathways (Figure 6.7) (Cho & Klemke 2002; Gustavsson *et al.* 2004). The p130Cas-mediated activation of Rac, and possibly of other GTPases that regulate the cytoskeleton, occurs through interaction with the adaptor protein Crk (Bouton *et al.* 2001; O'Neill *et al.* 2000). Crk, which binds to p130Cas via its SH2 domain, can bind via its SH3 domain to DOCK180, which is a GEF known to activate Rho-family GTPases such as Rac1 and TC10 (Chiang *et al.* 2001; Cho & Klemke 2002; Cote & Vuori 2002; Gual *et al.* 2002). Rac1 has a central role in cell migration, where it regulates lamellipodia protrusion, and has also been shown to be important in internalization of *Yersinia* (McGee *et al.* 2001; Weidow *et al.* 2000). In accordance, integrin-induced phosphorylation of p130Cas induces recruitment of DOCK180 to CrkII during phagocytosis (Albert *et al.* 2000). p130Cas can also participate

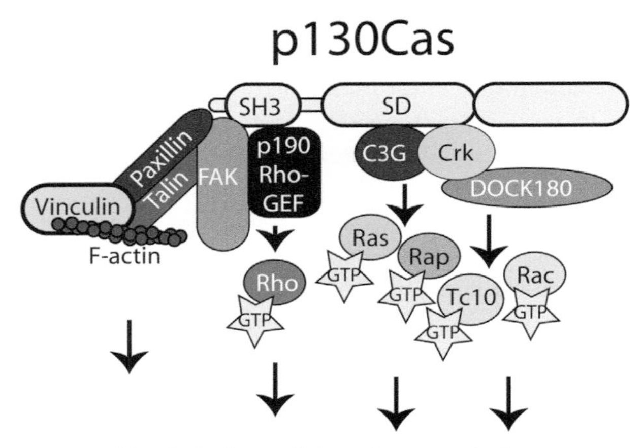

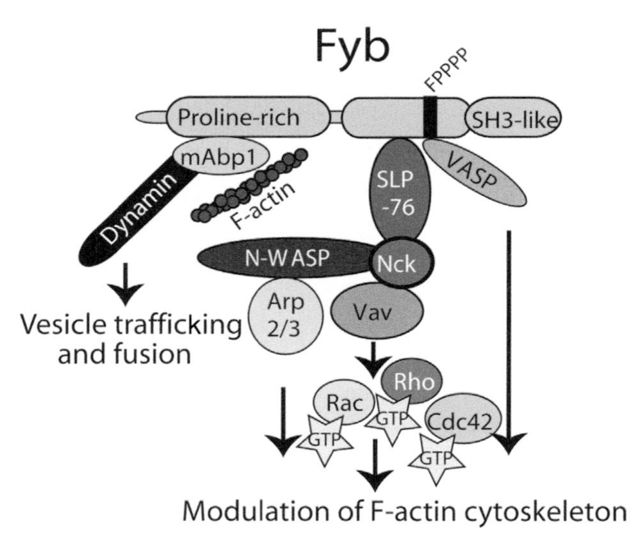

Figure 6.7 Schematic illustration of the potential ways in which p130Cas and Fyb can influence cellular F-actin. See text for reviews and references.

in signaling to the cytoskeleton in other ways, such as through binding to FAK, which in turn associates with and signals through actin-binding proteins (Schlaepfer & Mitra 2004). In addition, both p130Cas and Crk can bind C3G, a GEF for the GTPases Ras and Rap, which also can influence cytoskeletal regulation (Kirsch *et al.* 1998). Taken together, the central role for p130Cas in integrin-mediated adhesive and protrusive events makes this docking protein an opportune target for a virulence effector that is assigned to silence host cell responses induced by the binding of the bacterium to integrin receptors.

However, somewhat divergent results were obtained when the role of p130Cas in bacterial uptake was addressed by utilizing p130Cas−/− fibroblast cells compared with cells expressing a dominant negative variant of p130Cas. Bear in mind that these experiments were done in non-professional phagocytes. The p130Cas−/− fibroblasts were not affected in uptake of *Yersinia*, but a reduction in uptake was seen when a dominant-negative variant of p130Cas was overexpressed in this cell line or in cells expressing wild-type p130Cas (Weidow *et al.* 2000; Mogemark *et al.* 2005). It is possible that the results obtained with p130Cas−/− cells reflects contribution of other proteins, which act as functional homolog and compensate for the absence of p130Cas. Potential candidates could be the Cas family members HEF-1/Cas-L and Sin/Efs, which like p130Cas localize to focal adhesions, bind to and are phosphorylated by FAK and Src-family members, and contribute to assembly of signaling complexes downstream of the integrin receptor (O'Neill *et al.* 2000). It could also be that FAK contributes to the phagocytic capacity of p130Cas−/− cells, since it has been suggested that p130Cas and FAK function in diverse signaling pathways that mediate bacterial uptake and that these pathways can compensate for each other (Bruce-Staskal *et al.* 2002). Bacterial uptake is totally abolished in FAK−/− cells, so it could also be that FAK, and not p130Cas, is a key mediator of internalization, and that YopH, in order to block bacterial uptake, gains access to FAK through its interaction with p130Cas. However, we find this possibility less probable because FAK is not a substrate of YopH in macrophages, which are assumed to be the primary target cells for the activity of this effector (Hamid *et al.* 1999).

In experiments using p130Cas−/− cells it was obvious that the knockout cells were clearly less affected with respect to the Yop-mediated cytotoxic effect (Mogemark *et al.* unpublished). This result proves that this docking protein is the cellular target that directs YopH to cellular adhesion points, and which greatly facilitates the YopH-mediated disruptive effect of these structures, resulting in cell detachment. Whether the cytotoxic effect and blockage of phagocytosis reflects similar mechanisms that are affected by YopH is, however, still elusive. It could also be that the phagocytic blocking, which requires YopH to act close to the intracellular face of the bacterial adhesion site, is a two-step mechanism, which in addition to the binding to focal complex structures requires interaction with p130Cas and Fyb for full effect. The later-appearing cell detachment could then reflect a more global effect of YopH, which is seen when this effector is distributed all over the cell and taken to adhesion complexes via its binding to p130Cas.

The exact role of Fyb in macrophages is not known, but its specific expression only in cells of hematopoetic origin suggests an immunological function for this docking protein. In addition, the finding that this adaptor protein is rapidly dephosphorylated by the *Yersinia* antiphagocytic effector, YopH, implies a role for this protein in macrophage phagocytosis or other antimicrobial functions. Interestingly, both Fyb and the other YopH target in macrophages, p130Cas, constitute specific substrates for the Fes tyrosine kinase in these cells (Jucker *et al.* 1997); like p130Cas, Fyb has been implicated to participate in β1 integrin signaling, although not as a focal adhesion protein. Instead, there are many studies that imply a role for Fyb in cytoskeletal regulation (Figure 6.7), which is a feature in line with a role in phagocytic events.

Fyb is prominent in areas where actin dynamics are high; in spreading platelets this adaptor is confined to lamellipodia, and in mast cells it co-localizes with F-actin in membrane ruffles, adhesion plaques, or podosomes (Geng *et al.* 2001; Krause *et al.* 2000). It is noteworthy that motility is enhanced in integrin-stimulated T cells overexpressing Fyb (Hunter *et al.* 2000). Fyb becomes tyrosine phosphorylated upon stimulation of the T cell receptor, and forms a complex with Wiskott–Aldrich syndrome protein (WASP), VASP, Nck, and SLP-76 (Bachmann *et al.* 1999; Krause *et al.* 2000). A similar analysis in macrophages, following stimulation of Fcγ receptors, identified the same multimolecular complex at sites of phagosome formation (Coppolino *et al.* 2001). WASP is an activator of the F-actin nucleator Arp 2/3 (Takenawa & Miki 2001) and it is likely that WASP, via Arp2/3, contributes to local actin rearrangements at this site. Fyb does not bind directly to Nck or WASP, but it is likely that these are recruited via the SLP-76 protein that binds Nck and Vav, a guanine exchange factor that promotes activation of Rho GTPases, of which WASP is a downstream effector. Another link from Fyb to the cytoskeleton is through its binding to VASP, a protein that decreases F-actin branching and increases the rate of actin polymerization, leading to the formation of long non-branched actin filaments (Baba *et al.* 1999; Bear *et al.* 2002; Samarin *et al.* 2003). Moreover, a recent study showing that Fyb interacts with a F-actin binding protein, mAbp1, in macrophages (Yuan *et al.*, 2005), suggests that Fyb–mAbp1 acts as a possible linker of the WASP-containing complex to the F-actin network.

In addition to influences on F-actin dynamics, signaling downstream of Fyb in macrophages might involve modulation of vesicle trafficking and fusion. The Fyb-interacting mAbp1 protein, which is a ubiquitously expressed

F-actin binding protein, has dual properties in that, in addition to interacting with F-actin, it can connect directly with dynamin, the GTPase that regulates fission of endocytic vesicles (Kessels *et al.* 2000, 2001; Larbolette *et al.* 1999). Endocytosis shares features with and is closely connected to phagocytosis, and noteworthy here is that overexpression of the mAbp1 dynamin-binding domain blocks endocytic uptake (Kessels *et al.* 2001). Given these dual properties, it is anticipated that mAbp1 plays a role in the coordination of endocytic and cytoskeletal activities and could potentially be involved in the F-actin-dependent events involved in membrane invagination and vesicle trafficking. Both these events are essential steps in phagocytosis, and it is intriguing that Fyb, a target for an antiphagocytic factor in macrophages, binds to a protein that has the potential to regulate F-actin-dependent vesicle uptake. Taken together, there are several ways by which Fyb can mediate signaling to the cytoskeleton: through binding to the cytoskeletal regulator VASP, and/or to SLP-76, which in turn interacts with Vav, a GEF for Rho-family GTPases, and/or to mAbp1, which has the potential to modulate F-actin and vesicle dynamics. Hence, although the studies with Fyb-deficient T cells indicated no role for Fyb in T-cell-receptor-induced gross actin polymerization, a function in more subtle actin organization, such as takes place in the vicinity of phagocytic cups under formation, is highly possible (Griffiths & Penninger 2002). Hence this, together with the participation of Fyb in integrin signaling, makes Fyb a suitable target for an antiphagocytic effector, such as YopH.

CONCLUSION

Pathogenic *Yersinia* species direct specific virulence effectors into host cells to prevent phagocytosis. This is an essential virulence mechanism for these pathogens which strongly contributes to their extracellular replication during infection. It appears that *Yersinia* has developed a strategy where key molecules that control uptake of bacteria become the foci of action. YopH, a very potent PTPase, interferes with immediate early host cell signaling, which appears to be critical to activate/deactivate proteins in signal transduction. The proteins targeted by YopH, p130Cas, and Fyb are both docking proteins that participate in integrin-associated signaling by recruiting effector proteins to multi-protein complexes. p130Cas is implicated in the regulation of cellular adhesion and F-actin-associated protrusive effects. Fyb, which is an immune-cell-specific protein, participates in regulation of receptor-stimulated actin polymerization and maybe also in regulating vesicle dynamics. Other virulence effectors that contribute to the extracellular lifestyle of *Yersinia* are

YopE, YopT, and YpkA. These effectors target host cell Rho-family GTPases that have important roles in regulating the host cell cytoskeleton. It is also apparent that *Yersinia* has evolved to optimize swift delivery of certain effector proteins, since an efficient counteractive measure against a rapid phagocytic process requires that the delivery of blocking effectors occurs instantly after bacterium–cell contact.

ACKNOWLEDGEMENTS

We thank our technical associates, students, fellows, and colleagues for support, effort, and direct contributions to work cited. This work was supported by the Swedish Medical Research Council, the Swedish Cancer Foundation, and the King Gustaf Vth 80 year Foundation, and the Medical Faculty Research Foundation at Umeå University.

REFERENCES

Abassi, Y. A., Rehn, M., Ekman, N., Alitalo, K. and Vuori, K. (2003). p130Cas couples the tyrosine kinase Bmx/Etk with regulation of the actin cytoskeleton and cell migration. *J Biol Chem* **278**, 35636–43.

Aepfelbacher, M., Trasak, C., Wilharm, G. *et al.* (2003). Characterization of YopT effects on Rho GTPases in *Yersinia enterocolitica* infected cells. *J Biol Chem* **278**, 33217–23.

Aili, M., Hallberg, B., Wolf-Watz, H. and Rosqvist, R. (2002). GAP activity of *Yersinia* YopE. *Methods Enzymol* **358**, 359–70.

Aili, M., Telepnev, M., Hallberg, B., Wolf-Watz, H. and Rosqvist, R. (2003). *In vitro* GAP activity towards RhoA, Rac1 and Cdc42 is not a prerequisite for YopE induced HeLa cell cytotoxicity. *Microb Pathogen* **34**, 297–308.

Albert, M. L., Kim, J. I. and Birge, R. B. (2000). alphavbeta5 Integrin recruits the CrkII-Dock180-rac1 complex for phagocytosis of apoptotic cells. *Nat Cell Biol* **2**, 899–905.

Andersson, K., Carballeira, N., Magnusson, K. E. *et al.* (1996). YopH of *Yersinia pseudotuberculosis* interrupts early phosphotyrosine signalling associated with phagocytosis. *Molec Microbiol* **20**, 1057–69.

Andersson, K., Magnusson, K. E., Majeed, M., Stendahl, O. and Fällman, M. (1999). *Yersinia pseudotuberculosis*-induced calcium signaling in neutrophils is blocked by the virulence effector YopH. *Infect Immun* **67**, 2567–74.

Andor, A., Trulzsch, K., Essler, M. *et al.* (2001). YopE of *Yersinia*, a GAP for Rho GTPases, selectively modulates Rac-dependent actin structures in endothelial cells. *Cell Microbiol* **3**, 301–10.

Aspenstrom, P., Fransson, A. and Saras, J. (2004). Rho GTPases have diverse effects on the organization of the actin filament system. *Biochem J* **377**, 327–37.

Baba, Y., Nonoyama, S., Matsushita, M. *et al.* (1999). Involvement of Wiskott-Aldrich syndrome protein in B-cell cytoplasmic tyrosine kinase pathway. *Blood* **93**, 2003–12.

Bachmann, C., Fischer, L., Walter, U. and Reinhard, M. (1999). The EVH2 domain of the vasodilator-stimulated phosphoprotein mediates tetramerization, F-actin binding, and actin bundle formation. *J Biol Chem* **274**, 23549–57.

Barz, C., Abahji, T. N., Trulzsch, K. and Heesemann, J. (2000). The *Yersinia* Ser/Thr protein kinase YpkA/YopO directly interacts with the small GTPases RhoA and Rac-1. *FEBS Lett* **482**, 139–43.

Bear, J. E., Svitkina, T. M., Krause, M. *et al.* (2002). Antagonism between Ena/VASP proteins and actin filament capping regulates fibroblast motility. *Cell* **109**, 509–21.

Black, D. S. and Bliska, J. B. (1997). Identification of p130Cas as a substrate of *Yersinia* YopH (Yop51), a bacterial protein tyrosine phosphatase that translocates into mammalian cells and targets focal adhesions. *EMBO J* **16**, 2730–44.

Black, D. S. and Bliska, J. B. (2000). The RhoGAP activity of the *Yersinia* pseudotuberculosis cytotoxin YopE is required for antiphagocytic function and virulence. *Molec Microbiol* **37**, 515–27.

Black, D. S., Montagna, L. G., Zitsmann, S. and Bliska, J. B. (1998). Identification of an amino-terminal substrate-binding domain in the *Yersinia* tyrosine phosphatase that is required for efficient recognition of focal adhesion targets. *Molec Microbiol* **29**, 1263–74.

Black, D. S., Marie-Cardine, A., Schraven, B. and Bliska, J. B. (2000). The *Yersinia* tyrosine phosphatase YopH targets a novel adhesion- regulated signalling complex in macrophages. *Cell Microbiol* **2**, 401–14.

Bliska, J. B. and Black, D. S. (1995). Inhibition of the Fc receptor-mediated oxidative burst in macrophages by the *Yersinia pseudotuberculosis* tyrosine phosphatase. *Infect Immun* **63**, 681–5.

Bliska, J. B. and Falkow, S. (1992). Bacterial resistance to complement killing mediated by the Ail protein of *Yersinia enterocolitica. Proc Natl Acad Sci USA* **89**, 3561–5.

Bölin, I. and Wolf-Watz, H. (1988). The plasmid-encoded Yop2b protein of *Yersinia pseudotuberculosis* is a virulence determinant regulated by calcium and temperature at the level of transcription. *Molec Microbiol* **2**, 237–45.

Bölin, I., Norlander, L. and Wolf-Watz, H. (1982). Temperature-inducible outer membrane protein of *Yersinia pseudotuberculosis* and *Yersinia enterocolitica* is associated with the virulence plasmid. *Infect Immun* **37**, 506–12.

Booth, J. W., Trimble, W. S. and Grinstein, S. (2001). Membrane dynamics in phagocytosis. *Semin Immunol* **13**, 357–64.

Botelho, R. J., Tapper, H., Furuya, W., Mojdami, D. and Grinstein, S. (2002). Fc gamma R-mediated phagocytosis stimulates localized pinocytosis in human neutrophils. *J Immunol* **169**, 4423–9.

Bottone, E. J. (1999). *Yersinia enterocolitica*: overview and epidemiologic correlates. *Microbes Infect* **1**, 323–33.

Bouton, A. H., Riggins, R. B. and Bruce-Staskal, P. J. (2001). Functions of the adapter protein Cas: signal convergence and the determination of cellular responses. *Oncogene* **20**, 6448–58.

Bovallius, A. and Nilsson, G. (1975). Ingestion and survival of *Y. pseudotuberculosis* in HeLa cells. *Can J Microbiol* **21**, 1997–2007.

Bröms, J. E., Sundin, C., Francis, M. S. and Forsberg, Å. (2003). Comparative analysis of type III effector translocation by *Yersinia pseudotuberculosis* expressing native LcrV or PcrV from *Pseudomonas aeruginosa*. *J Infect Dis* **188**, 239–49.

Bruce-Staskal, P. J., Weidow, C. L., Gibson, J. J. and Bouton, A. H. (2002). Cas, Fak and Pyk2 function in diverse signaling cascades to promote *Yersinia* uptake. *J Cell Sci* **115**, 2689–700.

Brunet S., Thibault P., Gagnon E., Kearney, P., Bergeron J. J. and Desjardins M. (2003). Organelle proteomics: looking at less to see more. *Trends Cell Biol* **13**, 629–38.

Burrows, T. and Bacon, G. A. (1956). The basis of virulence in *Pasteurella pestis*: an antigen determining virulence. *Br J Exp Pathol* **37**, 481–93.

Caron, E. and Hall, A. (1998). Identification of two distinct mechanisms of phagocytosis controlled by different Rho GTPases. *Science* **282**, 1717–21.

Carter, R. S., Pennington, K. N., Ungurait, B. J., Arrate, P. and Ballard, D. W. (2003). Signal-induced ubiquitination of I kappaB Kinase-beta. *J Biol Chem* **278**, 48903–6.

Casamassima, A. and Rozengurt, E. (1998). Insulin-like growth factor I stimulates tyrosine phosphorylation of p130(Cas), focal adhesion kinase, and paxillin. Role of phosphatidylinositol 3′-kinase and formation of a p130(Cas).Crk complex. *J Biol Chem* **273**, 26149–56.

Chavrier, P. (2001). Molecular basis of phagocytosis. *Semin Immunol* **13**, 337–8.

Chiang, S. H., Baumann, C. A., Kanzaki, M. *et al.* (2001). Insulin-stimulated GLUT4 translocation requires the CAP-dependent activation of TC10. *Nature* **410**, 944–8.

Chimini, G. and Chavrier, P. (2000). Function of Rho family proteins in actin dynamics during phagocytosis and engulfment. *Nat Cell Biol* **2**, E191–6.

Cho, S. Y. and Klemke, R. L. (2002). Purification of pseudopodia from polarized cells reveals redistribution and activation of Rac through assembly of a CAS/Crk scaffold. *J Cell Biol* **156**, 725–36.

Clark, M. A., Hirst, B. H. and Jepson, M. A. (1998). M-cell surface beta1 integrin expression and invasin-mediated targeting of *Yersinia pseudotuberculosis* to mouse Peyer's patch M cells. *Infect Immun* **66**, 1237–43.

Coppolino, M. G., Krause, M., Hagendorff, P. *et al.* (2001). Evidence for a molecular complex consisting of Fyb/SLAP, SLP-76, Nck, VASP and WASP that links the actin cytoskeleton to Fcgamma receptor signalling during phagocytosis. *J Cell Sci* **114**, 4307–18.

Cornelis, G. R. (1998). The *Yersinia* Yop virulon, a bacterial system to subvert cells of the primary host defense. *Folia Microbiol* **43**, 253–61.

Cornelis, G. R. (2002). The *Yersinia* Ysc-Yop 'type III' weaponry. *Nat Rev Molec Cell Biol* **3**, 742–52.

Cornelis, G. R. and Wolf-Watz, H. (1997). The *Yersinia* Yop virulon: a bacterial system for subverting eukaryotic cells. *Molec Microbiol* **23**, 861–7.

Cote, J. F. and Vuori, K. (2002). Identification of an evolutionarily conserved superfamily of DOCK180- related proteins with guanine nucleotide exchange activity. *J Cell Sci* **115**, 4901–13.

da Silva, A. J., Janssen, O. and Rudd, C. E. (1993). T cell receptor zeta/CD3-p59fyn(T)-associated p120/130 binds to the SH2 domain of p59fyn(T). *J Exp Med* **178**, 2107–13.

da Silva, A. J., Li, Z., de Vera, C., Canto, E., Findell, P. and Rudd, C. E. (1997a). Cloning of a novel T-cell protein FYB that binds FYN and SH2-domain-containing leukocyte protein 76 and modulates interleukin 2 production. *Proc Natl Acad Sci USA* **94**, 7493–8.

da Silva, A. J., Raab, M., Li, Z. and Rudd, C. E. (1997b). TcR zeta/CD3 signal transduction in T-cells: downstream signalling via ZAP-70, SLP-76 and FYB. *Biochem Soc Trans* **25**, 361–6.

da Silva, A. J., Rosenfield, J. M., Mueller, I., Bouton, A., Hirai, H. and Rudd, C. E. (1997c). Biochemical analysis of p120/130: a protein-tyrosine kinase substrate restricted to T and myeloid cells. *J Immunol* **158**, 2007–16.

Deleuil, F., Mogemark, L., Francis, M. S., Wolf-Watz, H. and Fällman, M. (2003). Interaction between the *Yersinia* protein tyrosine phosphatase YopH and eukaryotic Cas/Fyb is an important virulence mechanism. *Cell Microbiol* **5**, 53–64.

Denu, J. M., Lohse, D. L., Vijayalakshmi, J., Saper, M. A. and Dixon, J. E. (1996). Visualization of intermediate and transition-state structures in protein-tyrosine phosphatase catalysis. *Proc Natl Acad Sci USA* **93**, 2493–8.

Dersch, P. (2003). Molecular and cellular mechanisms of bacterial entry into host cells. *Contrib Microbiol* **10**, 183–209.

Desjardins, M. (2003). ER-mediated phagocytosis: a new membrane for new functions. *Nat Rev Immunol* **3**, 280–91.

Desjardins, M., Huber, L. A., Parton, R. G. and Griffiths, G. (1994). Biogenesis of phagolysosomes proceeds through a sequential series of interactions with the endocytic apparatus. *J Cell Biol* **124**, 677–88.

Dolfi, F., Garcia-Guzman, M., Ojaniemi, M., Nakamura, H., Matsuda, M. and Vuori, K. (1998). The adaptor protein Crk connects multiple cellular stimuli to the JNK signaling pathway. *Proc Natl Acad Sci USA* **95**, 15394–9.

Dukuzumuremyi, J. M., Rosqvist, R., Hallberg, B., Akerstrom, B., Wolf-Watz, H. and Schesser, K. (2000). The *Yersinia* protein kinase A is a host factor inducible RhoA/Rac-binding virulence factor. *J Biol Chem* **275**, 35281–90.

Eitel, J. and Dersch, P. (2002). The YadA protein of *Yersinia pseudotuberculosis* mediates high-efficiency uptake into human cells under environmental conditions in which invasin is repressed. *Infect Immun* **70**, 4880–91.

El Tahir, Y. and Skurnik, M. (2001). YadA, the multifaceted *Yersinia* adhesin. *Int J Med Microbiol* **291**, 209–18.

Ernst, J. D. (2000). Bacterial inhibition of phagocytosis. *Cell Microbiol* **2**, 379–86.

Evdokimov, A. G., Tropea, J. E., Routzahn, K. M., Copeland, T. D. and Waugh, D. S. (2001). Structure of the N-terminal domain of *Yersinia pestis* YopH at 2.0 Å resolution. *Acta Crystallogr D Biol Crystallogr* **57**, 793–9.

Fällman, M., Andersson, K., Håkansson, S., Magnusson, K. E., Stendahl, O. and Wolf-Watz, H. (1995). *Yersinia pseudotuberculosis* inhibits Fc receptor-mediated phagocytosis in J774 cells. *Infect Immun* **63**, 3117–24.

Fällman, M., Deleuil, F. and McGee, K. (2002). Resistance to phagocytosis by *Yersinia*. *Int J Med Microbiol* **291**, 501–9.

Forsberg, A. and Wolf-Watz, H. (1988). The virulence protein Yop5 of *Yersinia pseudotuberculosis* is regulated at transcriptional level by plasmid-plB1-encoded trans-acting elements controlled by temperature and calcium. *Molec Microbiol* **2**, 121–33.

Fujii, Y., Wakahara, S., Nakao, T. *et al.* (2003). Targeting of MIST to Src-family kinases via SKAP55-SLAP-130 adaptor complex in mast cells. *FEBS Lett* **540**, 111–16.

Gagnon, E., Duclos, S., Rondeau, C. *et al.* (2002). Endoplasmic reticulum-mediated phagocytosis is a mechanism of entry into macrophages. *Cell* **110**, 119–31.

Galyov, E. E., Håkansson, S., Forsberg, Å. and Wolf-Watz, H. (1993). A secreted protein kinase of *Yersinia pseudotuberculosis* is an indispensable virulence determinant. *Nature* **361**, 730–2.

Galyov, E. E., Håkansson, S. and Wolf-Watz, H. (1994). Characterization of the operon encoding the YpkA Ser/Thr protein kinase and the YopJ protein of *Yersinia pseudotuberculosis*. *J Bact* **176**, 4543–8.

Garin, J., Diez, R., Kieffer, S. *et al.* (2001). The phagosome proteome: insight into phagosome functions. *J Cell Biol* **152**, 165–80.

Geng, L. and Rudd, C. E. (2002). Signalling scaffolds and adaptors in T-cell immunity. *Br J Haematol* **116**, 19–27.

Geng, L., Pfister, S., Kraeft, S. K. and Rudd, C. E. (2001). Adaptor FYB (Fyn-binding protein) regulates integrin-mediated adhesion and mediator release: Differential involvement of the FYB SH3 domain. *Proc Natl Acad Sci USA* **98**, 11527–32.

Grassl, G. A., Bohn, E., Muller, Y., Buhler, O. T. and Autenrieth, I. B. (2003). Interaction of *Yersinia enterocolitica* with epithelial cells: invasin beyond invasion. *Int J Med Microbiol* **293**, 41–54.

Greenberg, S. and Grinstein, S. (2002). Phagocytosis and innate immunity. *Curr Opin Immunol* **14**, 136–45.

Greenberg, S., Chang, P. and Silverstein, S. C. (1993). Tyrosine phosphorylation is required for Fc receptor-mediated phagocytosis in mouse macrophages. *J Exp Med* **177**, 529–34.

Griffiths, E. K. and Penninger, J. M. (2002). Communication between the TCR and integrins: role of the molecular adapter ADAP/Fyb/Slap. *Curr Opin Immunol* **14**, 317–22.

Griffiths, E. K., Krawczyk, C., Kong, Y. Y. *et al.* (2001). Positive regulation of T cell activation and integrin adhesion by the adapter Fyb/Slap. *Science* **293**, 2260–3.

Grosdent, N., Maridonneau-Parini, I., Sory, M. P. and Cornelis, G. R. (2002). Role of Yops and adhesins in resistance of *Yersinia enterocolitica* to phagocytosis. *Infect Immun* **70**, 4165–76.

Gual, P., Shigematsu, S., Kanzaki, M. *et al.* (2002). A Crk-II/TC10 signaling pathway is required for osmotic shock-stimulated glucose transport. *J Biol Chem* **277**, 43980–6.

Guan, K. L. and Dixon, J. E. (1990). Protein tyrosine phosphatase activity of an essential virulence determinant in *Yersinia*. *Science* **249**, 553–6.

Gustavsson, A., Yuan, M. and Fällman, M. (2004). Temporal dissection of beta1-integrin signaling indicates a role for p130Cas-Crk in filopodia formation. *J Biol Chem* **279**, 22893–901.

Håkansson, S., Galyov, E. E., Rosqvist, R. and Wolf-Watz, H. (1996a). The *Yersinia* YpkA Ser/Thr kinase is translocated and subsequently targeted to the inner surface of the HeLa cell plasma membrane. *Molec Microbiol* **20**, 593–603.

Håkansson, S., Schesser, K., Persson, C. *et al.* (1996b). The YopB protein of *Yersinia pseudotuberculosis* is essential for the translocation of Yop effector proteins across the target cell plasma membrane and displays a contact-dependent membrane disrupting activity. *EMBO J* **15**, 5812–23.

Hall, A. (1998). Rho GTPases and the actin cytoskeleton. *Science* **279**, 509–14.

Hall, A. and Nobes, C. D. (2000). Rho GTPases: molecular switches that control the organization and dynamics of the actin cytoskeleton. *Phil Trans R Soc Lond B* **355**, 965–70.

Hamburger, Z. A., Brown, M. S., Isberg, R. R. and Bjorkman, P. J. (1999). Crystal structure of invasin: a bacterial integrin-binding protein. *Science* **286**, 291–5.

Hamid, N., Gustavsson, A., Andersson, K. *et al.* (1999). YopH dephosphorylates Cas and Fyn-binding protein in macrophages. *Microb Pathogen* **27**, 231–42.

Hampton, M. B., Kettle, A. J. and Winterbourn, C. C. (1998). Inside the neutrophil phagosome: oxidants, myeloperoxidase, and bacterial killing. *Blood* **92**, 3007–17.

Hanski, C., Kutschka, U., Schmoranzer, H. P. *et al* (1989). Immunohistochemical and electron microscopic study of interaction of *Yersinia enterocolitica* serotype O8 with intestinal mucosa during experimental enteritis. *Infect Immun* **57**, 673–8.

Harte, M. T., Macklem, M., Weidow, C. L., Parsons, J. T. and Bouton, A. H. (2000). Identification of two focal adhesion targeting sequences in the adapter molecule p130(Cas). *Biochim Biophys Acta* **1499**, 34–48.

Holmström, A., Rosqvist, R., Wolf-Watz, H. and Forsberg, Å. (1995). Virulence plasmid-encoded YopK is essential for *Yersinia pseudotuberculosis* to cause systemic infection in mice. *Infect Immun* **63**, 2269–76.

Holmström, A., Olsson, J., Cherepanov, P. *et al.* (2001). LcrV is a channel size-determining component of the Yop effector translocon of *Yersinia*. *Molec Microbiol* **39**, 620–32.

Honda, H., Oda, H., Nakamoto, T. *et al.* (1998). Cardiovascular anomaly, impaired actin bundling and resistance to Src- induced transformation in mice lacking p130Cas [see comments]. *Nat Genet* **19**, 361–5.

Hueck, C. J. (1998). Type III protein secretion systems in bacterial pathogens of animals and plants. *Microbiol Molec Biol Rev* **62**, 379–433.

Hunter, A. J., Ottoson, N., Boerth, N., Koretzky, G. A. and Shimizu, Y. (2000). Cutting edge: a novel function for the SLAP-130/FYB adapter protein in beta 1 integrin signaling and T lymphocyte migration. *J Immunol* **164**, 1143–7.

Iriarte, M. and Cornelis, G. R. (1998). YopT, a new *Yersinia* Yop effector protein, affects the cytoskeleton of host cells. *Molec Microbiol* **29**, 915–29.

Isberg, R. R. (1989). Mammalian cell adhesion functions and cellular penetration of enteropathogenic *Yersinia* species. *Molec Microbiol* **3**, 1449–53.

Isberg, R. R. and Barnes, P. (2001). Subversion of integrins by enteropathogenic *Yersinia*. *J Cell Sci* **114**, 21–8.

Isberg, R. R. and Leong, J. M. (1990). Multiple beta 1 chain integrins are receptors for invasin, a protein that promotes bacterial penetration into mammalian cells. *Cell* **60**, 861–71.

Isberg, R. R. and Tran Van Nhieu, G. (1994). Binding and internalization of microorganisms by integrin receptors. *Trends Microbiol* **2**, 10–4.

Isberg, R. R., Voorhis, D. L. and Falkow, S. (1987). Identification of invasin: a protein that allows enteric bacteria to penetrate cultured mammalian cells. *Cell* **50**, 769–78.

Isberg, R. R., Swain, A. and Falkow, S. (1988). Analysis of expression and thermoregulation of the *Yersinia pseudotuberculosis* inv gene with hybrid proteins. *Infect Immun* **56**, 2133–8.

Isberg, R. R., Hamburger, Z. and Dersch, P. (2000). Signaling and invasin-promoted uptake via integrin receptors. *Microbes Infect* **2**, 793–801.

Jucker, M., McKenna, K., da Silva, A. J., Rudd, C. E. and Feldman, R. A. (1997). The Fes protein-tyrosine kinase phosphorylates a subset of macrophage proteins that are involved in cell adhesion and cell-cell signaling. *J Biol Chem* **272**, 2104–9.

Juris, S. J., Rudolph, A. E., Huddler, D., Orth, K. and Dixon, J. E. (2000). A distinctive role for the *Yersinia* protein kinase: actin binding, kinase activation, and cytoskeleton disruption. *Proc Natl Acad Sci USA* **97**, 9431–6.

Kanner, S. B., Reynolds, A. B., Wang, H. C., Vines, R. R. and Parsons, J. T. (1991). The SH2 and SH3 domains of pp60src direct stable association with tyrosine phosphorylated proteins p130 and p110. *EMBO J* **10**, 1689–98.

Kessels, M. M., Engqvist-Goldstein, A. E. and Drubin, D. G. (2000). Association of mouse actin-binding protein 1 (mAbp1/SH3P7), an Src kinase target, with dynamic regions of the cortical actin cytoskeleton in response to Rac1 activation. *Molec Biol Cell* **11**, 393–412.

Kessels, M. M., Engqvist-Goldstein, A. E., Drubin, D. G. and Qualmann, B. (2001). Mammalian Abp1, a signal-responsive F-actin-binding protein, links the actin cytoskeleton to endocytosis via the GTPase dynamin. *J Cell Biol* **153**, 351–66.

Kirsch, K. H., Georgescu, M. M. and Hanafusa, H. (1998). Direct binding of p130(Cas) to the guanine nucleotide exchange factor C3G. *J Biol Chem* **273**, 25673–9.

Kozma, R., Ahmed, S., Best, A. and Lim, L. (1995). The Ras-related protein Cdc42Hs and bradykinin promote formation of peripheral actin microspikes and filopodia in Swiss 3T3 fibroblasts. *Molec Cell Biol* **15**, 1942–52.

Krause, M., Sechi, A. S., Konradt, M., Monner, D., Gertler, F. B. and Wehland, J. (2000). Fyn-binding protein (Fyb)/SLP-76-associated protein (SLAP),

Ena/vasodilator-stimulated phosphoprotein (VASP) proteins and the Arp2/3 complex link T cell receptor (TCR) signaling to the actin cytoskeleton. *J Cell Biol* **149**, 181–94.

Larbolette, O., Wollscheid, B., Schweikert, J., Nielsen, P. J. and Wienands, J. (1999). SH3P7 is a cytoskeleton adapter protein and is coupled to signal transduction from lymphocyte antigen receptors. *Molec Cell Biol* **19**, 1539–46.

Law, S. F., Zhang, Y. Z., Fashena, S. J., Toby, G., Estojak, J. and Golemis, E. A. (1999). Dimerization of the docking/adaptor protein HEF1 via a carboxy-terminal helix-loop-helix domain. *Exp Cell Res* **252**, 224–35.

Leong, J. M., Morrissey, P. E., Marra, A. and Isberg, R. R. (1995). An aspartate residue of the *Yersinia* pseudotuberculosis invasin protein that is critical for integrin binding. *EMBO J* **14**, 422–31.

Leung, K. Y., Reisner, B. S. and Straley, S. C. (1990). YopM inhibits platelet aggregation and is necessary for virulence of *Yersinia pestis* in mice. *Infect Immun* **58**, 3262–71.

Lindler, L. E. and Tall, B. D. (1993). *Yersinia pestis* pH 6 antigen forms fimbriae and is induced by intracellular association with macrophages. *Molec Microbiol* **8**, 311–24.

Lindler, L. E., Klempner, M. S. and Straley, S. C. (1990). *Yersinia pestis* pH 6 antigen: genetic, biochemical, and virulence characterization of a protein involved in the pathogenesis of bubonic plague. *Infect Immun* **58**, 2569–77.

Liu, J., Kang, H., Raab, M., da Silva, A. J., Kraeft, S. K. and Rudd, C. E. (1998). FYB (FYN binding protein) serves as a binding partner for lymphoid protein and FYN kinase substrate SKAP55 and a SKAP55-related protein in T cells. *Proc Natl Acad Sci USA* **95**, 8779–84.

Magae, J., Nagi, T., Takaku, K. *et al.* (1994). Screening for specific inhibitors of phagocytosis of thioglycollate-elicited macrophages. *Biosci Biotechnol Biochem* **58**, 104–7.

Marra, A. and Isberg, R. R. (1997). Invasin-dependent and invasin-independent pathways for translocation of *Yersinia pseudotuberculosis* across the Peyer's patch intestinal epithelium. *Infect Immun* **65**, 3412–21.

McDonald, C., Vacratsis, P. O., Bliska, J. B. and Dixon, J. E. (2003). The *Yersinia* virulence factor YopM forms a novel protein complex with two cellular kinases. *J Biol Chem* **278**, 18514–23.

McGee, K., Zettl, M., Way, M. and Fällman, M. (2001). A role for N-WASP in invasin-promoted internalisation. *FEBS Lett* **509**, 59–65.

Meresse, S., Steele-Mortimer, O., Moreno, E., Desjardins, M., Finlay, B. and Gorvel, J. P. (1999). Controlling the maturation of pathogen-containing vacuoles: a matter of life and death. *Nat Cell Biol* **1**, E183–8.

Miller, V. L., Farmer, J. J. III, Hill, W. E. and Falkow, S. (1989). The ail locus is found uniquely in *Yersinia enterocolitica* serotypes commonly associated with disease. *Infect Immun* **57**, 121–31.

Mogemark, L., McGee, K., Yuan, M., Deleuil, F. and Fällman, M. (2005). Disruption of target cell adhesion structures by the *Yersinia* effector YopH requires interaction with the substrate domain of p130Cas. *Eur J Cell Biol* **84** (4), 447–89.

Montagna, L. G., Ivanov, M. I. and Bliska, J. B. (2001). Identification of residues in the N-terminal domain of the *Yersinia* tyrosine phosphatase that are critical for substrate recognition. *J Biol Chem* **276**, 5005–11.

Musci, M. A., Hendricks-Taylor, L. R., Motto, D. G. *et al.* (1997). Molecular cloning of SLAP-130, an SLP-76-associated substrate of the T cell antigen receptor-stimulated protein tyrosine kinases. *J Biol Chem* **272**, 11674–7.

Nakamoto, T., Sakai, R., Ozawa, K., Yazaki, Y. and Hirai, H. (1996). Direct binding of C-terminal region of p130Cas to SH2 and SH3 domains of Src kinase. *J Biol Chem* **271**, 8959–65.

Neyt, C. and Cornelis, G. R. (1999). Role of SycD, the chaperone of the *Yersinia* Yop translocators YopB and YopD. *Molec Microbiol* **31**, 143–56.

Nobes, C. D. and Hall, A. (1995). Rho, rac, and cdc42 GTPases regulate the assembly of multimolecular focal complexes associated with actin stress fibers, lamellipodia, and filopodia. *Cell* **81**, 53–62.

Nojima, Y., Morino, N., Mimura, T. *et al.* (1995). Integrin-mediated cell adhesion promotes tyrosine phosphorylation of p130Cas, a Src homology 3-containing molecule having multiple Src homology 2-binding motifs. *J Biol Chem* **270**, 15398–402.

Ojaniemi, M. and Vuori, K. (1997). Epidermal growth factor modulates tyrosine phosphorylation of p130Cas. Involvement of phosphatidylinositol 3′-kinase and actin cytoskeleton. *J Biol Chem* **272**, 25993–8.

Oktay, M., Wary, K. K., Dans, M., Birge, R. B. and Giancotti, F. G. (1999). Integrin-mediated activation of focal adhesion kinase is required for signaling to Jun NH2-terminal kinase and progression through the G1 phase of the cell cycle. *J Cell Biol* **145**, 1461–9.

O'Neill, G. M., Fashena, S. J. and Golemis, E. A. (2000). Integrin signalling: a new Cas(t) of characters enters the stage. *Trends Cell Biol* **10**, 111–19.

Orth, K., Xu, Z., Mudgett, M. B. *et al.* (2000). Disruption of signaling by *Yersinia* effector YopJ, a ubiquitin-like protein protease. *Science* **290**, 1594–7.

Palmer, L. E., Hobbie, S., Galan, J. E. and Bliska, J. B. (1998). YopJ of *Yersinia pseudotuberculosis* is required for the inhibition of macrophage TNF-alpha production and downregulation of the MAP kinases p38 and JNK. *Molec Microbiol* **27**, 953–65.

Panetti, T. S. (2002). Tyrosine phosphorylation of paxillin, FAK, and p130CAS: effects on cell spreading and migration. *Front Biosci* **7**, d143–50.

Pepe, J. C. and Miller, V. L. (1993a). The biological role of invasin during a *Yersinia enterocolitica* infection. *Infect Agents Disease* **2**, 236–41.

Pepe, J. C. and Miller, V. L. (1993b). *Yersinia enterocolitica* invasin: a primary role in the initiation of infection. *Proc Natl Acad Sci USA* **90**, 6473–7.

Pepe, J. C., Badger, J. L. and Miller, V. L. (1994). Growth phase and low pH affect the thermal regulation of the *Yersinia enterocolitica inv* gene. *Molec Microbiol* **11**, 123–35.

Persson, C., Carballeira, N., Wolf-Watz, H. and Fällman, M. (1997). The PTPase YopH inhibits uptake of *Yersinia*, tyrosine phosphorylation of p130Cas and FAK, and the associated accumulation of these proteins in peripheral focal adhesions. *EMBO J* **16**, 2307–18.

Persson, C., Nordfelth, R., Andersson, K., Forsberg, Å., Wolf-Watz, H. and Fällman, M. (1999). Localization of the *Yersinia* PTPase to focal complexes is an important virulence mechanism. *Molec Microbiol* **33**, 828–38.

Peterson, E. J. (2003). The TCR ADAPts to integrin-mediated cell adhesion. *Immunol Rev* **192**, 113–21.

Peterson, E. J., Woods, M. L., Dmowski, S. A. *et al.* (2001). Coupling of the TCR to integrin activation by Slap-130/Fyb. *Science* **293**, 2263–5.

Pettersson, J., Nordfelth, R., Dubinina, E. *et al.* (1996). Modulation of virulence factor expression by pathogen target cell contact [see comments]. *Science* **273**, 1231–3.

Pettersson, J., Holmström, A., Hill, J. *et al.* (1999). The V-antigen of *Yersinia* is surface exposed before target cell contact and involved in virulence protein translocation. *Molec Microbiol* **32**, 961–76.

Pierson, D. E. and Falkow, S. (1993). The *ail* gene of *Yersinia enterocolitica* has a role in the ability of the organism to survive serum killing. *Infect Immun* **61**, 1846–52.

Plow, E. F., Haas, T. A., Zhang, L., Loftus, J. and Smith, J. W. (2000). Ligand binding to integrins. *J Biol Chem* **275**, 21785–8.

Polte, T. R. and Hanks, S. K. (1995). Interaction between focal adhesion kinase and Crk-associated tyrosine kinase substrate p130Cas. *Proc Natl Acad Sci USA* **92**, 10678–82.

Portnoy, D. A., Moseley, S. L. and Falkow, S. (1981). Characterization of plasmids and plasmid-associated determinants of *Yersinia enterocolitica* pathogenesis. *Infect Immun* **31**, 775–82.

Ridley, A. J. and Hall, A. (1992a). Distinct patterns of actin organization regulated by the small GTP-binding proteins Rac and Rho. *Cold Spring Harb Symp Quant Biol* **57**, 661–71.

Ridley, A. J. and Hall, A. (1992b). The small GTP-binding protein rho regulates the assembly of focal adhesions and actin stress fibers in response to growth factors. *Cell* **70**, 389–99.

Ridley, A. J., Paterson, H. F., Johnston, C. L., Diekmann, D. and Hall, A. (1992). The small GTP-binding protein rac regulates growth factor-induced membrane ruffling. *Cell* **70**, 401–10.

Rosenshine, I., Duronio, V. and Finlay, B. B. (1992). Tyrosine protein kinase inhibitors block invasin-promoted bacterial uptake by epithelial cells. *Infect Immun* **60**, 2211–17.

Rosqvist, R. and Wolf-Watz, H. (1986). Virulence plasmid-associated HeLa cell induced cytotoxicity of *Yersinia pseudotuberculosis*. *Microb Pathogen* **1**, 229–40.

Rosqvist, R., Bolin, I. and Wolf-Watz, H. (1988a). Inhibition of phagocytosis in *Yersinia pseudotuberculosis*: a virulence plasmid-encoded ability involving the Yop2b protein. *Infect Immun* **56**, 2139–43.

Rosqvist, R., Skurnik, M. and Wolf-Watz, H. (1988b). Increased virulence of *Yersinia pseudotuberculosis* by two independent mutations. *Nature* **334**, 522–4.

Rosqvist, R., Forsberg, Å., Rimpilainen, M., Bergman, T. and Wolf-Watz, H. (1990). The cytotoxic protein YopE of *Yersinia* obstructs the primary host defence. *Molec Microbiol* **4**, 657–67.

Rosqvist, R., Forsberg, Å. and Wolf-Watz, H. (1991). Intracellular targeting of the *Yersinia* YopE cytotoxin in mammalian cells induces actin microfilament disruption. *Infect Immun* **59**, 4562–9.

Rosqvist, R., Magnusson, K. E. and Wolf-Watz, H. (1994). Target cell contact triggers expression and polarized transfer of *Yersinia* YopE cytotoxin into mammalian cells. *EMBO J* **13**, 964–72.

Ruckdeschel, K., Roggenkamp, A., Schubert, S. and Heesemann, J. (1996). Differential contribution of *Yersinia enterocolitica* virulence factors to evasion of microbicidal action of neutrophils. *Infect Immun* **64**, 724–33.

Sakai, R., Iwamatsu, A., Hirano, N. *et al.* (1994). A novel signaling molecule, p130, forms stable complexes *in vivo* with v-Crk and v-Src in a tyrosine phosphorylation-dependent manner. *EMBO J* **13**, 3748–56.

Saltman, L. H., Lu, Y., Zaharias, E. M. and Isberg, R. R. (1996). A region of the *Yersinia pseudotuberculosis* invasin protein that contributes to high affinity binding to integrin receptors. *J Biol Chem* **271**, 23438–44.

Samarin, S., Romero, S., Kocks, C., Didry, D., Pantaloni, D. and Carlier, M. F. (2003). How VASP enhances actin-based motility. *J Cell Biol* **163**, 131–42.

Schesser, K., Spiik, A. K., Dukuzumuremyi, J. M., Neurath, M. F., Pettersson, S. and Wolf-Watz, H. (1998). The *yopJ* locus is required for *Yersinia*-mediated inhibition of NF-kappaB activation and cytokine expression: YopJ contains a eukaryotic SH2-like domain that is essential for its repressive activity. *Molec Microbiol* **28**, 1067–79.

Schlaepfer, D. D. and Mitra, S. K. (2004). Multiple connections link FAK to cell motility and invasion. *Curr Opin Genet Dev* **14**, 92–101.

Shao, F., Merritt, P. M., Bao, Z., Innes, R. W. and Dixon, J. E. (2002). A Yersinia effector and a *Pseudomonas* avirulence protein define a family of cysteine proteases functioning in bacterial pathogenesis. *Cell* **109**, 575–88.

Simonet, M., Richard, S. and Berche, P. (1990). Electron microscopic evidence for in vivo extracellular localization of *Yersinia pseudotuberculosis* harboring the pYV plasmid. *Infect Immun* **58**, 841–5.

Skrzypek, E., Cowan, C. and Straley, S. C. (1998). Targeting of the *Yersinia pestis* YopM protein into HeLa cells and intracellular trafficking to the nucleus. *Molec Microbiol* **30**, 1051–65.

Small, J. V., Stradal, T., Vignal, E. and Rottner, K. (2002). The lamellipodium: where motility begins. *Trends Cell Biol* **12**, 112–20.

Smego, R. A., Frean, J. and Koornhof, H. J. (1999). Yersiniosis I: microbiological and clinicoepidemiological aspects of plague and non-plague *Yersinia* infections. *Eur J Clin Microbiol Infect Dis* **18**, 1–15.

Straley, S. C. and Bowmer, W. S. (1986). Virulence genes regulated at the transcriptional level by Ca^{2+} in *Yersinia pestis* include structural genes for outer membrane proteins. *Infect Immun* **51**, 445–54.

Sulakvelidze, A. (2000). Yersiniae other than *Y. enterocolitica*, *Y. pseudotuberculosis*, and *Y. pestis*: the ignored species. *Microbes Infect* **2**, 497–513.

Tachibana, K., Urano, T., Fujita, H. *et al.* (1997). Tyrosine phosphorylation of Crk-associated substrates by focal adhesion kinase. A putative mechanism for the integrin-mediated tyrosine phosphorylation of Crk-associated substrates. *J Biol Chem* **272**, 29083–90.

Takenawa, T. and Miki, H. (2001). WASP and WAVE family proteins: key molecules for rapid rearrangement of cortical actin filaments and cell movement. *J Cell Sci* **114**, 1801–9.

Tapper, H. (1996). The secretion of preformed granules by macrophages and neutrophils. *J Leukoc Biol* **59**, 613–22.

Tardy, F., Homble, F., Neyt, C. *et al.* (1999). *Yersinia enterocolitica* type III secretion-translocation system: channel formation by secreted Yops. *EMBO J* **18**, 6793–9.

Timms, J. F., Swanson, K. D., Marie-Cardine, A. *et al.* (1999). SHPS-1 is a scaffold for assembling distinct adhesion-regulated multi-protein complexes in macrophages. *Curr Biol* **9**, 927–30.

Tjelle, T. E., Lovdal, T. and Berg, T. (2000). Phagosome dynamics and function. *Bioessays* **22**, 255–63.

Tonks, N. K. and Neel, B. G. (1996). From form to function: signaling by protein tyrosine phosphatases. *Cell* **87**, 365–8.

Tonks, N. K. and Neel, B. G. (2001). Combinatorial control of the specificity of protein tyrosine phosphatases. *Curr Opin Cell Biol* **13**, 182–95.

Tran van Nhieu, G. and Isberg, R. R. (1991). The Yersinia pseudotuberculosis invasin protein and human fibronectin bind to mutually exclusive sites on the alpha 5 beta 1 integrin receptor. *J Biol Chem* **266**, 24367–75.

Underhill, D. M. and Ozinsky, A. (2002). Phagocytosis of microbes: complexity in action. *A Rev Immunol* **20**, 825–52.

Veale, M., Raab, M., Li, Z. *et al.* (1999). Novel isoform of lymphoid adaptor FYN-T-binding protein (FYB-130) interacts with SLP-76 and up-regulates interleukin 2 production. *J Biol Chem* **274**, 28427–35.

Vieira, O. V., Botelho, R. J. and Grinstein, S. (2002). Phagosome maturation: aging gracefully. *Biochem J* **366**, 689–704.

Visser, L. G., Annema, A. and van Furth, R. (1995). Role of Yops in inhibition of phagocytosis and killing of opsonized *Yersinia enterocolitica* by human granulocytes. *Infect Immun* **63**, 2570–5.

Von Pawel-Rammingen, U., Telepnev, M. V., Schmidt, G., Aktories, K., Wolf-Watz, H. and Rosqvist, R. (2000). GAP activity of the *Yersinia* YopE cytotoxin specifically targets the Rho pathway: a mechanism for disruption of actin microfilament structure. *Molec Microbiol* **36**, 737–48.

Wachtel, M. R. and Miller, V. L. (1995). *In vitro* and *in vivo* characterization of an *ail* mutant of *Yersinia enterocolitica. Infect Immun* **63**, 2541–8.

Weidow, C. L., Black, D. S., Bliska, J. B. and Bouton, A. H. (2000). CAS/Crk signalling mediates uptake of *Yersinia* into human epithelial cells. *Cell Microbiol* **2**, 549–60.

Wu, L., Fu, J. and Shen, S. H. (2002a). SKAP55 coupled with CD45 positively regulates T-cell receptor-mediated gene transcription. *Molec Cell Biol* **22**, 2673–86.

Wu, L., Yu, Z. and Shen, S. H. (2002b). SKAP55 recruits to lipid rafts and positively mediates the MAPK pathway upon T cell receptor activation. *J Biol Chem* **277**, 40420–7.

Yang, Y. and Isberg, R. R. (1993). Cellular internalization in the absence of invasin expression is promoted by the *Yersinia pseudotuberculosis* yadA product. *Infect Immun* **61**, 3907–13.

Yang, Y., Merriam, J. J., Mueller, J. P. and Isberg, R. R. (1996). The psa locus is responsible for thermoinducible binding of *Yersinia pseudotuberculosis* to cultured cells. *Infect Immun* **64**, 2483–9.

Yuan, M., Mogemark, L. and Fällman, M. (2005). Fyn binding protein, Fyb, interacts with mammalian actin binding protein, mAbp1. *FEBS Lett* **579** (11), 2339–47.

Zhang, Z. Y., Clemens, J. C., Schubert, H. L. *et al.* (1992). Expression, purification, and physicochemical characterization of a recombinant *Yersinia* protein tyrosine phosphatase. *J Biol Chem* **267**, 23759–66.

Zhou, L., Tan, A. and Hershenson, M. B. (2004). *Yersinia* YopJ inhibits proinflammatory molecule expression in human bronchial epithelial cells. *Respir Physiol Neurobiol* **140**, 89–97.

CHAPTER 7

Listeria invasion and spread in non-professional phagocytes

Frederick S. Southwick

INTRODUCTION

The intracellular pathogen *Listeria monocytogenes* exploits a number of normal host cell functions to survive and spread; its intracellular lifestyle explains many of the unique clinical characteristics of this deadly food-borne pathogen (Southwick & Purich 1996). *Listeria* is able to grow on refrigerated foods and also multiplies readily at room temperature. Humans with defects in cell-mediated immunity who ingest foods stored for prolonged periods in the refrigerator are at risk of contracting *Listeria*. When ingested in high numbers, this bacterium can quietly enter through the gastrointestinal tract, seed the bloodstream, and subsequently invade the meninges, causing serious and often fatal bacterial meningitis. Based on epidemiologic studies, defects in humoral immunity are not associated with an increased risk of contracting *Listeria*; however, defects in cell-mediated immunity (particularly in patients with CD4 counts below 200) confer an increased predisposition for listeriosis. Pregnant women, neonates, patients receiving corticosteroids and other immunosuppressants to prevent the rejection of organ transplants or to treat connective tissue disease, and patients with AIDS, are all at increased risk of developing *Listeria* infection (Lorber 1997). Because *Listeria* can grow within the cytoplasm of host cells and spread from cell to cell without ever coming in contact with the extracellular milieu, this pathogen is able to avoid antibodies as well as extracellular antibiotics, and can only be killed by cell-mediated immune mechanisms.

This bacterium's ability to invade non-professional phagocytes and spread from cell to cell allows it to gain entry into areas of the body not

Phagocytosis and Bacterial Pathogenicity, ed. J. D. Ernst and O. Stendahl. Published by Cambridge University Press.

normally reached by other species of bacterium. *Listeria* can breach the blood–brain barrier and actually invade the brain tissue, most commonly attacking the brainstem and causing rhomboencephalitis. *Listeria* can also invade the placenta and cause an infection *in utero* called granulomatous infantiseptica.

Unlike other forms of bacterial meningitis, which induce a massive neutrophil response in the cerebrospinal fluid spinal fluid, *Listeria* meningitis and encephalitis often results in an increased percentage of cerebrospinal monocytes, a reflection of its intracellular growth. A monocytic cerebrospinal cell response is also seen in another intracellular bacterial pathogen, *Mycobacterium tuberculosis*, and *Listeria monocytogenes* on occasion has been misdiagnosed as tuberculosis, in addition to being mistaken for viral meningitis. The ability of *Listeria* to hide in the protective environment of the cytoplasm renders antibiotics less effective. In fact, antibiotics that are unable to penetrate cells, such as aminoglycosides, are unable to kill intracellular *Listeria*. A conventional course of ten days of antibiotics is often associated with relapse of *Listeria* infection; a minimum course of three weeks of antibiotics is recommended to assure cure (Lorber 1997).

The unique clinical characteristics described above are primarily explained by the ability of *Listeria* to gain entry into the cytoplasm of individual host cells, actively multiply there, and then subsequently spread to adjacent cells without coming in contact with the extracellular environment (Southwick & Purich 1996). This chapter will review the mechanisms by which *Listeria* is able to accomplish these tasks. There are four major steps involved *Listeria monocytogenes* pathogenesis (Figure 7.1): (1) entry into host cells by phagocytosis; (2) escape from the phagolysosome and growth in the host cell cytoplasm; (3) actin-based motility within the cytoplasm of host cells; and (4) spread from cell to cell.

Induction of phagocytosis in non-professional phagocytes

Unlike many conventional bacteria that are only ingested by professional phagocytes (macrophages and neutrophils), *Listeria* is able to fool cells such as epithelial and endothelial cells that normally do not undergo phagocytosis to form pseudopods and internalize this pathogen. This Gram-positive rod expresses a family of proteins on its surface called internalins to accomplish this task. Nineteen members have been identified in *Listeria* (Cabanes *et al.* 2002), Internalin A (often called internalin or abbreviated InlA), and Internalin B (InlB) being the most completely characterized members. There are two classes of internalin. Class I members, including InlA, contain a specific sorting sequence LPXTG (where X = any amino acid) at their

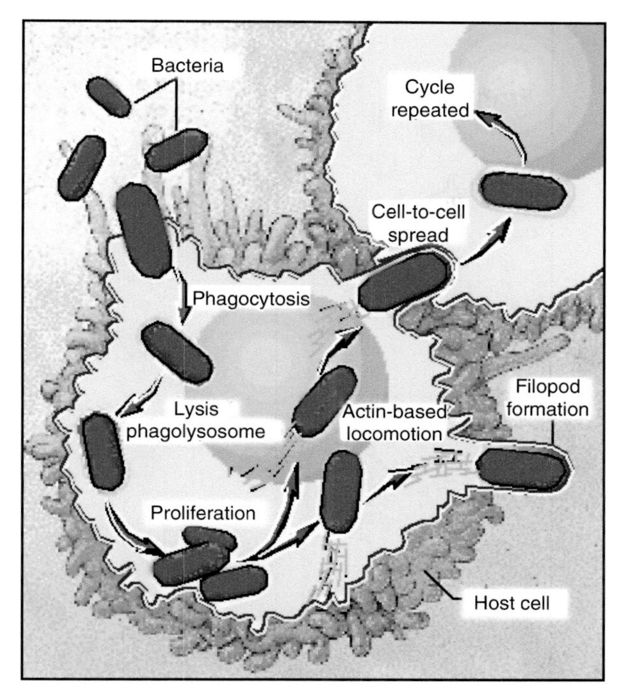

Figure 7.1 Schematic drawing of the different stages of *Listeria monocytogenes* invasion and spread within non-professional phagocytes.

carboxy-terminus. This sorting signal leads to retention of the protein in the cell-wall compartment where one or more transpeptidases (called sortases) catalyze cleavage between the threonine and glycine, and linkage to a peptide within the bacterial cell wall. Class II is exemplified by InlB. Unlike InlA, this protein is not covalently linked, but instead links to the cell wall via a series of three highly conserved approximately 80 amino acid modules that contain glycine (G)-tryptophan (W) dipeptides (termed GW modules). These carboxy-terminal regions interact with lipoteichoic acids on the bacterial surface and glycosaminoglycans on mammalian cells and represent a novel cell-wall adhesion motif. Class II surface proteins can also be secreted.

The internalin proteins all share a common characteristic. All members contain a series of leucine-rich repeat (LRR) domains, the number of repeats varying from 6 in InlG to 16 in InlA. These repeats consist of approximately 22 amino acids, and over half the residues in these repeats face outward (Cossart *et al.* 2003). The LRR sequence forms a concave face containing a preponderance of hydrophobic and charged amino acids (Marino *et al.* 2002;

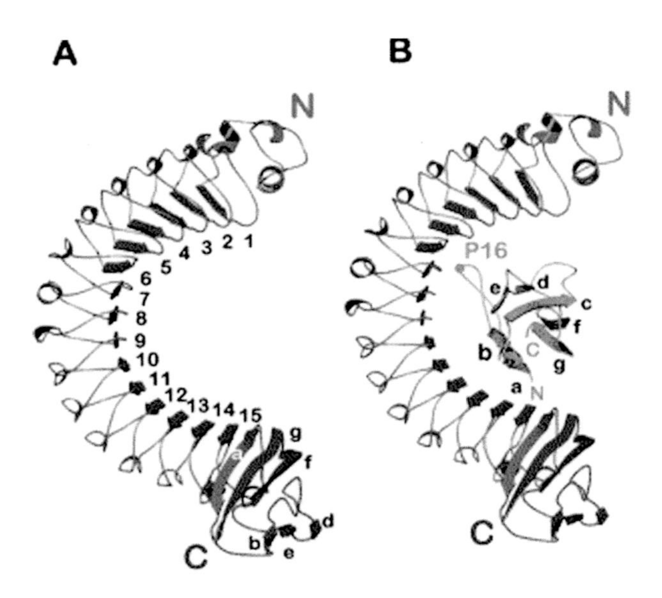

Figure 7.2 Structure of InlA. (A) Uncomplexed InlA′: cap domain, pink; LRR-domain, violet; Ig-like inter-repeat domain, blue. Strands of the LRR are numbered; strands of the Ig-like domain are indicated by letters. (B) The complex InlA′/hEC1 viewed as in (A). hEC1 is rendered in green; strands indicated by letters. (From Schubert, W. D. *et al.* 2002. *Cell* **111**: 825–36.) (See Plate 6.)

Schubert *et al.* 2002) (Figure 7.2). This region also contains two calcium-binding sites that could serve as metal ion bridges to enhance binding to other proteins; however, recent evidence suggests that this is not the case for InlB (Marino *et al.* 2004). The above structural determinants suggest that the concave face of each LRR repeat is likely to be involved in protein–protein interactions between the surface of the bacterium and the host cell outer membrane. Functional support for this conclusion is derived from the finding that the amino-terminal 213 amino acids of InlB, containing all 8 LRR repeats, is sufficient to induce entry of *Listeria* into mammalian cells (Braun *et al.* 1999). The region containing the LRR repeats and inter-repeat regions of InlA also supports entry into mammalian cells. Unlike Inl A and B, the other internalins by themselves are unable to support bacterial entry. However, InlE, G, and H enhance the ability of InlA to induce phagocytosis (Bergmann *et al.* 2002).

Specific mammalian receptors that participate in protein–protein interactions with InlA and InlB have been identified. Internalin A binds to

E-cadherin, a cell membrane receptor found on epithelial cells that is critical for cell–cell junction formation. E-cadherin receptors on adjacent cells bind to each other, forming tight adherens junctions between epithelial cells. This receptor has five domains (EC-1–EC-5) that closely resemble the structure of immunoglobulins, and also contains a long cytoplasmic tail that binds to the actin-binding protein beta-catenin. The affinity of InlA for E-cadherin is low compared with E-cadherin's binding affinity for itself; however, *Listeria* contains high concentrations of InlA on its surface, allowing the bacterial protein to effectively compete for and bind to E-cadherin in the adherens junctions. The concave LRR face of InlA forms a cup-like region where the first two extracellular domains of E-cadherin, EC1 and EC2, bind by hydrophobic and charged amino acid interactions (Schubert *et al.* 2002) (Figure 7.2B). A single amino acid sequence difference between human and murine E-cadherin identifies proline 16 as a critical amino acid for proper binding (Figure 7.2B). The murine protein has a glutamate at position 16 and fails to bind InlA (Lecuit *et al.* 1999). *Listeria* is unable to migrate across murine gastrointestinal enterocytes and oral ingestion of *Listeria* does not cause disease in wild-type mice. A transgenic mouse expressing human E-cadherin bypasses this defect and permits *Listeria* entry through the gastrointestinal enterocytes (Lecuit *et al.* 2001). InlA–E-cadherin interactions have also been shown to play a critical role in the passage of *Listeria* through the maternofetal barrier, E-cadherin being found on the basal and apical plasma membranes of syncytiotrophoblasts. InlA+, compared with InlA− *Listeria* strains, were able to pass through the syncytiotrophoblast barrier and replicate within the placenta, producing lesions typical of those found in pregnant women suffering from listeriosis (Lecuit *et al.* 2004).

Unlike *Salmonella* and *Shigella*, which utilize a type III secretion system to induce membrane ruffling for their own internalization, binding of *Listeria* InlA to E-cadherin induces internalization by the zipper mechanism of phagocytosis (Figure 7.3). Binding to E-cadherin stimulates actin filaments to assemble in discrete areas, causing the formation of pseudopods that progressively surround and internalize the bacterium into a phagolysosome. The E-cadherin–catenin adherence junction also concentrates vezatin, a transmembrane protein that attracts the unconventional myosin VIIa; this myosin may play a role in generating the directional forces required for internalizing *Listeria* (Kussel-Andermann *et al.* 2000; Dussurget *et al.* 2004; Sousa *et al.* 2004). The importance of InlA for the entry of *Listeria* into cells and across the intestinal, blood–brain and placental barriers is emphasized by a recent molecular epidemiology survey: 100% of the *Listeria* strains associated with infection in pregnancy, 98% of strains causing central nervous system

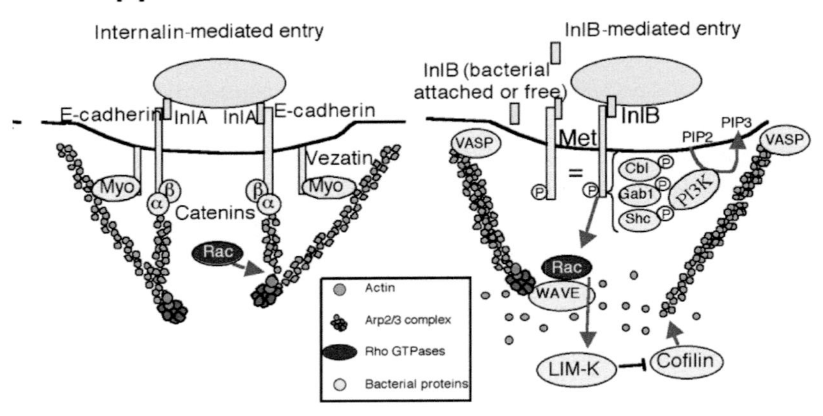

Figure 7.3 Zipper mechanisms of phagocytosis induced by InlA (internalin) and InlB.

infection, and 93% of those causing bacteremia expressed internalin A on their surface, compared with 66% of food isolates (Jacquet *et al.* 2004).

Internalin B mediates *Listeria* entry into hepatocytes, epithelial and non-epithelial cells through the hepatocyte growth factor/scatter factor receptor, or Met (Figure 7.3). Met is a transmembrane tyrosine kinase receptor that normally binds hepatocyte growth factor, binding being associated with dimerization and phosphorylation of the receptor. As observed with InlA, InlB binds to Met by its concave LRR repeats. InlB-mediated phagocytosis is enhanced by the presence of glycosaminoglycans on the host cell's surface. These molecules are thought to concentrate hepatocyte growth factor on the cell surface and prevent its proteolysis by extracellular proteases. In addition, when *Listeria* releases InlB, glycosaminoglycans are able to concentrate soluble InlB, and binding is inhibited by heparin. Curiously, soluble InlB induces a ruffling response and dramatic actin rearrangements, identical to the effects of hepatocyte growth factor; it has been suggested that soluble InlB may potentiate surface-bound InlB-mediated entry into cells (Cossart & Sansonetti 2004). The complement receptor gC1qR serves as a second membrane binding site for InlB (Braun *et al.* 2000). Unlike Met, this receptor lacks a classical transmembrane domain or a glycosyl-phosphatidylinositol (GPI) anchor and to date it is unknown how this receptor mediates phagocytosis.

InlB-mediated phagocytosis is associated with a localization of the actin nucleating protein complex Arp2/3, as well as actin filaments around the phagocytic cup (Figure 7.3). The actin filament recycling protein ADF/cofilin

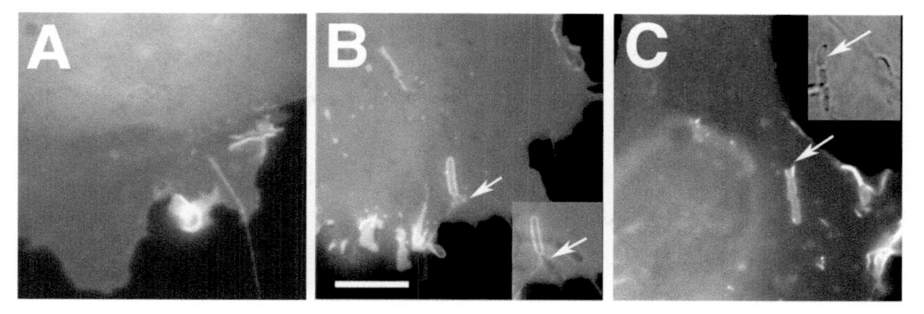

Figure 7.4 Localization of Akt-PH-GFP and PLCδ-PH-GFP during entry of *Listeria monocytogenes* into Ptk2 cells. Fluorescence images of Ptk2 cells transfected with the PH-GFP constructs 24 h before infection with *Listeria*. (A) Time 0, prior to infection, showing the Akt-PH domain localizing in active membrane ruffles. (B) Time 20–40 min after addition of *Listeria* to a culture dish containing cells transfected with Akt-PH-GFP. Two bacteria are being incorporated into the host cell by phagocytosis. At the site pointed to by an arrow, one bacterium has already been ingested and another is just starting to enter the phagocytic cup. The first bacterium has been internalized and is fully surrounded by Akt-PH-GFP. The arrow points to the edge of the membrane where the second bacterium has begun to undergo phagocytosis. The insert, bottom right, shows a dual-phase and fluorescent image of the same bacteria. The phase image shows the extracellular segment of the second bacterium not seen in the fluorescence image. Bar = 10 μm. (C) Time 20–40 min after addition of *Listeria* to a culture dish containing cells transfected with PLCδ-PH-GFP. The images are indistinguishable from Akt-PH-GFP. Two bacteria have been ingested and are outlined by the probe. The arrow points to the edge of the membrane where the third bacterium is undergoing phagocytosis. The insert, upper right, shows a phase image, and the arrow points to the same location as the fluorescence image.

is also observed to concentrate in this region. Impairment of either Arp2/3 or ADF/cofilin function results in a marked reduction in phagocytosis, indicating that actin nucleating activity and the recycling of actin filaments play major roles in InlB-mediated internalization by non-professional phagocytes. Dominant negative constructs of the small G-proteins Rac, CDC42, and Rho reveal that Rac is also of functional importance in this process (Bierne *et al.* 2001). InlB binding to Met results in the transient phosphorylation of Met and recruitment and phosphorylation of the adaptor proteins Gab1, Cbl, and Shc. These adaptor proteins also associate with the p85 subunit of phoshoinositide-3-kinase (PI-3-kinase) resulting in a transient rise of PI-3-kinase lipid products PI(3, 4)P_2 and PI(3,4,5)P_3 (Ireton *et al.* 1996, 1999), and PI(3,4,5)P_3 and its precursor PI(4,5)P_2 have been shown to concentrate around *Listeria* as it is ingested (Sidhu *et al.* 2006) (Figure 7.4). Inhibition of PI-3-kinase by wortmannin or LY94002 markedly impairs phagocytosis (Ireton *et al.* 1996, 1999; Bierne *et al.* 2000). InlB binding also activates phospholipase C gamma1, resulting in a rise in inositol (1,4,5) phosphate, IP_3, which in turn releases intracellular stores of Ca^{2+}, causing a transient rise in cytosplasmic Ca^{2+}. However, activation of phospholipase C and calcium signaling are not required for phagocytosis, but may play a role in post-phagocytic events (Bierne *et al.* 2000).

The proper organization of the host cell peripheral membrane is also critical for both InlA- and InlB-mediated phagocytosis. Cholesterol-laden lipid rafts accumulate at the site of bacterial entry; disruption of these rafts impairs bacterial internalization. Depletion of cholesterol from lipid rafts interferes with proper surface presentation of E-cadherin and impairs signal transduction by Met (Seveau *et al.* 2004).

In addition to the internalins, the exotoxin listeriolysin O (LLO) (see below) facilitates *Listeria* entry, and facilitation is mediated by calcium. LLO is able to form pores in host cells that result in oscillations of intracellular ionized calcium concentration (Wadsworth & Goldfine 1999; Repp *et al.* 2002). Inhibition of these fluxes by chelating extracellular calcium significantly impairs the phagocytosis of *Listeria* (Wadsworth & Goldfine 1999; Dramsi & Cossart 2003).

Escape from the phagolysosome, and growth in the host cell cytoplasm

Once *Listeria* are ingested, each bacterium is surrounded by membrane within the phagosome. Within 30 minutes the bacterium lyses this confining membrane and escapes into the cytoplasm, thus avoiding the potential toxic products usually released into this closed compartment. The exotoxin primarily responsible for escape from both primary and secondary vacuoles is listeriolysin O (LLO). *Listeria* mutants lacking LLO are incapable of escaping from vacuoles and are avirulent (Kathariou *et al.* 1987; Portnoy *et al.* 1988; Cossart *et al.* 1989). LLO is a cholesterol-dependent cytolysin similar to streptolysin O (SLO) of *Streptococcus pyogenes* and perfringolysin O (PFO) of *Clostridium perfringens.* These proteins bind to cholesterol and form oligomers containing 20–80 monomers that form membrane pores. LLO is unique among this family of proteins in being pH-sensitive (Geoffroy *et al.* 1987). Optimum lytic activity is observed at pH 5.0, the usual pH found in early phagosomes (Glomski *et al.* 2002). Once the phagosome membrane is lysed, the pH quickly rises to 7.0 and LLO is inactivated, preventing this hemolysin from indiscriminately damaging other membrane structures in the cell. LLO also contains an NH2-terminal PEST-like sequence that is thought to target the protein for phosphorylation and degradation. Mutations in the LLO PEST-like sequence result in increased concentrations of LLO in the cytoplasm (Decatur & Portnoy 2000). Mutations in LLO that alter pH-sensitivity or inactivate PEST-like sequence-mediated protein degradation result in greater host cell cytotoxicity and lower *Listeria* virulence in mice, emphasizing the importance of these two mechanisms for confining LLO activity to the phagosome (Glomski *et al.* 2003).

Listeria also secretes two different phospholipases C (PLC): phosphatidylinositol-specific PLC (PLCA) and a broad-range PLC (PLCB) (Mengaud *et al.* 1991; Smith *et al.* 1995; Vazquez-Boland *et al.* 1992; Goldfine & Wadsworth 2002). These proteins work in concert with LLO to lyse the phagosome. PLCA specifically acts on phosphatidylinositol (PI); unlike eucaryotic forms, the *Listeria* protein does not hydrolyze phosphorylated forms of PI, such as PI(4)P and PI(4,5)P_2. Knockout of the gene encoding PLCA (*plcA*) reduces the efficiency of *Listeria* escape from the phagosome of macrophages and modestly reduces cell-to-cell spread of the bacterium (Camilli *et al.* 1991). Loss of this protein also reduces the ability of *Listeria* to multiply in the liver, but not the spleen, of infected mice (Camilli *et al.* 1993). PLCA also weakly acts on (GPI)-anchored proteins (Mengaud *et al.* 1991); however, the *Listeria* form has $\frac{1}{50}$ to $\frac{1}{100}$ the activity of *Bacillus* PI-PLC (Gandhi *et al.* 1993). PLCB is similar in structure and activity to zinc-dependent enzymes secreted by other Gram-positive bacilli, including *Bacillus* and *Clostridium* activated enzyme (Goldfine *et al.* 1993; Vazquez-Boland *et al.* 1992). This enzyme can cleave a broad range of lipids including sphingomyelin as well as phosphoglycerolipids. Knockout of the gene encoding for PLCB (*plcB*) primarily impairs escape from vacuoles containing double membranes. These structures, sometimes termed secondary vacuoles, are the consequence of ingestion by uninfected cells of bacteria within membrane structures of adjacent infected cells, and these double membranes or secondary vacuoles are a manifestation of the cell-to-cell spread of *Listeria* (see Figure 7.1). The formation of a tissue culture plaque represents a zone of cell destruction and is another manifestation of the ability of *Listeria* to spread from cell to cell. PLCB-deficient *Listeria* show a decrease in plaque diameter compared with wild-type *Listeria*. Mutation experiments indicate that loss of the ability to cleave sphingomyelin is of particular importance for escape from secondary vacuoles (Zuckert *et al.* 1998). Deletion of the genes expressing both PLCA and PLCB causes greater reduction in plaque diameter, as well as a more profound reduction in virulence in mice, indicating that these two proteins work in concert to lyse vacuoles and allow

Listeria to spread from host cell to host cell (Smith *et al.* 1995). However, unlike LLO these *Listeria* PLCs are not absolutely required to cause disease in animals.

On escaping from host cell vacuoles, *Listeria* demonstrates a remarkable ability to survive and grow in the cytoplasm. The doubling time within the cytoplasm is 40–60 min, closely approximating the growth of *Listeria* in media (Goetz *et al.* 2001). The ability of *Listeria* to replicate in the cytoplasm has been compared to that of other bacteria by directly microinjecting GFP-expressing strains and analyzing intracellular multiplication by fluorescence microscopy. *Bacillus subtilis*, *Legionella pneumophila*, and *Salmonella* were found to be incapable of multiplying in the cytoplasm, indicating that *Listeria* possesses a virulence factor that allows it to obtain nutrients from the host cell (Goetz *et al.* 2001). This virulence gene has been identified by gene deletion and complementation experiments to be a hexose phosphate transporter (*hpt*). The *hpt* gene is contained on the same PrfA-regulated operon as LLO, PI-PLC, and ActA. Loss of this gene markedly slows intracellular growth and lowers virulence in animal studies. The *hpt* gene encodes a permease (Hpt) that is similar in structure to the mammalian glucose-6-phosphate transporter that imports glucose-6-phosphate to the endoplasmic reticulum for conversion to glucose. Based on this structural similarity, it is predicted that Hpt transports hexose phosphates from the host cell cytoplasm into the bacterium for utilization as energy and carbon sources for growth. Likely candidate hexose phosphates for transport that are found in abundance in the host cell cytoplasm include the glycogen precursor glucose-1-phosphate, as well as the glycolytic intermediate fructose-6-phosphate (Chico-Calero *et al.* 2002).

Actin-based motility within the cytoplasm of host cells

Immediately after escape into the cytoplasm *Listeria* becomes surrounded with actin filaments, which have the appearance of halos on fluorescence micrographs after rhodamine-conjugated phalloidin staining for filamentous actin (F-actin) (Dabiri *et al.* 1990). Approximately two hours after the initiation of infection the morphology of the actin filaments begins to change, actin filaments taking on an asymmetric morphology and accumulating at one end of the bacterium (Dabiri *et al.* 1990; Tilney & Portnoy 1989) (Figure 7.5). Actin filament tails are observed as the bacterium begins to move through the cytoplasm at speeds of up to 1.4 $\mu m\ s^{-1}$. The actin filament tail remains fixed in the cytoplasm, progressively growing at the interface between the bacterium and constantly disassembling at the back of the tail (Sanger *et al.*

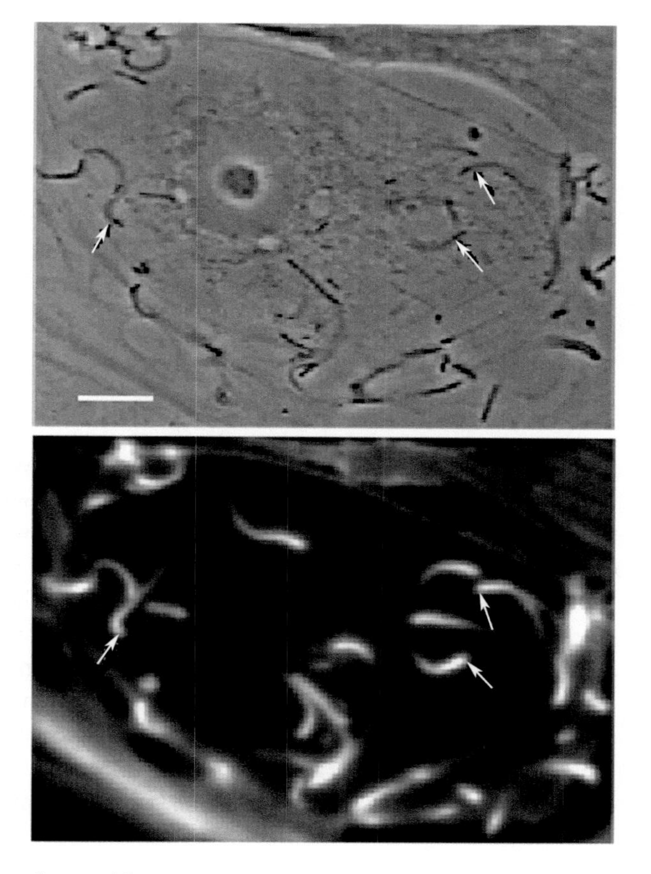

Figure 7.5 Phase and fluorescence images of PtK2 cultured cells, 4 h after the initiation of *Listeria* infection. Upper panel: The actin-rich tails appear as phase-dense structures trailing behind each small, dark bacterium. Each arrow points to the juncture between the actin tail and the bacterium. The bar represents 10 μm. Lower panel: The same microscopical field examined by fluorescence; each arrow points to the same juncture between the tail and the bacterium shown above. However, the bacteria cannot be seen in the fluorescent image. The slide has been treated with rhodamine-conjugated phalloidin and hence causes bright fluorescent staining of the tails that contain actin filaments. The fluorescent stain detects tails with greater sensitivity than phase-contrast microscopy, which accounts for the longer tails in the fluorescent image. (Adapted from Southwick and Purich, 1996, *N. Engl. J. Med.* **334**: 770–6.)

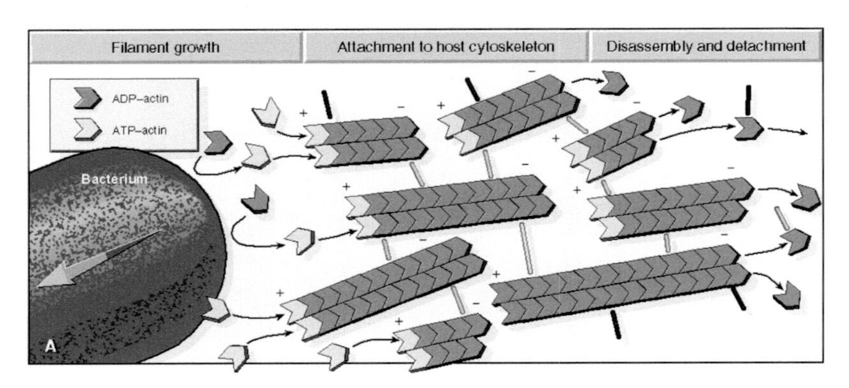

Figure 7.6 Schematic image of *Listeria*-induced actin rocket tail assembly. New ATP-actin monomers are added in the polymerization zone, located at the back surface of the motile bacteria. As the monomers are incorporated onto the barbed or fast-growing ends of actin filaments, ATP is hydrolyzed to ADP. The actin filaments in the tail steadily depolymerize over time and the regions of the tail farthest from the bacteria disappear as monomers steadily dissociate from the filaments as ADP-actiin. Actin filaments in the tails are linked to the host cell actin cytoskeleton by actin filament bundling-proteins such as alpha-actinin.

1992). The disassembly rate of the tails is constant, actin filaments having a half-life of 40–80 s (Sanger *et al.* 1992; Theriot *et al.* 1992). A polymerization zone is formed between the bacterium and the actin filament tail, and it is in this region that new actin monomers are added to the fast-growing or barbed ends of actin filaments (Sanger *et al.* 1992) (Figure 7.6). The speed at which the bacterium moves through the cytoplasm directly correlates with rate of tail assembly (Sanger *et al.* 1992; Theriot *et al.* 1992). Therefore, given the constant disassembly rate, the length of the tail is a direct reflection of the speed of bacterial movement.

The remarkable behavior of *Listeria* within the cytoplasm dramatically demonstrates that actin filament assembly can produce directional forces to generate movement (Dabiri *et al.* 1990). This model has laid to rest the old prevailing assumption that actin filaments represented a passive structural component that simply solidified lamellipod and pseudopod extensions, initially formed as a consequence of changes in osmotic pressure. *Listeria* actin-based motility has provided a simplified model system for dissecting the various steps required to regulate actin assembly and initiate directional forces during non-muscle cell motility. Chemotaxis and phagocytosis are initiated by receptor–ligand interactions leading to the activation of complex signal transduction pathways, and generation of complex multi-vectoral forces. *Listeria* actin-based motility bypasses these complexities. In many instances actin-regulatory pathways defined by using this model have proved applicable to the regulation of actin assembly in the periphery of motile cells during amoeboid movement. The discrete polymerization zone and actin filament tails associated with *Listeria* intracellular movements have allowed the identification of key actin regulatory proteins by immunofluorescence. The introduction of specific peptide inhibitors and other microfilament poisons, as well as the introduction of mutant proteins and infection with bacteria containing ActA mutations, have helped to define specific functional interactions. In addition, the ability to recapitulate *Listeria* actin-based motility in cytoplasmic extracts has allowed investigators to add and subtract various components and assess the functional consequences of these changes, as well as to analyze the forces generated by actin assembly. A multitude of investigators have embraced this system, and as a consequence the *Listeria* intracellular model has led to major advances in field of non-muscle cell motility (Marx 2003).

How does *Listeria* corrupt the host cell's actin-regulatory system to move within the cell? The first answer to this complex question was provided by the discovery of a 90 kDa protein on the surface of *Listeria*, called ActA. This protein stimulates actin tail formation and is solely responsible for *Listeria* actin-based motility in host cells (Domann *et al.* 1992; Kocks *et al.* 1992). Interference in the transcription of the *actA* gene by transposon mutation rendered the bacteria incapable of inducing actin filament tails, moving within the cytoplasm, or spreading efficiently from cell to cell. Reintroduction of the intact gene resulted in full recovery of these functions. ActA expression on the surface of bacteria moving by actin-based motility is polar, the protein concentrating at the back of the bacteria where actin filaments assemble. Polar expression is linked to bacterial division, low concentrations of ActA being noted at the site of division (Kocks *et al.* 1993). Further evidence for the exclusive ability of this single bacterial protein to mediate actin-based motility has been provided by three sets of experiments: (1) expression of ActA in the non-pathogenic species *L. innocua* converts this normally non-motile bacterium into a bacterium that can migrate through the cytoplasm by inducing actin assembly (Kocks *et al.* 1995); (2) incubation of *Streptococcus pneumoniae* with a recombinant ActA–LytA hybrid protein, attaching ActA to its surface, allows this bacterium to induce actin tail formation and move in *Xenopus* cytoplasmic extracts (Smith *et al.* 1995); and (3) ActA coated asymmetrically on latex beads readily induces the formation of actin filament tails in cell extracts

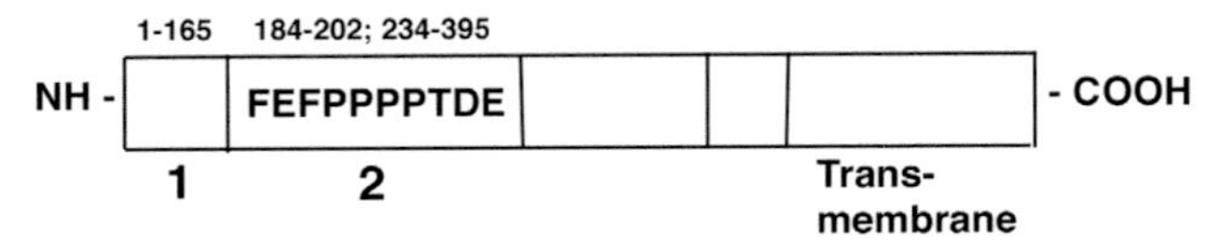

Figure 7.7 Schematic diagram of the functional regions of ActA.

(Cameron *et al.* 1999). Deletion of *actA* reduces the virulence of *Listeria* as assessed by intravenous infection of mice, highlighting the importance of bacterium-induced actin-based motile function for the spread of successful infection in the host.

A number of the functional domains in ActA have been defined by mutational analysis as well as by the introduction of inhibitory peptides by microinjection (Figure 7.7). This surface protein is anchored into the *Listeria* cell wall by its hydrophobic carboxy-terminus leaving the amino-terminus exposed to the outer surface. The immediate amino-terminal region (aa1–165) contains a cluster of acidic amino acid residues (aa31–58) as well as a series of basic residues (aa146–150) that are important for binding and activating the Arp2/3 nucleating complex (Lasa *et al.* 1997; Skoble *et al.* 2000). One segment of this basic region mimics a primary sequence in the actin-regulatory protein cofilin (aa145–156) (Skoble *et al.* 2000). This region also contains a binding site for monomeric actin (aa60–101) (Lasa *et al.* 1997; Skoble *et al.* 2000) as well as a region responsible for ActA dimer formation (aa97–126) (Mourrain *et al.* 1997). In the central region of the protein (aa234–395) there is a series of four oligoproline repeats of the type FEFPPPPTDE. This sequence contains an aromatic group and four prolines bordered on either end by acidic, negatively charged amino acid residues. This sequence type represents the ideal binding sequence for docking the ENA/VASP family of proteins (Purich & Southwick 1997; Machner *et al.* 2001). A discrete series of amino acids (aa184–202) can bind the phosphoinositide PI(4,5)P_2 (Cicchetti *et al.* 1999; Steffen *et al.* 2000). The functional importance of this binding domain has not been determined.

How does this multi-functional bacterial protein orchestrate directional actin assembly (see Figure 7.8)? The amino-terminal end of ActA binds to and activates the Arp2/3 complex to serve as a template that nucleates actin assembly. This region of ActA alone is capable of inducing the formation of actin tails; however, the rate of actin assembly is slower than in the presence of the intact molecule (Lasa *et al.* 1995). In host cells the adaptor

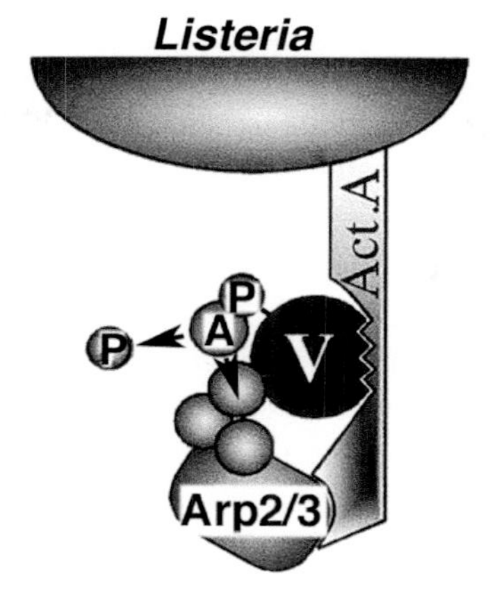

Figure 7.8 Schematic diagram of how ActA binds VASP and Arp2/3 to stimulate the addition of profilin–actin to the barbed ends of actin filaments. (Adapted from Laurent *et al.* (1999).)

proteins WASP and N-WASP activate the Arp2/3 complex after being activated by the G-proteins Cdc42 and Rac. However, ActA bypasses this step, and directly binds to and activates the Arp2/3 complex without a requirement for G-proteins. On binding to the Arp2/3 complex, WASP, N-WASP, and ActA change the conformation of the complex and allow it to interact with actin monomers at their pointed or slow-growing ends, stimulating elongation at their barbed or fast-growing ends. The activated Arp2/3 complex induces the formation of branched filaments that form 67° angles, allowing the application of force to a broader area of the cytoplasm.

Although the amino-terminal Arp2/3 binding region of ActA is absolutely required for the induction of actin tails, the addition of the oligoproline VASP-binding sequences to the ActA molecule markedly increases the rate of actin assembly and the speed of bacterial migration. Mutational analysis demonstrates that bacteria lacking the VASP-binding sequences have up to a 400-fold reduction in virulence (Auerbuch *et al.* 2003). ActA has four VASP binding sites; the progressive deletion of individual sequences incrementally decreases the concentration of VASP attracted to the polymerization zone, and progressively reduces the rate of actin tail assembly and intracellular speed (Smith *et al.* 1996). Introduction of the oligoproline FEFPPPTDE by microinjection halts *Listeria* intracellular motility, further emphasizing the functional importance of attracting VASP to the polymerization zone (Southwick & Purich 1994). VASP is a tetrameric protein that possesses 20 repeat sequences of the type GPPPPP. These oligoproline sites specifically bind profilin, a 15 kDa protein that carries actin monomers as a 1:1 complex to the barbed ends of actin filaments, where the actin monomers are then added to the filament. As the actin monomer binds to the filament, profilin is rapidly released, freeing profilin to bind and escort a new actin monomer to the barbed end. In this way profilin ushers actin monomers to the growing ends of actin filaments. Profilin also enhances the exchange of ATP for ADP in monomeric actin. When monomeric actin is added to a filament, ATP is quickly hydrolyzed; as the filament disassembles at the pointed or low-affinity end of the actin filament, monomers dissociate as ADP–actin. The hydrolysis of ATP is likely to provide the energy for force production by actin polymerization; ATP–actin is the form of monomeric actin required for efficient actin assembly. Profilin serves to recharge the ADP–actin monomer to the ATP form. Because there are four VASP binding sites on each ActA molecule and each VASP tetramer contains 20 potential profilin binding sites, a single ActA molecule could attract 80 profilin molecules. Assuming there are 40–50 ActA molecules at the back of a moving bacterium, the concentration of profilin on its surface could reach 1 mM (Kang *et al.* 1997). This would assure that *Listeria* could attract sufficient monomeric actin to maintain its rapid journey to the peripheral membrane of the host cell. In addition to profilin binding sites, VASP also contains an F-actin binding site. This site may help direct the trajectory of the newly forming actin filaments. It has been observed that VASP reduces the branching assembly of the Arp2/3 complex (Samarin *et al.* 2003; Skoble *et al.* 2001) and also maintains *Listeria* on a straighter path, allowing it to more quickly reach the periphery of the cell (Auerbuch *et al.* 2003).

In addition to Arp2/3 complex, VASP, and profilin, other host cell actin regulatory proteins have also been shown to play critical roles in actin tail formation. Immunofluorescence micrographs have demonstrated the presence of two actin recycling proteins, ADF/cofilin and gelsolin (Rosenblatt *et al.* 1997; Laine *et al.* 1998). Both proteins sever or break apart actin filaments. In addition ADF/cofilin increases the depolymerization rate of the pointed or slow-growing ends of actin filaments. Gelsolin requires free calcium to initiate severing, whereas ADF/cofilin is calcium-insensitive. Gelsolin is a more powerful severing protein, having a higher affinity for actin monomers in the actin filament than ADF/cofilin, and also is capable of severing newly forming actin filaments containing ATP–actin monomers, whereas ADF/cofilin is only capable of severing older filaments containing predominantly ADP–actin monomers. Gelsolin is able to disassemble *Listeria* actin tails at the low resting Ca^{2+} concentrations found in cytoplasm of infected cells (140 nM free calcium), suggesting that *Listeria* may lower gelsolin's threshold for calcium activation (Larson *et al.* 2005). In a reconstituted system *in vitro* the presence of ADF/cofilin is absolutely required for *Listeria* actin-based motility, emphasizing the importance of the recycling of actin filaments to provide new actin monomers for continued formation of new actin filament tails (Loisel *et al.* 1999).

Listeria also absolutely requires the presence of a barbed-end capping protein in order to generate directional force. By binding to the barbed or fast-growing ends of actin filaments these actin-regulatory proteins prevent the addition or removal of actin monomers. Because the barbed end is the primary end of actin filaments where actin monomers are added in the living cell, all cells contain significant concentrations of barbed-end capping proteins to prevent indiscriminant elongation of actin filaments. Two barbed-end capping proteins have been identified in *Listeria* actin tails (David *et al.* 1998; Laine *et al.* 1998). Gelsolin, in addition to recycling actin filaments, binds to the barbed ends and prevents actin monomer exchange. The second capping protein, CapZ, derives its name from the fact that it was first identified in the Z line of skeletal muscle, where it prevents actin filaments from growing beyond this anatomic structure (Caldwell *et al.* 1989). Unlike gelsolin, CapZ does not require calcium to function (Cooper & Pollard 1985). In the polymerization zone ActA is able to orchestrate the uncapping of the barbed ends, freeing these ends for the addition of actin monomers and allowing new actin assembly exclusively behind the motile bacteria. How ActA prevents capping in the polymerization zone remains to be clarified. Possible candidates for uncapping include VASP, which has been shown in high concentration to modestly block CapZ barbed-end capping activity (Bear *et al.* 2002) and profilin, which is able to compete with capping proteins for the barbed filament ends (Bubb *et al.* 2003).

Another vital protein for the formation of tails and the generation of directional force is the actin bundling protein alpha-actinin. Alpha-actinin forms a barrel-like structure consisting of two identical monomers that are linked together in an antiparallel fashion, resulting in an amino-terminal actin filament binding site at each end (Viel 1999). The two actin filament ends can link two actin filaments together. Mixing alpha-actinin and actin filaments results in the formation of multiple filament bundles, and alpha-actinin has been termed a bundling protein. This protein is found in high concentrations in the stress fibers of tissue culture cells. Immunofluorescence micrographs of live video microscopy following microinjection of fluorescently labeled alpha-actinin show that this protein is also highly concentrated in *Listeria* actin filament tails (Dabiri *et al.* 1990). This protein links the newly formed actin filaments together to form a tail structure and also links the actin tails to actin structures within the cytoplasm of the cell. These linkages are required to lock the tail in place and provide a stable platform so that the elongation of new filaments can drive the bacteria away from the tail. Microinjection of a 53 kDa fragment of alpha-actinin, which disrupts bundling, completely blocks *Listeria* actin-based motility (Dold *et al.* 1994).

Two other proteins of potential importance have been identified to be present in *Listeria* actin tails: coronin and the small G-protein Rac (David *et al.* 1998). Coronin was first identified as an actin-binding protein in *Dictyostelium* (de Hostos *et al.* 1991). In yeast this protein was found to stimulate actin polymerization and to also bind microtubules (Goode *et al.* 1999). Coronin is found at the leading edge of mammalian cells (Mishima & Nishida 1999) and interacts with multiple NADPH oxidase components from neutrophils (Grogan *et al.* 1997). Rac is intimately related to actin-based processes and plays a central role in the stimulation of lamellipod formation (Burridge & Wennerberg 2004). To date the functions of these two proteins in *Listeria* actin-based motility have not been clarified.

Although the primary constituents for *Listeria* actin-based motility have been defined, the mechanisms regulating their activation and cooperative interactions requires further investigation. G-proteins have not been shown to play a role in *Listeria* intracellular movement; however, recent evidence points to a likely role of another family of signal transduction mediators, the phosphoinositides. By using GFP-labeled PH domains that recognize $PI(4,5)P_2$ and $PI(3,4,5)P_3$, these phosphoinositides have been identified along the surface of intracellular *Listeria* and within the actin tails of motile bacteria. Treatment with the PI-3-kinase inhibitor LY290042 causes near-complete, reversible inhibition of *Listeria* actin-based motility and dissociation of PIP_3 from the surface and tails of *Listeria*, emphasizing the functional importance of this pathway (Sidhu *et al.* 2006). Deletion of the putative ActA phosphoinositide-binding site does not affect localization of either $PI(4,5)P_2$ or $PI(3,4,5)P_3$. The protein or proteins responsible for concentrating the phosphoinositides are presently being investigated as are mechanisms by which these phospholipids facilitate *Listeria* actin-based motility.

Spread from cell to cell

Using the multiple host cell components described above, *Listeria* are propelled through the cytoplasm by stimulating actin assembly. Their trajectories eventually lead the bacteria to the peripheral cytoplasm where each bacterium pushes outward, forming a long F-actin-rich projection or filopod with the bacterium at its tip. The projection pushes into an adjacent cell and induces phagocytosis. The bacterium is internalized and is surrounded by membrane from its original host cell as well as a second membrane from its new host cell. This mechanism of cell-to-cell spread allows *Listeria* to avoid the extracellular space. Once in the new host cell, the double membrane is lysed and the cycle begins anew. Although there has been considerable conjecture about the mechanisms underlying the ingestion of bacteria containing membrane from an adjacent cell (Robbins *et al.* 1999), the role of bacterial products, the possible expression of unique membrane receptors, the excitation of specific signal-transduction pathways, and the potential role of autophagy have not been investigated.

Other intracellular pathogens also utilize host cell actin

The first pathogen recognized to induce actin filament assembly in the cytoplasm was *Shigella flexneri* (Bernardini *et al.* 1989). *Shigella*'s life cycle is similar to that of *Listeria*; however, this Gram-negative rod uses its type III secretion system to inject proteins that induce massive ruffling and internalization of the bacteria into giant phagosomes (termed the "trigger mechanism" for phagocytosis) (Cossart & Sansonetti 2004). *Shigella* uses an unrelated surface protein, IcsA, to attract and activate N-WASP (Suzuki *et al.* 1998); N-WASP in turn activates Arp2/3 complex nucleation (Gouin *et al.* 2005). *Rickettsia* species, *Burkhodleria pseudomallei*, and *Mycobacterium marinum* also move through the host cell cytoplasm by actin-based motility. The mechanisms by which they accomplish this task are less well understood; however, essential components are progressively being defined. Vaccinia virus and *Escherichia coli* (EPEC) induce actin assembly polymerization at the plasma membrane. Like *Listeria*, each of these organisms utilizes host cell actin assembly to cause disease (Gouin *et al.* 2005).

CONCLUSIONS

The dissection of the molecular mechanisms underlying *Listeria* entry into non-professional phagocytes, escape from the phagosome, and induction of directional actin assembly have provided remarkable insights into the signal transduction pathways of phagocytosis, hemolysin and phospholipase function, and the elegant orchestration of actin-regulatory proteins to induce directional actin assembly. The ability of *Listeria* to exploit receptors such as E-cadherin and hepatocyte growth factor receptors explain this pathogen's tropism and ability to penetrate anatomic barriers that other bacterial pathogens are unable to breach. The ability of *Listeria* to hijack host cell Arp2/3 complexes, VASP, and profilin, as well as capping, recycling, and bundling proteins to move within cells and spread from cell to cell without coming in contact with the extracellular space, explains why this pathogen is most dangerous in humans and other mammals with defective cell-mediated immunity. The lessons learned by studying *Listeria* can also be applied to normal host processes, and this simplified model of cell motility promises to continue to significantly advance the fields of cell biology and bacterial pathogenesis.

ACKNOWLEDGEMENTS

I thank Dr Pascale Cossart for her many helpful suggestions. This chapter was made possible by NIH funding: RO1AI-23262 and RO1AI34276.

REFERENCES

Auerbuch, V., J. J. Loureiro, F. B. Gertler, J. A. Theriot, and D. A. Portnoy. 2003. Ena/VASP proteins contribute to Listeria monocytogenes pathogenesis by controlling temporal and spatial persistence of bacterial actin-based motility. *Mol Microbiol.* **49**: 1361–75.

Bear, J. E., T. M. Svitkina, M. Krause *et al.* 2002. Antagonism between Ena/VASP proteins and actin filament capping regulates fibroblast motility. *Cell.* **109**: 509–21.

Bergmann, B., D. Raffelsbauer, M. Kuhn, M. Goetz, S. Hom, and W. Goebel. 2002. InlA- but not InlB-mediated internalization of Listeria monocytogenes by non-phagocytic mammalian cells needs the support of other internalins. *Mol Microbiol.* **43**: 557–70.

Bernardini, M. L., J. Mounier, H. d'Hauteville, M. Coquis-Rondon, and P. J. Sansonetti. 1989. Identification of icsA, a plasmid locus of Shigella flexneri that governs bacterial intra- and intercellular spread through interaction with F-actin. *Proc Natl Acad Sci USA.* **86**: 3867–71.

Bierne, H., S. Dramsi, M. P. Gratacap *et al.* 2000. The invasion protein InIB from Listeria monocytogenes activates PLC-gamma1 downstream from PI 3-kinase. *Cell Microbiol.* **2**: 465–76.

Bierne, H., E. Gouin, P. Roux, P. Caroni, H. L. Yin, and P. Cossart. 2001. A role for cofilin and LIM kinase in Listeria-induced phagocytosis. *J Cell Biol.* **155**: 101–12.

Braun, L., F. Nato, B. Payrastre, J. C. Mazie, and P. Cossart. 1999. The 213-amino-acid leucine-rich repeat region of the Listeria monocytogenes InlB protein is sufficient for entry into mammalian cells, stimulation of PI 3-kinase and membrane ruffling. *Mol Microbiol.* **34**: 10–23.

Braun, L., B. Ghebrehiwet, and P. Cossart. 2000. gC1q-R/p32, a C1q-binding protein, is a receptor for the InlB invasion protein of Listeria monocytogenes. *EMBO J.* **19**: 1458–66.

Bubb, M. R., E. G. Yarmola, B. G. Gibson, and F. S. Southwick. 2003. Depolymerization of actin filaments by profilin. Effects of profilin on capping protein function. *J Biol Chem.* **278**: 24629–35.

Burridge, K., and K. Wennerberg. 2004. Rho and Rac take center stage. *Cell.* **116**: 167–79.

Cabanes, D., P. Dehoux, O. Dussurget, L. Frangeul, and P. Cossart. 2002. Surface proteins and the pathogenic potential of Listeria monocytogenes. *Trends Microbiol.* **10**: 238–45.

Caldwell, J. E., J. A. Waddle, J. A. Cooper, J. A. Hollands, S. J. Casella, and J. F. Casella. 1989. cDNAs encoding the beta subunit of cap Z, the actin-capping protein of the Z line of muscle. *J Biol Chem.* **264**: 12648–52.

Cameron, L. A., M. J. Footer, A. van Oudenaarden, and J. A. Theriot. 1999. Motility of ActA protein-coated microspheres driven by actin polymerization. *Proc Natl Acad Sci USA.* **96**: 4908–13.

Camilli, A., H. Goldfine, and D. A. Portnoy. 1991. Listeria monocytogenes mutants lacking phosphatidylinositol-specific phospholipase C are avirulent. *J Exp Med.* **173**: 751–4.

Camilli, A., L. G. Tilney, and D. A. Portnoy. 1993. Dual roles of plcA in Listeria monocytogenes pathogenesis. *Mol Microbiol.* **8**: 143–57.

Chico-Calero, I., M. Suarez, B. Gonzalez-Zorn *et al.* 2002. Hpt, a bacterial homolog of the microsomal glucose- 6-phosphate translocase, mediates rapid intracellular proliferation in Listeria. *Proc Natl Acad Sci USA.* **99**: 431–6.

Cicchetti, G., P. Maurer, P. Wagener, and C. Kocks. 1999. Actin and phosphoinositide binding by the ActA protein of the bacterial pathogen Listeria monocytogenes. *J Biol Chem.* **274**: 33616–26.

Cooper, J. A., and T. D. Pollard. 1985. Effect of capping protein on the kinetics of actin polymerization. *Biochemistry.* **24**: 793–9.

Cossart, P., and P. J. Sansonetti. 2004. Bacterial invasion: the paradigms of enteroinvasive pathogens. *Science.* **304**: 242–8.

Cossart, P., M. F. Vicente, J. Mengaud, F. Baquero, J. C. Perez-Diaz, and P. Berche. 1989. Listeriolysin O is essential for virulence of Listeria monocytogenes: direct evidence obtained by gene complementation. *Infect Immun.* **57**: 3629–36.

Cossart, P., J. Pizarro-Cerda, and M. Lecuit. 2003. Invasion of mammalian cells by Listeria monocytogenes: functional mimicry to subvert cellular functions. *Trends Cell Biol.* **13**: 23–31.

Dabiri, G. A., J. M. Sanger, D. A. Portnoy, and F. S. Southwick. 1990. Listeria monocytogenes moves rapidly through the host-cell cytoplasm by inducing directional actin assembly. *Proc Natl Acad Sci USA*. **87**: 6068–72.

David, V., E. Gouin, M. V. Troys *et al.* 1998. Identification of cofilin, coronin, Rac and capZ in actin tails using a Listeria affinity approach. *J Cell Sci*. **111** (19): 2877–84.

de Hostos, E. L., B. Bradtke, F. Lottspeich, R. Guggenheim, and G. Gerisch. 1991. Coronin, an actin binding protein of Dictyostelium discoideum localized to cell surface projections, has sequence similarities to G protein beta subunits. *EMBO J*. **10**: 4097–104.

Decatur, A. L., and D. A. Portnoy. 2000. A PEST-like sequence in listeriolysin O essential for Listeria monocytogenes pathogenicity. *Science*. **290**: 992–5.

Dold, F. G., J. M. Sanger, and J. W. Sanger. 1994. Intact alpha-actinin molecules are needed for both the assembly of actin into the tails and the locomotion of Listeria monocytogenes inside infected cells. *Cell Motil Cytoskeleton*. **28**: 97–107.

Domann, E., J. Wehland, M. Rohde *et al.* 1992. A novel bacterial virulence gene in Listeria monocytogenes required for host cell microfilament interaction with homology to the proline-rich region of vinculin. *EMBO J*. **11**: 1981–90.

Dramsi, S., and P. Cossart. 2003. Listeriolysin O-mediated calcium influx potentiates entry of Listeria monocytogenes into the human Hep-2 epithelial cell line. *Infect Immun*. **71**: 3614–18.

Dussurget, O., J. Pizarro-Cerda, and P. Cossart. 2004. Molecular determinants of Listeria monocytogenes virulence. *A Rev Microbiol*. **58**: 587–610.

Gandhi, A. J., B. Perussia, and H. Goldfine. 1993. Listeria monocytogenes phosphatidylinositol (PI)-specific phospholipase C has low activity on glycosyl-PI-anchored proteins. *J Bacteriol*. **175**: 8014–17.

Geoffroy, C., J. L. Gaillard, J. E. Alouf, and P. Berche. 1987. Purification, characterization, and toxicity of the sulfhydryl-activated hemolysin listeriolysin O from Listeria monocytogenes. *Infect Immun*. **55**: 1641–6.

Glomski, I. J., M. M. Gedde, A. W. Tsang, J. A. Swanson, and D. A. Portnoy. 2002. The Listeria monocytogenes hemolysin has an acidic pH optimum to compartmentalize activity and prevent damage to infected host cells. *J Cell Biol*. **156**: 1029–38.

Glomski, I. J., A. L. Decatur, and D. A. Portnoy. 2003. Listeria monocytogenes mutants that fail to compartmentalize listerolysin O activity are cytotoxic, avirulent, and unable to evade host extracellular defenses. *Infect Immun*. **71**: 6754–65.

Goetz, M., A. Bubert, G. Wang *et al.* 2001. Microinjection and growth of bacteria in the cytosol of mammalian host cells. *Proc Natl Acad Sci USA.* **98**: 12221–6.

Goldfine, H., and S. J. Wadsworth. 2002. Macrophage intracellular signaling induced by Listeria monocytogenes. *Microbes Infect.* **4**: 1335–43.

Goldfine, H., N. C. Johnston, and C. Knob. 1993. Nonspecific phospholipase C of Listeria monocytogenes: activity on phospholipids in Triton X-100-mixed micelles and in biological membranes. *J Bacteriol.* **175**: 4298–306.

Goode, B. L., J. J. Wong, A. C. Butty *et al.* 1999. Coronin promotes the rapid assembly and cross-linking of actin filaments and may link the actin and microtubule cytoskeletons in yeast. *J Cell Biol.* **144**: 83–98.

Gouin, E., M. D. Welch, and P. Cossart. 2005. Actin-based motility of intracellular pathogens. *Curr Op Microbiol.* (In press.)

Grogan, A., E. Reeves, N. Keep *et al.* 1997. Cytosolic phox proteins interact with and regulate the assembly of coronin in neutrophils. *J Cell Sci.* **110** (24): 3071–81.

Ireton, K., B. Payrastre, H. Chap *et al.* 1996. A role for phosphoinositide 3-kinase in bacterial invasion. *Science.* **274**: 780–2.

Ireton, K., B. Payrastre, and P. Cossart. 1999. The Listeria monocytogenes protein InlB is an agonist of mammalian phosphoinositide 3-kinase. *J Biol Chem.* **274**: 17025–32.

Jacquet, C., M. Doumith, J. I. Gordon, P. M. Martin, P. Cossart, and M. Lecuit. 2004. A molecular marker for evaluating the pathogenic potential of foodborne Listeria monocytogenes. *J Infect Dis.* **189**: 2094–100.

Kang, F., R. O. Laine, M. R. Bubb, F. S. Southwick, and D. L. Purich. 1997. Profilin interacts with the Gly-Pro-Pro-Pro-Pro-Pro sequences of vasodilator-stimulated phosphoprotein (VASP): implications for actin-based Listeria motility. *Biochemistry.* **36**: 8384–92.

Kathariou, S., P. Metz, H. Hof, and W. Goebel. 1987. Tn916-induced mutations in the hemolysin determinant affecting virulence of Listeria monocytogenes. *J Bacteriol.* **169**: 1291–7.

Kocks, C., E. Gouin, M. Tabouret, P. Berche, H. Ohayon, and P. Cossart. 1992. L. monocytogenes-induced actin assembly requires the actA gene product, a surface protein. *Cell.* **68**: 521–31.

Kocks, C., R. Hellio, P. Gounon, H. Ohayon, and P. Cossart. 1993. Polarized distribution of Listeria monocytogenes surface protein ActA at the site of directional actin assembly. *J Cell Sci.* **105** (3): 699–710.

Kocks, C., J. B. Marchand, E. Gouin *et al.* 1995. The unrelated surface proteins ActA of Listeria monocytogenes and IcsA of Shigella flexneri are sufficient to confer actin-based motility on Listeria innocua and Escherichia coli respectively. *Mol Microbiol.* **18**: 413–23.

Kussel-Andermann, P., A. El-Amraoui, S. Safuddine *et al.* 2000. Vezatin, a novel transmembrane protein, bridges myosin VIIA to the cadherin-catenins complex. *EMBO J.* **19**: 6020–9.

Laine, R. O., K. L. Phaneuf, C. C. Cunningham, D. Kwiatkowski, T. Azuma, and F. S. Southwick. 1998. Gelsolin, a protein that caps the barbed ends and severs actin filaments, enhances the actin-based motility of Listeria monocytogenes in host cells. *Infect Immun.* **66**: 3775–82.

Larson, L., S. Arnaudeau, B. Gibson *et al.* (2005). Gelsolin mediates calcium-dependent disassembly of Listeria actin tails. Submitted.

Lasa, I., V. David, E. Gouin, J. B. Marchand, and P. Cossart. 1995. The amino-terminal part of ActA is critical for the actin-based motility of Listeria monocytogenes; the central proline-rich region acts as a stimulator. *Mol Microbiol.* **18**: 425–36.

Lasa, I., E. Gouin, M. Goethals *et al.* 1997. Identification of two regions in the N-terminal domain of ActA involved in the actin comet tail formation by Listeria monocytogenes. *EMBO J.* **16**: 1531–40.

Laurent, V., T. P. Loisel, B. Harbeck *et al.* 1999. Role of proteins of the Ena/VASP family in actin-based motility of Listeria monocytogenes. *J Cell Biol.* **144**: 1245–58.

Lecuit, M., S. Dramsi, C. Gottardi, M. Fedor-Chaiken, B. Gumbiner, and P. Cossart. 1999. A single amino acid in E-cadherin responsible for host specificity towards the human pathogen Listeria monocytogenes. *EMBO J.* **18**: 3956–63.

Lecuit, M., S. Vandormael-Pournin, J. Lefort *et al.* 2001. A transgenic model for listeriosis: role of internalin in crossing the intestinal barrier. *Science.* **292**: 1722–5.

Lecuit, M., D. M. Nelson, S. D. Smith *et al.* 2004. Targeting and crossing of the human maternofetal barrier by Listeria monocytogenes: role of internalin interaction with trophoblast E-cadherin. *Proc Natl Acad Sci USA.* **101**: 6152–7.

Loisel, T. P., R. Boujemaa, D. Pantaloni, and M. F. Carlier. 1999. Reconstitution of actin-based motility of Listeria and Shigella using pure proteins. *Nature.* **401**: 613–16.

Lorber, B. 1997. Listeriosis. *Clin Infect Dis.* **24**: 1–9; [quiz] 10–1.

Machner, M. P., C. Urbanke, M. Barzik *et al.* 2001. ActA from Listeria monocytogenes can interact with up to four Ena/VASP homology 1 domains simultaneously. *J Biol Chem.* **276**: 40096–103.

Marino, M., M. Banerjee, R. Jonquieres, P. Cossart, and P. Ghosh. 2002. GW domains of the Listeria monocytogenes invasion protein InlB are SH3-like and mediate binding to host ligands. *EMBO J.* **21**: 5623–34.

Marino, M., M. Banerjee, J. Copp *et al.* 2004. Characterization of the calcium-binding sites of Listeria monocytogenes InlB. *Biochem Biophys Res Commun.* **316**: 379–86.

Marx, J. 2003. Cell biology. How cells step out. *Science.* **302**: 214–16.

Mengaud, J., C. Braun-Breton, and P. Cossart. 1991. Identification of phosphatidylinositol-specific phospholipase C activity in Listeria monocytogenes: a novel type of virulence factor? *Mol Microbiol.* **5**: 367–72.

Mishima, M., and E. Nishida. 1999. Coronin localizes to leading edges and is involved in cell spreading and lamellipodium extension in vertebrate cells. *J Cell Sci.* **112** (17): 2833–42.

Mourrain, P., I. Lasa, A. Gautreau, E. Gouin, A. Pugsley, and P. Cossart. 1997. ActA is a dimer. *Proc Natl Acad Sci USA.* **94**: 10034–9.

Portnoy, D. A., P. S. Jacks, and D. J. Hinrichs. 1988. Role of hemolysin for the intracellular growth of Listeria monocytogenes. *J Exp Med.* **167**: 1459–71.

Purich, D. L., and F. S. Southwick. 1997. ABM-1 and ABM-2 homology sequences: consensus docking sites for actin-based motility defined by oligoproline regions in Listeria ActA surface protein and human VASP. *Biochem Biophys Res Commun.* **231**: 686–91.

Repp, H., Z. Pamukci, A. Koschinski *et al.* 2002. Listeriolysin of Listeria monocytogenes forms Ca^{2+}-permeable pores leading to intracellular Ca^{2+} oscillations. *Cell Microbiol.* **4**: 483–91.

Robbins, J. R., A. I. Barth, H. Marquis, E. L. de Hostos, W. J. Nelson, and J. A. Theriot. 1999. Listeria monocytogenes exploits normal host cell processes to spread from cell to cell. *J Cell Biol.* **146**: 1333–50.

Rosenblatt, J., B. J. Agnew, H. Abe, J. R. Bamburg, and T. J. Mitchison. 1997. Xenopus actin depolymerizing factor/cofilin (XAC) is responsible for the turnover of actin filaments in Listeria monocytogenes tails. *J Cell Biol.* **136**: 1323–32.

Samarin, S., S. Romero, C. Kocks, D. Didry, D. Pantaloni, and M. F. Carlier. 2003. How VASP enhances actin-based motility. *J Cell Biol.* **163**: 131–42.

Sanger, J. M., J. W. Sanger, and F. S. Southwick. 1992. Host cell actin assembly is necessary and likely to provide the propulsive force for intracellular movement of Listeria monocytogenes. *Infect Immun.* **60**: 3609–19.

Schubert, W. D., C. Urbanke, T. Ziehm *et al.* 2002. Structure of internalin, a major invasion protein of Listeria monocytogenes, in complex with its human receptor E-cadherin. *Cell.* **111**: 825–36.

Seveau, S., H. Bierne, S. Giroux, M. C. Prevost, and P. Cossart. 2004. Role of lipid rafts in E-cadherin- and HGF-R/Met-mediated entry of Listeria monocytogenes into host cells. *J Cell Biol.* **166**: 743–53.

Sidhu, G., W. Li, E. Bishai, N. Laryngakis, T. Balla, and F. S. Southwick. 2006. Phosphoinositide-3-kinase is required for intracellular Listeria monocytogenes actin-based motility and filopod formation. Submitted.

Skoble, J., D. A. Portnoy, and M. D. Welch. 2000. Three regions within ActA promote Arp2/3 complex-mediated actin nucleation and Listeria monocytogenes motility. *J Cell Biol.* **150**: 527–38.

Skoble, J., V. Auerbuch, E. D. Goley, M. D. Welch, and D. A. Portnoy. 2001. Pivotal role of VASP in Arp2/3 complex-mediated actin nucleation, actin branch-formation, and Listeria monocytogenes motility. *J Cell Biol.* **155**: 89–100.

Smith, G. A., H. Marquis, S. Jones, N. C. Johnston, D. A. Portnoy, and H. Goldfine. 1995. The two distinct phospholipases C of Listeria monocytogenes have overlapping roles in escape from a vacuole and cell-to-cell spread. *Infect Immun.* **63**: 4231–7.

Smith, G. A., J. A. Theriot, and D. A. Portnoy. 1996. The tandem repeat domain in the Listeria monocytogenes ActA protein controls the rate of actin-based motility, the percentage of moving bacteria, and the localization of vasodilator-stimulated phosphoprotein and profilin. *J Cell Biol.* **135**: 647–60.

Sousa, S., D. Cabanes, A. El-Amraoui, C. Petit, M. Lecuit, and P. Cossart. 2004. Unconventional myosin VIIa and vezatin, two proteins crucial for Listeria entry into epithelial cells. *J Cell Sci.* **117**: 2121–30.

Southwick, F. S., and D. L. Purich. 1994. Arrest of Listeria movement in host cells by a bacterial ActA analogue: implications for actin-based motility. *Proc Natl Acad Sci USA.* **91**: 5168–72.

Southwick, F. S., and D. L. Purich. 1996. Intracellular pathogenesis of listeriosis. *N Engl J Med.* **334**: 770–6.

Steffen, P., D. A. Schafer, V. David, E. Gouin, J. A. Cooper, and P. Cossart. 2000. Listeria monocytogenes ActA protein interacts with phosphatidylinositol 4,5-bisphosphate in vitro. *Cell Motil Cytoskeleton.* **45**: 58–66.

Suzuki, T., H. Miki, T. Takenawa, and C. Sasakawa. 1998. Neural Wiskott-Aldrich syndrome protein is implicated in the actin-based motility of Shigella flexneri. *EMBO J.* **17**: 2767–76.

Theriot, J. A., T. J. Mitchison, L. G. Tilney, and D. A. Portnoy. 1992. The rate of actin-based motility of intracellular Listeria monocytogenes equals the rate of actin polymerization. *Nature.* **357**: 257–60.

Tilney, L. G., and D. A. Portnoy. 1989. Actin filaments and the growth, movement, and spread of the intracellular bacterial parasite, Listeria monocytogenes. *J Cell Biol.* **109**: 1597–608.

Vazquez-Boland, J. A., C. Kocks, S. Dramsi *et al.* 1992. Nucleotide sequence of the lecithinase operon of Listeria monocytogenes and possible role of lecithinase in cell-to-cell spread. *Infect Immun.* **60**: 219–30.

Viel, A. 1999. Alpha-actinin and spectrin structures: an unfolding family story. *FEBS Lett.* **460**: 391–4.

Wadsworth, S. J., and H. Goldfine. 1999. Listeria monocytogenes phospholipase C-dependent calcium signaling modulates bacterial entry into J774 macrophage-like cells. *Infect Immun.* **67**: 1770–8.

Zuckert, W. R., H. Marquis, and H. Goldfine. 1998. Modulation of enzymatic activity and biological function of Listeria monocytogenes broad-range phospholipase C by amino acid substitutions and by replacement with the Bacillus cereus ortholog. *Infect Immun.* **66**: 4823–31.

CHAPTER 8

Mycobacterium tuberculosis: mechanisms of phagocytosis and intracellular survival

Joel D. Ernst and Andrea Wolf

INTRODUCTION

Mycobacterium tuberculosis, the cause of tuberculosis, has infected an estimated one-third of the world's human population and causes more deaths per year than any other single bacterial pathogen (Corbett *et al.* 2003). Although tuberculosis is most frequently an infection of the lungs, it can affect virtually any organ of the body (Raviglione & O'Brien 2004). In most individuals the infection remains latent without symptoms or transmission, but in approximately 10% the infection progresses to active disease and kills at least half of these. Untreated, active disease provides the opportunity for transmission of *M. tuberculosis* to other individuals through coughing up of the bacteria by an infected person, which provides droplet nuclei that are inhaled into the lung alveoli and establish a new infection. Tuberculosis is most common in developing countries; because T-lymphocyte-mediated cellular immunity is essential for control of the infection, the ongoing epidemic of HIV infection in regions with a high prevalence of tuberculosis is worsening an already severe problem. Moreover, the development of multiple drug resistance in *M. tuberculosis* has amplified the problems of treatment of tuberculosis in many parts of the world.

LIFE CYCLE OF *M. TUBERCULOSIS*

Although bacteria are not classically considered to have morphologically distinct stages representing phases of their life cycle as eucaryotic parasites

Phagocytosis and Bacterial Pathogenicity, ed. J. D. Ernst and O. Stendahl. Published by Cambridge University Press.

do, it is clear that pathogenic bacteria such as *M. tuberculosis* adapt to distinct environmental niches by major alterations in their patterns of gene expression (Schnappinger *et al.* 2003). *M. tuberculosis* is transmitted by the airborne route, in small (10 μm diameter) droplet nuclei that can be inhaled into the lung alveoli. In order to generate droplet nuclei of this size, larger droplets that have been expectorated by an individual with active tuberculosis must lose water by evaporation. During this process, the bacteria require mechanisms to survive progressive increases in osmolality and the ion concentrations of the droplets. Moreover, effective prolonged survival may require mechanisms for avoiding or repairing DNA damage that may occur from exposure to UV irradiation in the environment. After being inhaled into lung alveoli, *M. tuberculosis* are believed to be rapidly phagocytosed by alveolar macrophages, which are quiescent but nevertheless equipped to kill other inhaled bacteria effectively. Therefore, within the phagosome environment, *M. tuberculosis* must respond with mechanisms for acquiring nutrients, acquiring and retaining divalent cations including iron, and defending against other antimicrobial components of the macrophage. Studies of *M. tuberculosis* gene expression have indicated that 540 genes, representing approximately 11% of the genome, are differentially expressed within macrophage phagosomes compared with broth culture (Schnappinger *et al.* 2003); many of the differentially expressed genes have clear roles in adapting to known conditions within phagosomes.

After phagocytosis, *M. tuberculosis* multiplies within macrophages and the expanding population of bacteria spreads to newly recruited macrophages and dendritic cells (Peters *et al.* 2004; Peters & Ernst 2003; Peters *et al.* 2001). A subset of the bacteria also transit from the lung to the local draining lymph node (Chackerian *et al.* 2002), where they likely initiate an adaptive immune response by antigen presentation to naïve T lymphocytes. In addition, some of the bacterial population is believed to disseminate by the bloodstream to other peripheral organs, where the bacteria may remain latent for decades before reactivating to cause extrapulmonary tuberculosis. Nevertheless, the lungs are the site of replication of the major bulk of the *M. tuberculosis* population, and replication may occur intracellularly as well as extracellularly, if the response to infection results in tissue necrosis and formation of large cavities. After development of an adaptive cellular immune response (approximately 3 weeks after initial infection), which is dominated by $CD4^+$ T lymphocytes that exhibit a Th1 pattern of differentiation (Flynn & Chan 2001), growth of *M. tuberculosis* in the lung is restricted, and progression of disease is halted, at least temporarily. The arrest of bacterial replication within macrophages upon the activation and recruitment of *M. tuberculosis*-specific $CD4^+$ effector T lymphocytes is mediated by secretion of interferon gamma (IFNγ), as well as through an incompletely characterized mechanism that requires cell–cell contact between the T lymphocytes and infected macrophages (Cowley & Elkins 2003). In the majority of infected people, the adaptive immune response to *M. tuberculosis* is sufficient to prevent active disease throughout life. Because very few, if any, *M. tuberculosis*-infected humans clear the bacteria, the state of infection without symptoms or signs of active disease is referred to as "latent tuberculosis," although it is currently unclear whether the bacteria are actually latent and not replicating, or whether they are replicating and dying at equivalent rates without expressing genes that result in tissue damage. In a significant minority of people, latent infection may reactivate and cause progressive disease. Although it is known that immunodeficient states, such as those imposed by HIV infection or immunosuppressive drugs, may allow reactivation and progressive disease to develop, it is currently unknown whether the bacteria play an active or passive role during the transition from latency to reactivated disease. In those people that develop active disease, bacterial replication may be intracellular or extracellular, and it is those people with active disease in the lungs that cough and expel infectious droplet nuclei to start the infectious life cycle anew in a previously naïve host.

M. TUBERCULOSIS AS A FACULTATIVE INTRACELLULAR PATHOGEN OF PHAGOCYTES

Considerable evidence indicates that *M. tuberculosis* is a facultative intracellular pathogen, mainly of macrophages and other professional phagocytes. Histopathologic studies of tissues from individuals with tuberculosis reveal that the bacteria are closely associated with cells with the microscopic features and tissue distribution of macrophages (Rich 1944). A recent study confirmed that, in humans with pulmonary tuberculosis, the bacteria reside in cells with the ultrastructural, endocytic, and phagocytic properties of macrophages (Mwandumba *et al.* 2004). In that study, tubercle bacilli were located in phagosomes that did not accumulate endocytosed markers and that were non-acidified; this result confirms the observations that *M. tuberculosis* survives in phagosomes sequestered from late endosomes and lysosomes in cultured macrophages (Armstrong & Hart 1971; Sturgill-Koszycki *et al.* 1994; Clemens & Horwitz 1995; Russell *et al.* 1996). In some cases, however, local tissue necrosis (and/or proteolytic degradation of extracellular matrix) causes tuberculous cavities to form, and the bacteria can reach high extracellular densities (*c*.10^9 bacteria per milliliter). Although recent studies have also demonstrated that *M. tuberculosis* can reside in dendritic cells *in vivo* (Tailleux *et al.* 2003b), it is not yet clear what proportion of the bacteria reside in dendritic cells rather than in macrophages, nor is it known whether the bacteria inhabit distinct cell types at different phases of infection. *M. tuberculosis* has also been found to survive and replicate in human and murine macrophages *in vitro* (Armstrong & Hart 1971; Shepard 1957; Zimmerli *et al.* 1996a). Replication of *M. tuberculosis* in macrophages is consistently demonstrable, but the fate of *M. tuberculosis* in human and murine dendritic cells depends on experimental conditions: some studies show that *M. tuberculosis* replicates in cultured dendritic cells, whereas others reveal that *M. tuberculosis* survives, but does not replicate, in cultured dendritic cells (Henderson *et al.* 1997; Bodnar *et al.* 2001; Jiao *et al.* 2002; Tailleux *et al.* 2003a).

Additional evidence that *M. tuberculosis* is an intracellular pathogen of professional phagocytes *in vivo* is provided by the observation that the quantitatively most important mechanisms of protective immunity in tuberculosis are conferred by MHC class II-restricted $CD4^+$ T lymphocytes (Mogues *et al.* 2001). Since the MHC class II antigen-processing and presentation molecules reside in vesicular compartments, and since MHC class II is expressed predominantly by professional antigen-presenting cells (macrophages, dendritic cells, and B lymphocytes), the importance of this mechanism in control of progressive tuberculosis provides additional (if indirect) evidence that *M. tuberculosis* resides predominantly in vesicular compartments (i.e. phagosomes) in macrophages (and possibly dendritic cells).

Molecular genetic evidence also indicates that *M. tuberculosis* genes that are essential for survival in cultured macrophages are also essential *in vivo*, in animal model systems. For example, the gene encoding isocitrate lyase, the initial enzyme of the glyoxylate shunt, is induced when *M. tuberculosis* enters macrophages, is essential for bacterial replication in cultured macrophages, and is also essential for persistence of *M. tuberculosis* in immunocompetent mice (McKinney *et al.* 2000). Likewise, null mutants of phoP, a component of a major signal transduction pathway in *M. tuberculosis*, fail to replicate normally in cultured macrophages and are markedly attenuated in mice (Perez *et al.* 2001).

In summary, considerable evidence indicates that *M. tuberculosis* not only survives within professional phagocytes, but actually exploits them as a common site of residence, replication, and long-term persistence. These observations imply that *M. tuberculosis* employs phagocytosis as an efficient mechanism to enter the cells that normally function to kill and degrade invading bacteria.

PHAGOCYTIC RECEPTORS THAT MEDIATE PHAGOCYTOSIS OF *M. TUBERCULOSIS*

Consistent with the observation that *M. tuberculosis* is an intracellular pathogen, multiple receptors have been identified that recognize *M. tuberculosis* surface components and mediate phagocytosis of the bacteria. Although expression of some of these receptors is widespread, others are restricted to a single cell type. The broad range of receptors used by *M. tuberculosis* and other pathogenic mycobacteria to enter phagocytes indicates that the bacteria benefit from residence in phagocytic cells, at least for a portion of their infectious life cycle.

Macrophages

Complement receptor 3

Complement receptor 3 (CR3; also termed $\alpha_M\beta_2$ integrin, or CD11b/CD18) is a heterodimeric protein that binds the iC3b fragment of complement, serves as a cell adhesion molecule by recognizing fibrinogen and ICAM-1, and recognizes non-opsonized bacteria and yeast that possess beta-glucan-containing carbohydrates. CR3 is broadly expressed on macrophages, neutrophils, and myeloid dendritic cells. Although surface CR3 can bind particulate targets, it requires activation in order to mediate phagocytosis in macrophages (for additional information on CR3, see Chapters 2 and 3, this volume).

Non-opsonized *M. tuberculosis* can bind to murine macrophages by interactions that are inhibitable by mycobacterial lipoarabinomannan, lipomannan, and phosphatidylinositol mannan (Stokes & Speert 1995). Moreover, the same mycobacterial components inhibit binding of non-opsonized *M. tuberculosis* to CR3-transfected CHO cells (Cywes *et al.* 1997), indicating that CR3 contributes to interaction of non-opsonized *M. tuberculosis* with macrophages. This interaction is likely through a lectin-like domain located near the C-terminus of CR3 (Xia & Ross 1999). The mycobacterial protein known as "antigen 85C" can also interact directly with the I domain of CR3 (Hetland & Wiker 1994), although the quantitative contribution of this binding to interactions of whole bacteria *in vitro* or *in vivo* has not been determined. In certain subsets of macrophages, some of the binding of non-opsonized *M. tuberculosis* may be mediated by interactions with mannose receptors (see below).

Whereas non-opsonized *M. tuberculosis* binds readily to macrophages, opsonization with serum markedly enhances binding and phagocytosis of the bacteria, and the third component of complement (C3) provides most, if not all of the opsonic activity for *M. tuberculosis* in non-immune serum (Schlesinger *et al.* 1990). *M. tuberculosis* and other virulent mycobacteria have evolved multiple mechanisms for activating the complement system and acquiring surface-bound iC3b. Like other bacteria, *M. tuberculosis* activates the alternative pathway of complement activation in non-immune human serum (Ramanathan *et al.* 1980; Ferguson *et al.* 2004), and this is sufficient to markedly increase phagocytosis of *M. tuberculosis*. Several *M. tuberculosis* molecules have been found to be C3 acceptor molecules, including a heparin-binding hemagglutinin (HBHA) (Mueller-Ortiz *et al.* 2001), although deletion of the gene encoding HBHA did not decrease the amount of C3 bound by *M. tuberculosis* (Mueller-Ortiz *et al.* 2002). In individuals that have been infected with *M. tuberculosis* and that have circulating antibodies to *M. tuberculosis* surface antigens, and/or in the presence of low concentrations of complement proteins (such as in lung fluid), the classical pathway can also contribute to complement opsonization (Ferguson *et al.* 2004; Hetland *et al.* 1998). Finally, serum mannose-binding lectin binds to *M. tuberculosis* and to its abundant mannosylated glycolipids (Garred *et al.* 1994). Although it is likely that this can result in complement activation by the mannose-binding lectin/MASP pathway, direct evidence for this has not yet been reported.

In addition to the mechanisms of complement activation and opsonization shared by other bacteria, pathogenic mycobacteria possess an additional mechanism for complement activation that appears to be unique. *M. tuberculosis* (but not non-pathogenic mycobacteria) can bind the C2a fragment of the second component of complement, and the mycobacteria–C2a complex can serve as a C3 convertase in the absence of other components of the classical pathway, such as C4 (Schorey *et al.* 1997). This results in deposition of iC3b on the surface of the mycobacteria, and enhanced phagocytosis.

Thus, *M. tuberculosis* possesses four known distinct mechanisms for activating complement, to promote its opsonization with iC3b for recognition by CR3 (CD11b/CD18) and CR4 (CD11c/CD18), which predominates on alveolar macrophages (Hirsch *et al.* 1994). This suggests that *M. tuberculosis* may benefit from complement opsonization and complement-receptor-mediated entry into phagocytic cells.

Because CR3-mediated entry of *Leishmania major* promastigotes promotes intracellular survival of the parasites by allowing entry without activation of an oxidative burst (Mosser & Edelson 1985, 1987), a similar mechanism for *M. tuberculosis* has been examined. Complement-opsonized *M. tuberculosis* has been found to enter cultured human macrophages without inducing an increase in intracellular ionized calcium (Malik *et al.* 2000). Moreover, this avoidance of a calcium signal may contribute to the intracellular survival of virulent *M. tuberculosis* in macrophages: pharmacological elevation of intramacrophage calcium during phagocytosis of *M. tuberculosis* led to enhanced phagosome maturation and decreased survival of the bacteria (Malik *et al.* 2003). While these findings are of considerable interest, it is unclear whether these phenomena are restricted to CR3-mediated *M. tuberculosis* phagocytosis.

Although CR3 is clearly a major macrophage receptor for *M. tuberculosis*, there is currently no evidence that use of this receptor for entry is uniquely advantageous to the bacteria. In human monocyte-derived macrophages, monoclonal-antibody-mediated blockade of CR3-mediated entry had no measurable effect on the subsequent survival or intracellular growth of a virulent strain of *M. tuberculosis* (Zimmerli *et al.* 1996a). Similar results have been found by using murine macrophages. Whereas uptake of opsonized or non-opsonized *M. tuberculosis* was higher in wild-type macrophages than in CR3 (CD11b)-deficient macrophages, the rate of intracellular growth of the bacteria was indistinguishable in $CR3^{-/-}$ and wild-type macrophages (Melo *et al.* 2000). Finally, a thorough study of infection *in vivo* of $CD11b^{-/-}$ mice on either a C57BL/6 or a BALB/c background revealed no difference in the course of infection in the presence or absence of CR3. There were no differences in survival between the infected $CD11b^{-/-}$ and wild-type mice, and there were no differences in the rate of bacterial growth or in the ultimate bacterial burden in lungs, liver, or spleen in the presence or absence of CR3 (Hu *et al.* 2000). An additional study, using an independently constructed line of $CD11b^{-/-}$ mice, reached a similar conclusion after infection *in vivo* with *M. tuberculosis* (Melo *et al.* 2000). Taken together, these results indicate that CR3-mediated entry of *M. tuberculosis* is unlikely to confer a measurable advantage on the survival of *M. tuberculosis* compared with entry through other receptors.

Taken together, the data currently available indicate that *M. tuberculosis* can utilize CR3 to enter macrophages whether or not the bacteria are opsonized. Moreover, *M. tuberculosis* utilizes four distinct mechanisms for complement opsonization that may contribute to macrophage phagocytosis in different body compartments or at different stages of infection. However, despite the multiple mechanisms for interaction of *M. tuberculosis* and CR3, no clear advantage to the bacteria has been identified so far. For additional information on CR3 and *M. tuberculosis*, see Velasco-Velazquez *et al.* (2003).

Mannose receptor

The macrophage mannose receptor serves as an endocytic receptor as well as a receptor that can mediate phagocytosis of particles, including bacteria, with appropriate carbohydrate structures on their surface (Ezekowitz *et al.* 1990). *In vitro* studies with human monocyte-derived macrophages have demonstrated that *M. tuberculosis* binds to one or more receptors that can be inhibited by yeast mannan or mannosylated bovine serum albumin (Schlesinger 1993; Astarie-Dequeker *et al.* 1999). Since these substances have also been characterized as ligands for the macrophage mannose receptor (Taylor *et al.* 1990), it has been concluded that *M. tuberculosis* interacts with the previously characterized macrophage mannose receptor. Unlike CR3, in which transfection of CHO cells with CD18 and CD11b conferred the ability to bind and internalize *M. tuberculosis* (Cywes *et al.* 1996), studies have not yet been reported that definitively establish that the molecularly characterized mannose receptor mediates the reported interactions with *M. tuberculosis* and macrophages. One prominent mycobacterial molecule reported to interact with the mannan-inhibitable receptor is lipoarabinomannan (LAM), a lipoglycan that possesses terminal mannosyl residues, especially in pathogenic mycobacteria (Schlesinger *et al.* 1994, 1996). Since the mannose receptor exhibits no intrinsic signal transduction capability and there is no evidence for its interaction with accessory signalling molecules, it has been suggested that the mannose receptor may provide a mechanism for macrophage entry of *M. tuberculosis* without triggering cell activation (Beharka *et al.* 2002). Although studies in mannose receptor-knockout mice have not been reported, in studies using human monocyte-derived macrophages, mannose receptor blockade did not influence the intracellular survival or rate of intracellular replication of a virulent strain of *M. tuberculosis* (Zimmerli *et al.* 1996a).

Fc receptors

Receptors for the Fc domain of immunoglobulin G exist in several molecular forms, and mediate phagocytosis of IgG-opsonized particles by macrophages (see Chapters 2 and 3, this volume). Although in-depth analyses of the role(s) of Fc receptors in the biology of tuberculosis have not been undertaken, most studies indicate that anti-mycobacterial IgG plays little, if any, role in immunity to tuberculosis. This may be at least partly explained by the findings of a classic study by Armstrong & Hart (1975) that demonstrated that opsonization of *M. tuberculosis* (H37Rv) with serum from rabbits immunized with sonicated *M. bovis* BCG did not increase the ability of murine macrophages to kill ingested *M. tuberculosis*, even though immune opsonization promoted fusion of mycobacterial phagosomes with lysosomes. Therefore, despite the paucity of detailed information on Fc receptor-mediated phagocytosis of *M. tuberculosis*, it appears that the intracellular trafficking, but not the ultimate intracellular fate, of the bacteria is distinct from that mediated by other receptors.

Other macrophage receptors

In addition to the aforementioned macrophage receptors, additional receptors have been implicated in mediating phagocytosis of *M. tuberculosis*. CD44, which is commonly considered to be an adhesion protein, especially for activated T lymphocytes, is also expressed on macrophages and other hematopoietic cells. CD44$^{-/-}$ mice are more susceptible to progressive tuberculosis disease; this increased susceptibility is likely to be largely due to diminished recruitment of antigen-presenting cells and T lymphocytes to the site of infection (Leemans *et al.* 2003). In addition, resident peritoneal macrophages from CD44$^{-/-}$ mice were found to bind and ingest fewer *M. tuberculosis* than did cells from wild-type controls, and soluble recombinant CD44 bound directly to *M. tuberculosis*, as demonstrated by flow cytometry (Leemans *et al.* 2003). These findings imply that CD44 may contribute to macrophage phagocytosis of *M. tuberculosis*, but do not indicate whether the role of CD44 is to enhance binding of bacteria to macrophages, or whether CD44 itself can mediate internalization of the bacteria. Likewise, type A scavenger receptors (SR-A) have been implicated in binding and possibly internalization of virulent *M. tuberculosis*. Competitive inhibitors of SR-A diminish binding and phagocytosis of *M. tuberculosis* by human monocyte-derived macrophages (Zimmerli *et al.* 1996a), and sulfolipids from *M. tuberculosis* competitively inhibit binding of other ligands to soluble SR-A (Ernst 1998).

Non-receptor host modulators of *M. tuberculosis* phagocytosis

Pulmonary surfactant proteins A and D (Sp-A and Sp-D) are members of the collectin family of proteins. They possess C-type lectin domains and collagen-like domains, which assemble into triple helical structures. Both of these collectins are present in lung lining fluid, but they have divergent effects on macrophage phagocytosis of *M. tuberculosis*. Sp-A increases phagocytosis of *M. tuberculosis* by human monocyte-derived macrophages by an effect on macrophages, rather than by a direct interaction with the bacteria (Gaynor *et al.* 1995), although evidence for binding of Sp-A to *M. tuberculosis* has been reported (Downing *et al.* 1995). The current weight of evidence for the mechanism of Sp-A enhancement of phagocytosis of *M. tuberculosis* indicates that

Sp-A upregulates surface expression of macrophage mannose receptors, perhaps by affecting post-translational trafficking of mannose receptors (Beharka *et al.* 2002). In contrast, Sp-D decreases macrophage phagocytosis of *M. tuberculosis*, apparently through a direct interaction with the bacteria, rather than through a direct effect on macrophages (Ferguson *et al.* 1999). It is currently unclear how the apparently opposite effects of Sp-A and Sp-D balance or compete with each other, since both are present in the lung lining fluid, where inhaled *M. tuberculosis* are likely to encounter both of them.

Dendritic cells

Dendritic cells are important modulators of T lymphocyte activation and differentiation, therefore they are likely to play an important role in initiation and maintenance of the adaptive immune response to *M. tuberculosis*. Whereas "myeloid" dendritic cells (the subtype that predominates in non-lymphoid tissues such as the lung) resemble macrophages, they possess the unique ability to migrate from peripheral tissues to secondary lymphoid tissues, and they secrete distinct cytokines upon encountering *M. tuberculosis* (Hickman *et al.* 2002). While myeloid dendritic cells express CR3, CR4 (CD18/CD11c), and mannose receptors, a dendritic cell-specific receptor, DC-SIGN (dendritic cell-specific intercellular adhesion molecule 3-grabbing non-integrin), provides an additional dendritic cell receptor for *M. tuberculosis*. Studies with blocking antibodies reveal that antibodies to DC-SIGN inhibit human monocyte-derived dendritic cell binding and phagocytosis of *M. tuberculosis* by approximately 90% whether or not the bacteria are opsonized, but only minor inhibition was observed with antibodies to CR3 or mannose receptor (Tailleux *et al.* 2003b; Geijtenbeek *et al.* 2003). DC-SIGN interacts directly with mycobacteria, and does so by recognizing (at least) lipoarabinomannan. Utilization of DC-SIGN may also bias the cytokine response of cultured dendritic cells toward production of interleukin-10 (Geijtenbeek *et al.* 2003) and away from production of IL-12, suggesting that interaction of *M. tuberculosis* with DC-SIGN may result in generation of a suboptimal cellular immune response.

Neutrophils

The role of neutrophils in innate immunity to tuberculosis, or in its pathogenesis, has not been fully defined. Nevertheless, neutrophils are recruited to sites of infection with *M. tuberculosis* (Seiler *et al.* 2003) and are capable of phagocytosing live *M. tuberculosis* (Majeed *et al.* 1998). The major, if not the only, receptor for pathogenic mycobacteria on human neutrophils appears to be CR3 (Peyron *et al.* 2000).

Whereas *M. tuberculosis* has been reported to inhibit apoptosis of macrophages (Balcewicz-Sablinska *et al.* 1998; Keane *et al.* 2000, 2002; Riendeau & Kornfeld 2003), *M. tuberculosis* promotes apoptosis of human neutrophils by a mechanism that is dependent on reactive oxygen species and accompanied by activation of caspase 3, induction of proapoptotic Bax, and downregulation of antiapoptotic Bcl-x_L (Perskvist *et al.* 2002). Moreover, although it is widely accepted that ingestion of apoptotic cells by macrophages yields a non-inflammatory response of macrophages, ingestion of *M. tuberculosis*-infected apoptotic neutrophils is a prominent exception: neutrophils made apoptotic by exposure to UV light induce little TNFα and considerable TGFβ, but *M. tuberculosis*-apoptotic neutrophils induce macrophages to produce large amounts of TNFα and little TGFβ (Perskvist *et al.* 2002). Therefore, neutrophils may contribute to early innate immune responses to *M. tuberculosis* by amplifying responses of macrophages.

MYCOBACTERIAL LIGANDS FOR PHAGOCYTIC RECEPTORS

Given the observations that multiple phagocytic receptors recognize *M. tuberculosis* and related mycobacteria, it is not surprising that multiple bacterial ligands have been identified that mediate interactions with phagocytes.

Lipoarabinomannan (LAM) is one widely investigated component of the *M. tuberculosis* cell wall implicated in interacting with mannose receptors (Kang & Schlesinger 1998) and DC-SIGN (Tailleux *et al.* 2003b; Geijtenbeek *et al.* 2003). LAM is a peripherally exposed lipoglycan of mycobacteria that possesses terminal mannosyl residues on a backbone of core mannose and arabinose residues; the terminal mannose residues are crucial for mediating interactions of LAM with phagocytes (Schlesinger *et al.* 1994). While LAM has been shown to be sufficient for interactions with phagocytes, as demonstrated by enhanced uptake of latex beads upon coating with LAM (Schlesinger *et al.* 1994), it is less clear that LAM is either predominant or essential. Resolution of this question awaits genetic definition of the biosynthetic pathway of LAM, and preparation of strains of *M. tuberculosis* that express modified LAM molecules, or that are LAM-deficient.

Indirect observations suggest that additional mycobacterial molecules exist that mediate non-opsonic interactions with phagocytes. For example, non-opsonic binding of *M. tuberculosis* to the lectin-like domain of CR3 is competitively inhibited by laminarin (seaweed beta-glucan) and *N*-acetylglucosamine, indicating that LAM is not responsible for interacting with this domain of CR3.

As noted previously, the heparin-binding hemagglutinin (HBHA), which promotes interaction of *M. tuberculosis* with non-hematopoietic cells (Menozzi *et al.* 1998), is also an acceptor for covalent attachment of iC3b after opsonization through activation of the alternative pathway of complement activation (Mueller-Ortiz *et al.* 2001). Since deletion of the gene encoding HBHA did not diminish the binding of iC3b, other acceptor molecules clearly exist on the surface of *M. tuberculosis* (Mueller-Ortiz *et al.* 2002).

In summary, multiple phagocytic receptors and multiple bacterial ligands participate in the interaction of *M. tuberculosis* with phagocytes and lead to internalization of the bacteria. Given that other bacteria that have adapted to exist as extracellular pathogens have evolved multiple mechanisms for evading phagocytosis (Ernst 2000), the evidence that *M. tuberculosis* uses multiple ligand–receptor interactions to promote phagocytosis implies that *M. tuberculosis* is well adapted to intraphagocyte life, and may in fact require it as an obligate step in its pathogenic life cycle. Despite the observations described in the preceding sections, a recent publication has reported that sonication of mycobacteria, which is commmonly performed to disperse clumps of mycobacteria for use in assays of phagocytosis, may remove capsular materials that inhibit phagocytosis by certain subtypes of macrophages (Stokes *et al.* 2004). Further investigation is clearly necessary to determine whether this has biased the results of experiments *in vitro* with mycobacteria and phagocytes.

REQUIREMENT FOR CELLULAR CHOLESTEROL IN PHAGOCYTOSIS OF MYCOBACTERIA

Efficient phagocytosis of mycobacteria by neutrophils and macrophages is perturbed by depleting cholesterol from phagocyte plasma membranes. Treatment of neutrophils or macrophages with the cholesterol-sequestering reagents nystatin, filipin, or beta-cyclodextrin, with or without inhibition of cholesterol synthesis, significantly decreased CR3-mediated phagocytosis of *Mycobacterium kansasii* or *M. bovis* BCG (Peyron *et al.* 2000; Gatfield & Pieters 2000). Moreover, the membrane domains in contact with mycobacteria are enriched in cholesterol, as visualized with filipin and fluorescent microscopy. Although one report demonstrated that mycobacteria have a high capacity for binding cholesterol, and suggested that mycobacteria must interact with cholesterol in order to initiate phagocytosis (Gatfield & Pieters 2000), this model does not account for the likelihood that the lipid-rich cell wall of mycobacteria can bind abundant quantities of other lipids and non-polar molecules. Instead, it is much more likely that the role of cholesterol in phagocytosis of mycobacteria is in composing cholesterol-rich lipid rafts, which serve to promote association of raft-associated signal transduction molecules with CR3, as indicated by the finding that removal of glycosylphosphatidylinositol-linked proteins with phosphatidylinositol-specific phospholipase C, or treatment with *N*-acetyl-D-glucosamine, which disrupts interaction of CR3 with certain raft-associated proteins, decreased CR3-mediated phagocytosis of *M. kansasii* in a manner that resembled the effect of cholesterol depletion (Peyron *et al.* 2000). The latter model resembles results of studies of the role of cholesterol in CD48-mediated entry of fimH-expressing *E. coli* (Shin *et al.* 2000).

POSTPHAGOCYTIC FATE OF MYCOBACTERIA

Phagosome escape

Although the importance of MHC class II and $CD4^+$ T lymphocytes in protective immunity to tuberculosis supports a vesicular (i.e. phagosomal) location for the majority of an infecting population of pathogenic mycobacteria, several reports indicate that pathogenic mycobacteria can escape phagosomes and exist in the cytoplasm of host cells. An early report using electron microscopy to visualize *M. tuberculosis* within human and murine macrophages reported finding a subset of virulent mycobacteria in the cytoplasm, without a visible phagosome membrane (McDonough *et al.* 1993). Since this technique is subject to details of technique, and since other investigators did not observe extraphagosomal mycobacteria under similar circumstances, it has been widely believed that escape of mycobacteria from phagosomes occurs rarely, if at all. However, a recent report clearly demonstrated that a subset of *Mycobacterium marinum* can not only escape from phagosomes but can also use host cytoplasmic actin polymerization for intracellular motility (Stamm *et al.* 2003). Further investigation of these findings is warranted, and it should be worthwhile to determine whether the observations with *M. marinum* extend to other pathogenic mycobacteria, including *M. tuberculosis*.

Avoidance of phagosome maturation

Modulation of host cell pathways

Since the seminal observation by Armstrong and Hart (1971) that *M. tuberculosis* survives in phagosomes that do not acquire markers of lysosomes, a major focus of research in the cell biology of tuberculosis has been to understand the mechanism of this phenomenon and how it relates to the overall biology of tuberculosis. Subsequent to the initial report of defective interactions of *M. tuberculosis* phagosomes, mycobacterial phagosomes were also found to fail to become acidified (Crowle *et al.* 1991). This latter observation was found to be due to defective recruitment of vacuolar proton ATPase molecules to the mycobacterial phagosome membrane, a phenomenon restricted to phagosomes containing live, but not killed, mycobacteria, and that could be corrected by cytokine activation of the macrophages (Sturgill-Koszycki *et al.* 1994; Xu *et al.* 1994; Schaible *et al.* 1998). Together, these observations established that pathogenic mycobacteria survive in phagosomes that do not acquire the characteristics of a terminal phagolysosome.

Although the initial interpretation of the failure of mycobacterial phagosomes to mature to phagolysosomes was that mycobacterial phagosomes were resistant to fusion with other intracellular vesicles, further investigation revealed that, whereas they excluded membrane and luminal markers of late endosomes and lysosomes, mycobacterial phagosomes were able to fuse with early endosomes and markers derived from the trans-Golgi network (Clemens & Horwitz 1995; Russell *et al.* 1996; Sturgill-Koszycki *et al.* 1996). This indicated that mycobacterial phagosomes exhibit a selective trafficking defect and not a generalized resistance to membrane fusion.

Efforts to understand the mechanism underlying the selective ability of mycobacterial phagosomes to interact with early endosomes, but not late endosomes or lysosomes, revealed that phagosomes containing live mycobacteria are arrested at a step of maturation characterized by retention of the small GTPase rab5 and failure of acquision of rab7 (Via *et al.* 1997; Clemens *et al.* 2000) (Figure 8.1). Further investigation has revealed that, although several rab5 effectors are present, one rab5 effector, Early Endosomal Antigen-1 (EEA1) is deficient on phagosomes containing live mycobacteria (Fratti *et al.* 2001). EEA1 is transiently bound to phagosomes containing latex beads, through interactions with rab5 and with the membrane lipid phosphatidylinositol-3-phosphate (PI3P) (Patki *et al.* 1997; Simonsen *et al.* 1998). EEA1 itself acts as a tethering molecule for syntaxin 6, a protein that mediates membrane fusion and delivery of lysosomal membrane proteins and contents to model phagosomes (Simonsen *et al.* 1999). Therefore, a deficiency of EEA1 on mycobacterial phagosome membranes could be envisioned to account for the failure of acquisition of markers and contents of lysosomes.

Examination of proximal signaling events that might account for the observation that EEA1 is deficient on mycobacterial phagosomes has revealed evidence that live *M. tuberculosis* enters macrophages without inducing the transient increases in intracellular ionized calcium that characteristically

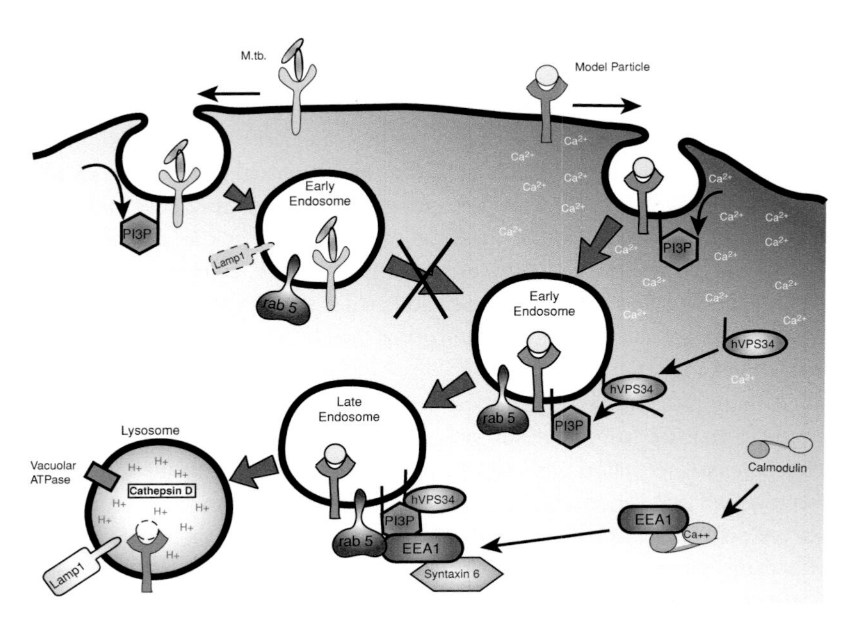

Figure 8.1 Maturation of phagosomes containing live *Mycobacterium tuberculosis* (M. tb.) compared with that of phagosomes containing a model particle. Phagosomes containing live M.tb. retain markers of early endosomes such as Rab5 and transferrin receptor, whereas phagosomes containing model particles acquire hVPS34, EEA1, and syntaxin 6, which promote phagosome acquisition of the vacuolar ATPase, LAMP-1, and mature cathepsin D. An M.tb.-associated defect in calcium signaling may be the most proximal determinant of the fate of distinct phagosomes. See the text for details.

accompany phagocytosis of other particles (or heat-killed *M. tuberculosis*) (Malik *et al.* 2000). This phenomenon is not restricted to macrophage phagocytosis, as phagocytosis of *M. tuberculosis* by neutrophils also occurs without increases in intracellular ionized calcium (Majeed *et al.* 1998). Further investigation has revealed that this proximal defect in generating a calcium signal results in failure of calcium- and calmodulin-dependent recruitment of EEA1 to mycobacterial phagosome membranes, at least in part due to defective recruitment of hVPS34, a phosphatidylinositol 3-kinase (Vergne *et al.* 2003). Moreover, live mycobacteria have been reported to perturb generation of PI-3-phosphate on phagosome membranes. Use of macrophages expressing fluorescent fusion proteins that bind specifically to PI3P (2xFYVE-EGFP and p40PX-EGFP), together with real-time confocal microscopy, revealed that phagocytosis of latex beads results in two phases of PI3P generation and recruitment of PI3P-binding proteins to phagosomes (Chua & Deretic 2004).

The first phase of PI3P generation begins 5–10 min after initial particle–cell contact, lasts approximately 5 min, and is followed by transient waves of PI3P, at intervals of approximately 20 min. In contrast, after phagocytosis of mycobacteria, the first wave appears abnormally early (simultaneous with contact between the bacteria and the macrophage) and is not followed by any secondary waves of PI3P generation. Because the secondary waves of PI3P depend on calcium and calmodulin, these observations further support the hypothesis that defective calcium signaling is the most proximal signaling abnormality that accompanies phagocytosis of mycobacteria, and indicate that defective PI3P generation and recruitment of EEA1 and syntaxin 6 are secondary to the calcium defect. Further investigation is required to define the mechanism(s) underlying the defective calcium signaling that accompanies phagocytosis of mycobacteria. Moreover, because macrophage phagosome maturation is not calcium-dependent in all contexts (Zimmerli *et al.* 1996b), whether the mechanism described above is the sole mechanism for *M. tuberculosis* inhibition of phagosome maturation remains to be established.

Bacterial molecules that modulate phagosome maturation

Several mycobacterial components have been reported to play key roles in mediating the defects in phagosome maturation. Lipoarabinomannan (LAM)-coated latex beads reproduce the defective calcium signaling and defective recruitment of EEA1 and hVPS34 characteristic of mycobacteria (Vergne *et al.* 2003; Fratti *et al.* 2003), but it is currently unclear whether LAM mediates these effects by interacting with host cell receptors or with other host cell proteins, or by perturbing the lipid structure of nascent and maturing phagosome membranes. It is also currently unclear how to reconcile these observations with LAM-coated beads with the observations that dead mycobacteria (which contain LAM, since it is constitutively produced) enter phagosomes that mature normally.

Protein kinase G (PknG) is one of 11 predicted serine/threonine kinases in the *M. tuberculosis* genome, and appears to be essential for survival of *M. tuberculosis* in a mouse model, as indicated by its under-representation in mutants isolated from the lungs of mice infected with a transposon mutant library (Sassetti & Rubin 2003). Characterization of a deletion mutant of *M. bovis* BCG lacking protein kinase G indicated that the mutant exhibited a growth defect in cultured macrophages, and was found in $LAMP^+$/β-$hexosaminidase^+$ vesicles with a greater frequency than was the wild-type strain of BCG (Wallburger *et al.* 2004). PknG is apparently absent from the avirulent mycobacterium *M. smegmatis*, which enters mature phagolysosomes and does not survive in macrophages. Expression of PknG in *M. smegmatis* allowed the bacteria to persist in $LAMP^-$/β-hexosaminidase$^-$ vesicles, and allowed them to survive longer in cells of the J774 murine macrophage cell line. PknG was detected in phagosomes and in the cytosol of macrophages infected with live, but not heat-killed, BCG, indicating that it has access to host proteins involved in signal transduction and phagosome maturation. Finally, a chemical inhibitor of PknG mimicked the effects of deletion of the gene, as reflected by poorer survival of BCG in macrophages, and by colocalization of the bacteria with lysosomal markers. Taken together, these results indicate that PknG may contribute to mycobacterial disruption of phagosome maturation. Further investigation is likely to reveal whether PknG functions with other mycobacterial components implicated in arresting phagosome maturation, and whether PknG intersects with the calcium- and PI3P-dependent macrophage pathways necessary for phagosome maturation.

In addition to the above approaches, in which candidate molecules were examined for a role in arresting phagosome maturation, the results of a genetic approach to identification of mycobacterial constituents that contribute to abnormal phagosome maturation have also recently been reported. By using a genetic screen that used iron dextran loading of macrophage lysosomes and magnetic selection of transposon mutants that resided in phagosomes that had acquired iron dextran, mutants of at least 15 genes were identified (Pethe *et al.* 2004). Several of these mutants were isolated multiple times and in independent screens, and the screen was validated by finding that when the mutants were examined individually 11 of them blocked acidification of their phagosomes, 4 of 5 examined occupied phagosomes that did not acquire normal quantities of preloaded colloidal gold, and 5 of the 8 mutants examined were attenuated for survival in bone-marrow-derived macrophages. The disrupted genes encoded products predicted to have diverse functions, including lipid metabolism and transport, isoprenol synthesis, and cyclization of terpenoids, and notably did not include PknG. Although the mutants do not clearly assort into an obvious single pathway, it is clear that further analysis of the disrupted genes and characterization of the mutants will provide valuable insight into the mechanisms used by *M. tuberculosis* to arrest maturation of its phagosome and survive within macrophages.

PATHOGENIC ADVANTAGES TO PROMOTING PHAGOCYTOSIS

In contrast to bacteria such as *Streptococcus pneumoniae*, *Helicobacter pylori*, and *Yersinia* species, which use diverse mechanisms to avoid being phagocytosed, *M. tuberculosis* has clearly adapted to cooperate with phagocytosis and thereby enter macrophages and other phagocytic cells. *M. tuberculosis* must have evolved these mechanisms in response to selective pressure, and must have done so after or simultaneously with evolving effective mechanisms for surviving the intramacrophage environment. What advantages might living inside of macrophage phagosomes confer on pathogenic mycobacteria?

First, by entering long-lived phagocytic cells, pathogenic mycobacteria may avoid removal by physical means, such as ciliary action in the airways or peristalsis in the gut, without expressing adhesins of their own. By entering macrophages, mycobacteria can establish a stable cellular niche; *M. tuberculosis* actually upregulates expression of macrophage adhesion proteins, especially ICAM-1 (Lopez Ramirez *et al.* 1994). Second, since *M. tuberculosis* has evolved effective mechanisms for inhibiting maturation of phagosomes, the mycobacterial phagosome environment may not be a hostile one, as it is not acidified and does not contain processed lysosomal hydrolases. Therefore, the biggest challenge for mycobacteria residing in immature phagosomes is to acquire nutrients and fuel. Because mycobacterial phagosomes exhibit properties of recycling endosomes, they may be ideally suited to acquire nutrients and fuel, since these organelles are used for sorting and trafficking of macromolecules that enter cells by receptor-mediated endocytosis, such as lipoproteins and transferrin-bound iron. Third, mycobacteria that are phagocytosed by motile phagocytic cells such as dendritic cells or neutrophils acquire the ability to migrate to distant tissues or to different portions of the lung without being motile themselves. Because mycobacteria do not contain flagella or other means of self-locomotion, their best approach to migrating is by transport within a phagocytic cell. While this may confer an advantage on mycobacteria, it is not obvious what *M. tuberculosis* gains by migrating from the lungs, since the lungs and aerosol transmission are the most effective ways for *M. tuberculosis* to be transmitted to a new host.

Another consideration in attempting to understand the advantage to mycobacteria of being phagocytosed is that the fate and therefore the advantage may vary, depending on the specific type of phagocytic cell. For example, phagocytosis by a tissue macrophage may offer a stable environment for one mycobacterium whereas phagocytosis by a motile dendritic cell may offer a very different environment and distinct advantages. For example, emerging evidence suggests that *M. tuberculosis* does not replicate as efficiently in dendritic cells as it does in macrophages (Tailleux *et al.* 2003a). Therefore, it is possible that *M. tuberculosis* has not yet perfected its ability to cause disease and be transmitted to a new host. In evolving mechanisms for efficient entry into macrophages, it has not been able to avoid being phagocytosed by dendritic cells. Although mycobacteria that enter dendritic cells may not thrive and may initiate an adaptive immune response that effectively contains the infection, the mechanisms used by *M. tuberculosis* to gain entry into cells that support replication and dissemination of the bacteria seem to be functioning well, since tuberculosis is a highly prevalent disease.

TRANSLATION OF KNOWLEDGE OF PHAGOCYTOSIS OF *M. TUBERCULOSIS*

How can we use what we know about phagocytosis of mycobacteria to improve control of the worldwide problem of tuberculosis? For one, as understanding of the mechanisms used by *M. tuberculosis* to inhibit phagosome maturation increases, it is likely that novel drug targets that disrupt these mechanisms will emerge. Since mutant mycobacteria that do not arrest phagosome maturation are attenuated for growth in macrophages, targeting this process is logical, although it is unclear whether disrupting any of these mechanisms after the mycobacteria are within immature phagosomes will force phagosome maturation and death of the bacteria.

Is it worth considering measures such as a vaccine that would inhibit phagocytosis of *M. tuberculosis* and thereby force them to remain extracellular? In evaluating the potential of this approach, it is worth while to note that, in a mouse model, depletion of alveolar macrophages results in lower bacterial burdens in the lungs and other organs, and longer survival of the animals (Leemans *et al.* 2001, 2005). These results imply that residence in macrophages is clearly beneficial to *M. tuberculosis*, but additional work is necessary to determine whether blocking phagocytosis of *M. tuberculosis* has a similar beneficial effect on the host. If LAM is indeed a crucial and essential molecule that promotes phagocytosis of *M. tuberculosis* through multiple receptors, then a vaccine approach to blocking LAM-mediated phagocytosis may well offer potential against tuberculosis.

SUMMARY AND CONCLUSIONS

M. tuberculosis, a highly successful human pathogen, uses multiple mechanisms that result in its efficient phagocytosis by macrophages, dendritic cells, and neutrophils. The bacteria use four distinct pathways for complement opsonization, and utilize multiple distinct phagocytic receptors to enter phagocytic cells. Whereas other bacteria avoid phagocytosis because they are sensitive to killing by the microbicidal mechanisms of professional phagocytes, *M. tuberculosis* has evolved the ability to survive within phagocytes by causing arrest of phagosome maturation. This results in residence of

M. *tuberculosis* and other pathogenic bacteria in an intracellular compartment that is much less hostile than that in mature phagosomes. The available knowledge currently implies that *M. tuberculosis* uses multiple mechanisms to accomplish its survival, persistence, and progression, which is not surprising for such a successful pathogen. Our challenge as scientists is to learn enough about these mechanisms to develop safe, effective, economical, and durable means of overcoming them and controlling the worldwide scourge of tuberculosis.

The space constraints and defined focus of this review necessitated the omission of numerous significant contributions to the understanding of the biology of tuberculosis, including certain aspects of phagocytosis of *M. tuberculosis* and postphagocytic events. We regret these omissions, and express our sincere regrets to the responsible authors.

REFERENCES

Armstrong, J. A., and P. D. Hart. 1971. Response of cultured macrophages to *Mycobacterium tuberculosis*, with observations on fusion of lysosomes with phagosomes. *J Exp Med* **134**: 713–40.

Armstrong, J. A., and P. D. Hart. 1975. Phagosome-lysosome interactions in cultured macrophages infected with virulent tubercle bacilli. Reversal of the usual nonfusion pattern and observations on bacterial survival. *J Exp Med* **142**: 1–16.

Astarie-Dequeker, C., E. N. N'Diaye, V. Le Cabec, M. G. Rittig, J. Prandi, and I. Maridonneau-Parini. 1999. The mannose receptor mediates uptake of pathogenic and nonpathogenic mycobacteria and bypasses bactericidal responses in human macrophages. *Infect Immun* **67**: 469–77.

Balcewicz-Sablinska, M. K., J. Keane, H. Kornfeld, and H. G. Remold. 1998. Pathogenic *Mycobacterium tuberculosis* evades apoptosis of host macrophages by release of TNF-R2, resulting in inactivation of TNF-alpha. *J Immunol* **161**: 2636–41.

Beharka, A. A., C. D. Gaynor, B. K. Kang, D. R. Voelker, F. X. McCormack, and L. S. Schlesinger. 2002. Pulmonary surfactant protein A up-regulates activity of the mannose receptor, a pattern recognition receptor expressed on human macrophages. *J Immunol* **169**: 3565–73.

Bodnar, K. A., N. V. Serbina, and J. L. Flynn. 2001. Fate of *Mycobacterium tuberculosis* within murine dendritic cells. *Infect Immun* **69**: 800–9.

Chackerian, A. A., J. M. Alt, T. V. Perera, C. C. Dascher, and S. M. Behar. 2002. Dissemination of *Mycobacterium tuberculosis* is influenced by host factors and precedes the initiation of T-cell immunity. *Infect Immun* **70**: 4501–9.

Chua, J., and V. Deretic. 2004. *Mycobacterium tuberculosis* reprograms waves of phosphatidylinositol 3-phosphate on phagosomal organelles. *J Biol Chem* **279**: 36982–92.

Clemens, D. L., and M. A. Horwitz. 1995. Characterization of the *Mycobacterium tuberculosis* phagosome and evidence that phagosomal maturation is inhibited. *J Exp Med* **181**: 257–70.

Clemens, D. L., B. Y. Lee, and M. A. Horwitz. 2000. Deviant expression of Rab5 on phagosomes containing the intracellular pathogens *Mycobacterium tuberculosis* and *Legionella pneumophila* is associated with altered phagosomal fate. *Infect Immun* **68**: 2671–84.

Corbett, E. L., C. J. Watt, N. Walker *et al.* 2003. The growing burden of tuberculosis: global trends and interactions with the HIV epidemic. *Arch Intern Med* **163**: 1009–21.

Cowley, S. C., and K. L. Elkins. 2003. CD4+ T cells mediate IFN-gamma-independent control of *Mycobacterium tuberculosis* infection both in vitro and in vivo. *J Immunol* **171**: 4689–99.

Crowle, A. J., R. Dahl, E. Ross, and M. H. May. 1991. Evidence that vesicles containing living, virulent *Mycobacterium tuberculosis* or *Mycobacterium avium* in cultured human macrophages are not acidic. *Infect Immun* **59**: 1823–31.

Cywes, C., N. L. Godenir, H. C. Hoppe *et al.* 1996. Nonopsonic binding of *Mycobacterium tuberculosis* to human complement receptor type 3 expressed in Chinese hamster ovary cells. *Infect Immun* **64**: 5373–83.

Cywes, C., H. C. Hoppe, M. Daffe, and M. R. Ehlers. 1997. Nonopsonic binding of *Mycobacterium tuberculosis* to complement receptor type 3 is mediated by capsular polysaccharides and is strain dependent. *Infect Immun* **65**: 4258–66.

Downing, J. F., R. Pasula, J. R. Wright, H. L. Twigg, III, and W. J. Martin, 2nd. 1995. Surfactant protein a promotes attachment of *Mycobacterium tuberculosis* to alveolar macrophages during infection with human immunodeficiency virus. *Proc Natl Acad Sci USA* **92**: 4848–52.

Ernst, J. D. 1998. Macrophage receptors for *Mycobacterium tuberculosis. Infect Immun* **66**: 1277–81.

Ernst, J. D. 2000. Bacterial inhibition of phagocytosis. *Cell Microbiol* **2**: 379–86.

Ezekowitz, R. A., K. Sastry, P. Bailly, and A. Warner. 1990. Molecular characterization of the human macrophage mannose receptor: demonstration of multiple carbohydrate recognition-like domains and phagocytosis of yeasts in Cos-1 cells. *J Exp Med* **172**: 1785–94.

Ferguson, J. S., D. R. Voelker, F. X. McCormack, and L. S. Schlesinger. 1999. Surfactant protein D binds to *Mycobacterium tuberculosis* bacilli and lipoarabinomannan via carbohydrate-lectin interactions resulting in reduced phagocytosis of the bacteria by macrophages. *J Immunol* **163**: 312–21.

Ferguson, J. S., J. J. Weis, J. L. Martin, and L. S. Schlesinger. 2004. Complement protein C3 binding to *Mycobacterium tuberculosis* is initiated by the classical pathway in human bronchoalveolar lavage fluid. *Infect Immun* **72**: 2564–73.

Flynn, J. L., and J. Chan. 2001. Immunology of tuberculosis. *A Rev Immunol* **19**: 93–129.

Fratti, R. A., J. M. Backer, J. Gruenberg, S. Corvera, and V. Deretic. 2001. Role of phosphatidylinositol 3-kinase and Rab5 effectors in phagosomal biogenesis and mycobacterial phagosome maturation arrest. *J Cell Biol* **154**: 631–44.

Fratti, R. A., J. Chua, I. Vergne, and V. Deretic. 2003. *Mycobacterium tuberculosis* glycosylated phosphatidylinositol causes phagosome maturation arrest. *Proc Natl Acad Sci USA* **100**: 5437–42.

Garred, P., M. Harboe, T. Oettinger, C. Koch, and A. Svejgaard. 1994. Dual role of mannan-binding protein in infections: another case of heterosis? *Eur J Immunogenet* **21**: 125–31.

Gatfield, J., and J. Pieters. 2000. Essential role for cholesterol in entry of mycobacteria into macrophages. *Science* **288**: 1647–50.

Gaynor, C. D., F. X. McCormack, D. R. Voelker, S. E. McGowan, and L. S. Schlesinger. 1995. Pulmonary surfactant protein A mediates enhanced phagocytosis of *Mycobacterium tuberculosis* by a direct interaction with human macrophages. *J Immunol* **155**: 5343–51.

Geijtenbeek, T. B., S. J. Van Vliet, E. A. Koppel *et al.* 2003. Mycobacteria target DC-SIGN to suppress dendritic cell function. *J Exp Med* **197**: 7–17.

Henderson, R. A., S. C. Watkins, and J. L. Flynn. 1997. Activation of human dendritic cells following infection with *Mycobacterium tuberculosis*. *J Immunol* **159**: 635–43.

Hetland, G., and H. G. Wiker. 1994. Antigen 85C on Mycobacterium bovis, BCG and *M. tuberculosis* promotes monocyte-CR3-mediated uptake of microbeads coated with mycobacterial products. *Immunology* **82**: 445–9.

Hetland, G., H. G. Wiker, K. Hogasen, B. Hamasur, S. B. Svenson, and M. Harboe. 1998. Involvement of antilipoarabinomannan antibodies in classical complement activation in tuberculosis. *Clin Diagn Lab Immunol* **5**: 211–18.

Hickman, S. P., J. Chan, and P. Salgame. 2002. *Mycobacterium tuberculosis* induces differential cytokine production from dendritic cells and macrophages with divergent effects on naive T cell polarization. *J Immunol* **168**: 4636–42.

Hirsch, C. S., J. J. Ellner, D. G. Russell, and E. A. Rich. 1994. Complement receptor-mediated uptake and tumor necrosis factor-alpha-mediated growth inhibition of *Mycobacterium tuberculosis* by human alveolar macrophages. *J Immunol* **152**: 743–53.

Hu, C., T. Mayadas-Norton, K. Tanaka, J. Chan, and P. Salgame. 2000. *Mycobacterium tuberculosis* infection in complement receptor 3-deficient mice. *J Immunol* **165**: 2596–602.

Jiao, X., R. Lo-Man, P. Guermonprez *et al.* 2002. Dendritic cells are host cells for mycobacteria in vivo that trigger innate and acquired immunity. *J Immunol* **168**: 1294–301.

Kang, B. K., and L. S. Schlesinger. 1998. Characterization of mannose receptor-dependent phagocytosis mediated by *Mycobacterium tuberculosis* lipoarabinomannan. *Infect Immun* **66**: 2769–77.

Keane, J., H. G. Remold, and H. Kornfeld. 2000. Virulent *Mycobacterium tuberculosis* strains evade apoptosis of infected alveolar macrophages. *J Immunol* **164**: 2016–20.

Keane, J., B. Shurtleff, and H. Kornfeld. 2002. TNF-dependent BALB/c murine macrophage apoptosis following *Mycobacterium tuberculosis* infection inhibits bacillary growth in an IFN-gamma independent manner. *Tuberculosis (Edinb)* **82**: 55–61.

Leemans, J. C., N. P. Juffermans, S. Florquin *et al.* 2001. Depletion of alveolar macrophages exerts protective effects in pulmonary tuberculosis in mice. *J Immunol* **166**: 4604–11.

Leemans, J. C., S. Florquin, M. Heikens, S. T. Pals, R. van der Neut, and T. Van Der Poll. 2003. CD44 is a macrophage binding site for *Mycobacterium tuberculosis* that mediates macrophage recruitment and protective immunity against tuberculosis. *J Clin Invest* **111**: 681–9.

Leemans, J. C., T. Thepen, S. Weijer *et al.* 2005. Macrophages play a dual role during pulmonary tuberculosis in mice. *J Infect Dis* **191**: 65–74.

Lopez Ramirez, G. M., W. N. Rom, C. Ciotoli *et al.* 1994. *Mycobacterium tuberculosis* alters expression of adhesion molecules on monocytic cells. *Infect Immun* **62**: 2515–20.

Majeed, M., N. Perskvist, J. D. Ernst, K. Orselius, and O. Stendahl. 1998. Roles of calcium and annexins in phagocytosis and elimination of an attenuated strain of *Mycobacterium tuberculosis* in human neutrophils. *Microb Pathog* **24**: 309–20.

Malik, Z. A., G. M. Denning, and D. J. Kusner. 2000. Inhibition of Ca(2+) signaling by *Mycobacterium tuberculosis* is associated with reduced phagosome-lysosome fusion and increased survival within human macrophages. *J Exp Med* **191**: 287–302.

Malik, Z. A., C. R. Thompson, S. Hashimi, B. Porter, S. S. Iyer, and D. J. Kusner. 2003. Cutting edge: *Mycobacterium tuberculosis* blocks Ca2+ signaling and phagosome maturation in human macrophages via specific inhibition of sphingosine kinase. *J Immunol* **170**: 2811–15.

McDonough, K. A., Y. Kress, and B. R. Bloom. 1993. Pathogenesis of tuberculosis: interaction of *Mycobacterium tuberculosis* with macrophages. *Infect Immun* **61**: 2763–73.

McKinney, J. D., K. Honer zu Bentrup, E. J. Munoz-Elias *et al.* 2000. Persistence of *Mycobacterium tuberculosis* in macrophages and mice requires the glyoxylate shunt enzyme isocitrate lyase. *Nature* **406**: 735–8.

Melo, M. D., I. R. Catchpole, G. Haggar, and R. W. Stokes. 2000. Utilization of CD11b knockout mice to characterize the role of complement receptor 3 (CR3, CD11b/CD18) in the growth of *Mycobacterium tuberculosis* in macrophages. *Cell Immunol* **205**: 13–23.

Menozzi, F. D., R. Bischoff, E. Fort, M. J. Brennan, and C. Locht. 1998. Molecular characterization of the mycobacterial heparin-binding hemagglutinin, a mycobacterial adhesin. *Proc Natl Acad Sci USA* **95**: 12625–30.

Mogues, T., M. E. Goodrich, L. Ryan, R. LaCourse, and R. J. North. 2001. The relative importance of T cell subsets in immunity and immunopathology of airborne *Mycobacterium tuberculosis* infection in mice. *J Exp Med* **193**: 271–80.

Mosser, D. M., and P. J. Edelson. 1985. The mouse macrophage receptor for C3bi (CR3) is a major mechanism in the phagocytosis of Leishmania promastigotes. *J Immunol* **135**: 2785–9.

Mosser, D. M., and P. J. Edelson. 1987. The third component of complement (C3) is responsible for the intracellular survival of Leishmania major. *Nature* **327**: 329–31.

Mueller-Ortiz, S. L., A. R. Wanger, and S. J. Norris. 2001. Mycobacterial protein HbhA binds human complement component C3. *Infect Immun* **69**: 7501–11.

Mueller-Ortiz, S. L., E. Sepulveda, M. R. Olsen, C. Jagannath, A. R. Wanger, and S. J. Norris. 2002. Decreased infectivity despite unaltered C3 binding by a DeltahbhA mutant of *Mycobacterium tuberculosis. Infect Immun* **70**: 6751–60.

Mwandumba, H. C., D. G. Russell, M. H. Nyirenda *et al.* 2004. *Mycobacterium tuberculosis* resides in nonacidified vacuoles in endocytically competent alveolar macrophages from patients with tuberculosis and HIV infection. *J Immunol* **172**: 4592–8.

Patki, V., J. Virbasius, W. S. Lane, B. H. Toh, H. S. Shpetner, and S. Corvera. 1997. Identification of an early endosomal protein regulated by phosphatidylinositol 3-kinase. *Proc Natl Acad Sci USA* **94**: 7326–30.

Perez, E., S. Samper, Y. Bordas, C. Guilhot, B. Gicquel, and C. Martin. 2001. An essential role for phoP in *Mycobacterium tuberculosis* virulence. *Mol Microbiol* **41**: 179–87.

Perskvist, N., M. Long, O. Stendahl, and L. Zheng. 2002. *Mycobacterium tuberculosis* promotes apoptosis in human neutrophils by activating caspase-3 and altering expression of Bax/Bcl-xL via an oxygen-dependent pathway. *J Immunol* **168**: 6358–65.

Peters, W., and J. D. Ernst. 2003. Mechanisms of cell recruitment in the immune response to *Mycobacterium tuberculosis. Microbes Infect* **5**: 151–8.

Peters, W., H. M. Scott, H. F. Chambers, J. L. Flynn, I. F. Charo, and J. D. Ernst. 2001. Chemokine receptor 2 serves an early and essential role in resistance to *Mycobacterium tuberculosis. Proc Natl Acad Sci USA* **98**: 7958–63.

Peters, W., J. G. Cyster, M. Mack *et al.* 2004. CCR2-dependent trafficking of F4/80dim macrophages and CD11cdim/intermediate dendritic cells is crucial for T cell recruitment to lungs infected with *Mycobacterium tuberculosis. J Immunol* **172**: 7647–53.

Pethe, K., D. L. Swenson, S. Alonso, J. Anderson, C. Wang, and D. G. Russell. 2004. Isolation of *Mycobacterium tuberculosis* mutants defective in the arrest of phagosome maturation. *Proc Natl Acad Sci USA* **101**: 13642–7.

Peyron, P., C. Bordier, E. N. N'Diaye, and I. Maridonneau-Parini. 2000. Nonopsonic phagocytosis of Mycobacterium kansasii by human neutrophils depends on cholesterol and is mediated by CR3 associated with glycosylphosphatidylinositol-anchored proteins. *J Immunol* **165**: 5186–91.

Ramanathan, V. D., J. Curtis, and J. L. Turk. 1980. Activation of the alternative pathway of complement by mycobacteria and cord factor. *Infect Immun* **29**: 30–5.

Raviglione, M. C., and R. C. O'Brien. 2004. Tuberculosis. In *Harrison's Principles of Internal Medicine.* D. L. Kasper, E. Braunwald, A. S. Fauci *et al.*, editors, pp. 953–66. New York: McGraw Hill.

Rich, A. R. 1944. *The Pathogenesis of Tuberculosis.* Springfield, IL: C. C. Thomas.

Riendeau, C. J., and H. Kornfeld. 2003. THP-1 cell apoptosis in response to Mycobacterial infection. *Infect Immun* **71**: 254–9.

Russell, D. G., J. Dant, and S. Sturgill-Koszycki. 1996. *Mycobacterium avium-* and *Mycobacterium tuberculosis*-containing vacuoles are dynamic, fusion-competent vesicles that are accessible to glycosphingolipids from the host cell plasmalemma. *J Immunol* **156**: 4764–73.

Sassetti, C. M., and E. J. Rubin. 2003. Genetic requirements for mycobacterial survival during infection. *Proc Natl Acad Sci USA* **100**: 12989–94.

Schaible, U. E., S. Sturgill-Koszycki, P. H. Schlesinger, and D. G. Russell. 1998. Cytokine activation leads to acidification and increases maturation of *Mycobacterium avium*-containing phagosomes in murine macrophages. *J Immunol* **160**: 1290–6.

Schlesinger, L. S. 1993. Macrophage phagocytosis of virulent but not attenuated strains of *Mycobacterium tuberculosis* is mediated by mannose receptors in addition to complement receptors. *J Immunol* **150**: 2920–30.

Schlesinger, L. S., C. G. Bellinger-Kawahara, N. R. Payne, and M. A. Horwitz. 1990. Phagocytosis of *Mycobacterium tuberculosis* is mediated by human monocyte complement receptors and complement component C3. *J Immunol* **144**: 2771–80.

Schlesinger, L. S., S. R. Hull, and T. M. Kaufman. 1994. Binding of the terminal mannosyl units of lipoarabinomannan from a virulent strain of *Mycobacterium tuberculosis* to human macrophages. *J Immunol* **152**: 4070–9.

Schlesinger, L. S., T. M. Kaufman, S. Iyer, S. R. Hull, and L. K. Marchiando. 1996. Differences in mannose receptor-mediated uptake of lipoarabinomannan from virulent and attenuated strains of *Mycobacterium tuberculosis* by human macrophages. *J Immunol* **157**: 4568–75.

Schnappinger, D., S. Ehrt, M. I. Voskuil *et al.* 2003. Transcriptional adaptation of *Mycobacterium tuberculosis* within macrophages: insights into the phagosomal environment. *J Exp Med* **198**: 693–704.

Schorey, J. S., M. C. Carroll, and E. J. Brown. 1997. A macrophage invasion mechanism of pathogenic mycobacteria. *Science* **277**: 1091–3.

Seiler, P., P. Aichele, S. Bandermann *et al.* 2003. Early granuloma formation after aerosol *Mycobacterium tuberculosis* infection is regulated by neutrophils via CXCR3-signaling chemokines. *Eur J Immunol* **33**: 2676–86.

Shepard, C. C. 1957. Growth characteristics of tubercle bacilli and certain other mycobacteria in HeLa cells. *J Exp Med* **105**: 39–55.

Shin, J. S., Z. Gao, and S. N. Abraham. 2000. Involvement of cellular caveolae in bacterial entry into mast cells. *Science* **289**: 785–8.

Simonsen, A., R. Lippe, S. Christoforidis *et al.* 1998. EEA1 links PI(3)K function to Rab5 regulation of endosome fusion. *Nature* **394**: 494–8.

Simonsen, A., J. M. Gaullier, A. D'Arrigo, and H. Stenmark. 1999. The Rab5 effector EEA1 interacts directly with syntaxin-6. *J Biol Chem* **274**: 28857–60.

Stamm, L. M., J. H. Morisaki, L. Y. Gao *et al.* 2003. Mycobacterium marinum escapes from phagosomes and is propelled by actin-based motility. *J Exp Med* **198**: 1361–8.

Stokes, R. W., and D. P. Speert. 1995. Lipoarabinomannan inhibits nonopsonic binding of *Mycobacterium tuberculosis* to murine macrophages. *J Immunol* **155**: 1361–9.

Stokes, R. W., R. Norris-Jones, D. E. Brooks, T. J. Beveridge, D. Doxsee, and L. M. Thorson. 2004. The glycan-rich outer layer of the cell wall of *Mycobacterium tuberculosis* acts as an antiphagocytic capsule limiting the association of the bacterium with macrophages. *Infect Immun* **72**: 5676–86.

Sturgill-Koszycki, S., P. H. Schlesinger, P. Chakraborty *et al.* 1994. Lack of acidification in Mycobacterium phagosomes produced by exclusion of the vesicular proton-ATPase. *Science* **263**: 678–81.

Sturgill-Koszycki, S., U. E. Schaible, and D. G. Russell. 1996. Mycobacterium-containing phagosomes are accessible to early endosomes and reflect a transitional state in normal phagosome biogenesis. *EMBO J* **15**: 6960–8.

Tailleux, L., O. Neyrolles, S. Honore-Bouakline *et al.* 2003a. Constrained intracellular survival of *Mycobacterium tuberculosis* in human dendritic cells. *J Immunol* **170**: 1939–48.

Tailleux, L., O. Schwartz, J. L. Herrmann *et al.* 2003b. DC-SIGN is the major *Mycobacterium tuberculosis* receptor on human dendritic cells. *J Exp Med* **197**: 121–7.

Taylor, M. E., J. T. Conary, M. R. Lennartz, P. D. Stahl, and K. Drickamer. 1990. Primary structure of the mannose receptor contains multiple motifs resembling carbohydrate-recognition domains. *J Biol Chem* **265**: 12156–62.

Velasco-Velazquez, M. A., D. Barrera, A. Gonzalez-Arenas, C. Rosales, and J. Agramonte-Hevia. 2003. Macrophage–*Mycobacterium tuberculosis* interactions: role of complement receptor 3. *Microb Pathog* **35**: 125–31.

Vergne, I., J. Chua, and V. Deretic. 2003. Tuberculosis toxin blocking phagosome maturation inhibits a novel Ca2+/calmodulin-PI3K hVPS34 cascade. *J Exp Med* **198**: 653–9.

Via, L. E., D. Deretic, R. J. Ulmer, N. S. Hibler, L. A. Huber, and V. Deretic. 1997. Arrest of mycobacterial phagosome maturation is caused by a block in vesicle fusion between stages controlled by rab5 and rab7. *J Biol Chem* **272**: 13326–31.

Walburger, A., A. Koul, G. Ferrari, L. Nguyen *et al.* 2004. Protein kinase G from pathogenic mycobacteria promotes survival within macrophages. *Science* **304**: 1800–4.

Xia, Y., and G. D. Ross. 1999. Generation of recombinant fragments of CD11b expressing the functional beta-glucan-binding lectin site of CR3 (CD11b/CD18). *J Immunol* **162**: 7285–93.

Xu, S., A. Cooper, S. Sturgill-Koszycki *et al.* 1994. Intracellular trafficking in *Mycobacterium tuberculosis* and *Mycobacterium avium*-infected macrophages. *J Immunol* **153**: 2568–78.

Zimmerli, S., S. Edwards, and J. D. Ernst. 1996a. Selective receptor blockade during phagocytosis does not alter the survival and growth of *Mycobacterium tuberculosis* in human macrophages. *Am J Respir Cell Mol Biol* **15**: 760–70.

Zimmerli, S., M. Majeed, M. Gustavsson, O. Stendahl, D. A. Sanan, and J. D. Ernst. 1996b. Phagosome-lysosome fusion is a calcium-independent event in macrophages. *J Cell Biol* **132**: 49–61.

Index

Printed in the United States
By Bookmasters